SSM+Vue.js 3 全栈开发实战

杨章伟　肖异骐　刘祥淼　编著

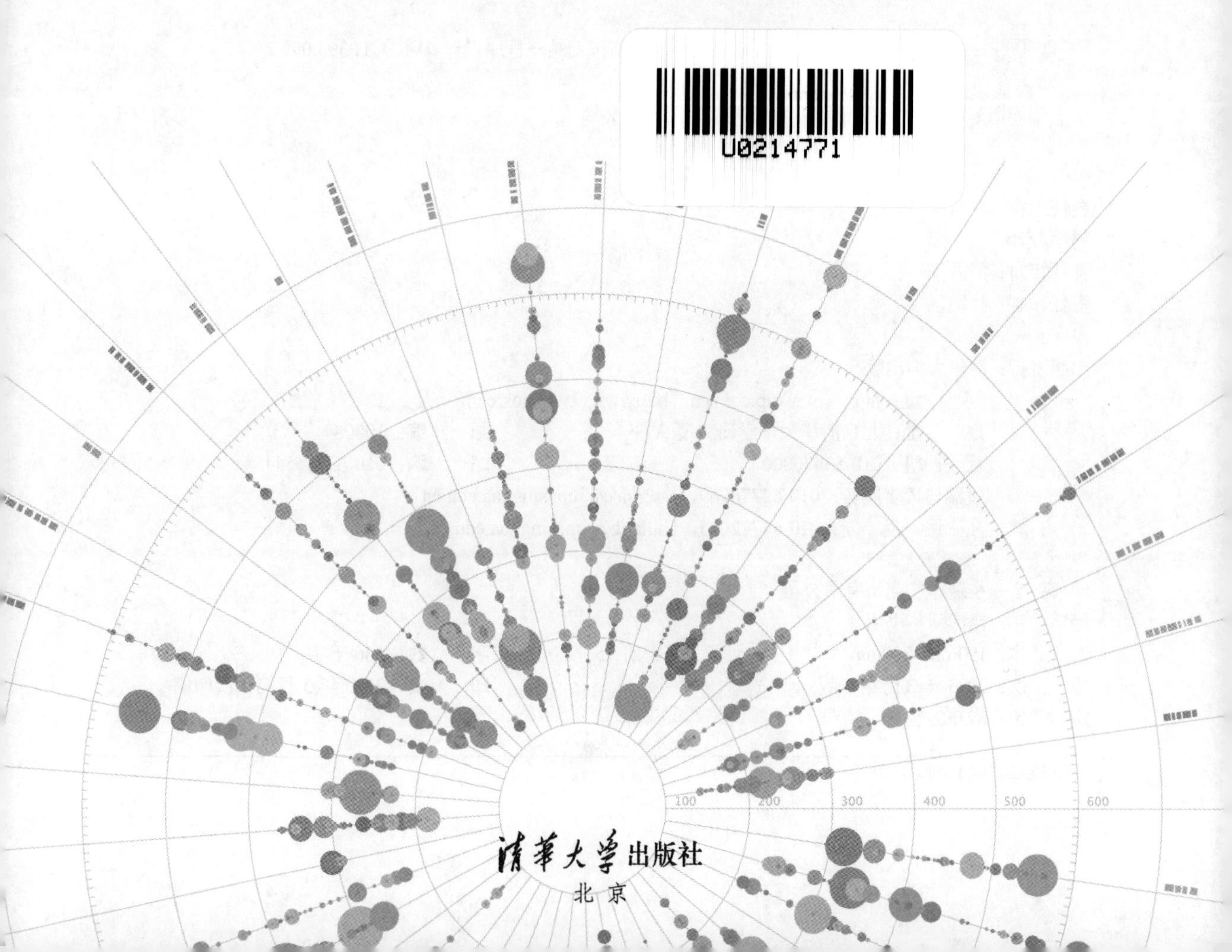

清華大學出版社
北京

内 容 简 介

SSM（Spring、Spring MVC 和 MyBatis）和 Vue.js 3 是当前使用广泛的前端和后端技术框架。本书由浅入深、循序渐进地讲解 SSM 和 Vue.js 框架的基础知识和应用，书中使用大量案例，可以很好地帮助读者学习和理解 SSM+Vue.js 前后端分离开发技术。本书配套示例源码、PPT 课件、作者答疑服务。

本书共分 18 章，内容包括 Spring 基础、Spring 中的 Bean、Spring AOP、Spring 的数据库开发、Spring 的事务管理、初识 MyBatis、MyBatis 的核心配置、动态 SQL、MyBatis 的关联映射、MyBatis 与 Spring 的整合、Vue.js 3 入门、Spring MVC 入门、Spring MVC 数据绑定、JSON 数据交互和 RESTful 支持、拦截器、SSM 框架整合、新闻发布管理系统实战、图书管理系统实战。读者通过对 SSM+Vue.js 框架知识的学习和对章节示例、实战案例的实践，可以很好地掌握 SSM+Vue.js 框架技术的基础知识，为开发大型项目打下坚实基础。

本书内容精练、重点突出、示例丰富，适合 SSM 框架初学者、SSM+Vue.js 框架前后端分离开发的项目开发人员阅读，可以作为 Java Web 开发人员的必备参考书，也可作为高等院校或高职高专计算机专业的教材使用。

图书在版编目（CIP）数据

SSM+Vue.js 3 全栈开发实战 / 杨章伟，肖异骐，刘祥淼编著. —北京：清华大学出版社，2023.1
ISBN 978-7-302-62446-2

I. ①S… II. ①杨… ②肖… ③刘… III. ①网页制作工具—程序设计 IV. ①TP393.092.2

中国国家版本馆 CIP 数据核字（2023）第 016839 号

责任编辑：夏毓彦
封面设计：王　翔
责任校对：闫秀华
责任印制：曹婉颖

出版发行：清华大学出版社
网　　址：http://www.tup.com.cn，http://www.wqbook.com
地　　址：北京清华大学学研大厦 A 座　　邮　　编：100084
社 总 机：010-83470000　　邮　　购：010-62786544
投稿与读者服务：010-62776969，c-service@tup.tsinghua.edu.cn
质 量 反 馈：010-62772015，zhiliang@tup.tsinghua.edu.cn

印 装 者：小森印刷霸州有限公司
经　　销：全国新华书店
开　　本：190mm×260mm　　**印　张：**21.75　　**字　　数：**586 千字
版　　次：2023 年 3 月第 1 版　　**印　　次：**2023 年 3 月第 1 次印刷
定　　价：89.00 元

产品编号：084294-01

前　言

为什么后台开发要用 SSM 框架

SSM 框架编写了大量的基础功能，使得程序员可将工作重心放到业务逻辑的实现上。SSM 框架具备良好的规范性和重用性，易扩展、易维护。换而言之，SSM 框架简化了烦琐的配置工作，让编写代码如丝绸一般顺滑。

为什么前端开发要用 Vue.js 框架

Vue.js 是一款友好的、多用途且高性能的 JavaScript 框架，使用 Vue.js 可以创建可维护性和可测试性更强的代码库。Vue.js 允许将一个网页分割成可复用的组件，每个组件都包含属于自己的 HTML、CSS、JavaScript，可以用来渲染网页中相应的地方，所以越来越多的前端开发者使用 Vue.js，与 SSM 框架整合，实现前后端分离开发。

SSM+Vue.js 框架有哪些优点

SSM+Vue.js 框架实现了前后端分离开发，SSM 负责后台开发，Vue.js 承担前端开发。在 Web 应用系统开发中，SSM+Vue.js 框架能够实现响应式数据绑定，会自动对页面中某些数据的变化做出同步响应，使数据的更改更为简单，并使代码具备良好的移植性和可维护性。

本书特点

（1）内容丰富，知识全面。全书共分 18 章，采用由浅入深、循序渐进的方式进行讲解，内容涉及 Spring、Spring MVC、MyBatis、Vue.js 及其整合。

（2）格式统一，讲解规范。书中案例基本上都采用了分步骤实现方法，使得读者可以清晰地掌握每个技术点的具体实现步骤，从而提高学习效率。

（3）案例精讲，注重实践。为了方便读者学习和理解，笔者根据自己多年的项目经验，在讲授知识点的同时，匹配了大量实例（含源码），从而让读者可以边学边实践。

（4）贴心提醒，轻松掌握。根据需要在各章使用了“注意”“说明”等提示信息，让读者可以在学习过程中更轻松地理解相关知识点及概念。

配套示例源码、PPT 课件下载

本书配套示例源码、PPT 课件等资源，可用微信扫描下面的二维码获取，也可按扫描后的页面提示把下载链接转发到邮箱中下载。如果有疑问或建议，请用电子邮件联系 booksaga@163.com，邮件主题写“SSM+Vue.js 3 全栈开发实战”。

本书读者

- SSM 框架开发初学者。
- Java Web 后端开发工程师。
- SSM 和 Vue.js 框架前后端分离开发的新手。
- Web 应用开发人员。
- 高性能前后端 Web 应用开发人员。
- 高等院校和培训学校相关课程的师生。

笔　者

2023 年 1 月

目　　录

第1章 Spring基础

Spring是目前非常流行的Java Web开发框架，可用于解决企业应用开发的复杂性问题。对于一个开发企业级应用的程序员来说，掌握Spring框架是必备技能之一。本章主要讲解Spring的基础知识。

本章主要涉及的知识点如下：

- Spring的基础知识。
- Spring的控制反转/依赖注入。

1.1 Spring概述

本节首先介绍Sping是一个什么样的框架，然后介绍Spring框架支持的包和相关文件的获取方式及其目录结构。学习这些内容的目的是为使用Spring框架打好基础。

1.1.1 什么是Spring

Spring是一个以IoC（Inversion of Control，控制反转）和AOP（Aspect Oriented Programming，面向切面编程）为内核的框架。IoC是Spring的基础，它实现的是一种控制，简单地说，就是以前是调用new构造方法来创建对象，现在变成使用Spring来创建对象。DI（Dependency Inject，依赖注入）与IoC的含义相同，它们从两个角度描述同一个概念。实际上DI就是对象的属性，已经被注入好相关值，直接使用即可。

IoC和DI将在第1.2节详细介绍，AOP将在第3章详细介绍。

注意：如果读者是第一次学习本框架，那么务必严格按照本书的指导，先进行模仿操作，直至看到实际效果，之后再进行改动和调整，从而进一步加深理解，直到熟练掌握。

1.1.2 Spring 的下载及目录结构

Spring 经过十多年的发展，版本不断升级。本书中的实例代码基于 Spring 5.2.9 编写。使用 Spring 框架进行开发需要用到 Spring 框架包和第三方依赖包，具体如下：

1. Spring 框架包

本书中的实例代码基于 Spring 5.2.9 编写，建议读者下载该版本（也可以根据实际情况下载最新的版本）。Spring 框架压缩包名称为 spring-framework-5.2.9.RELEASE-dist.zip，可以通过网址 http://repo.spring.io/simple/libs-release-local/org/springframework/spring/5.2.9.RELEASE/下载。下载完成后，将压缩包解压，最终目录结构如图 1.1 所示。

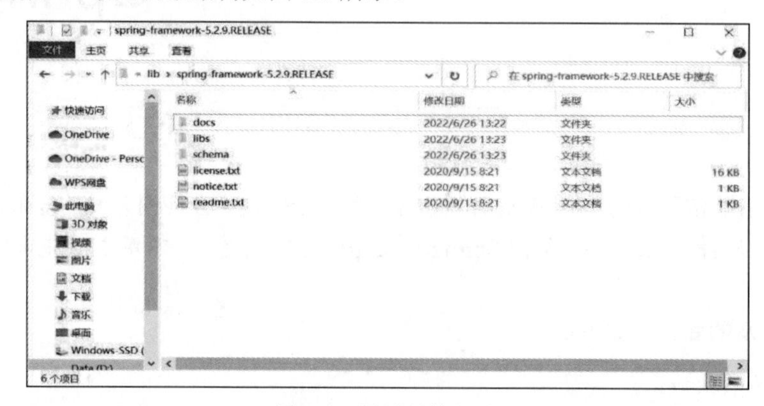

图 1.1　解压后的目录

其中，libs 目录下包含 63 个 JAR 文件，如图 1.2 所示。

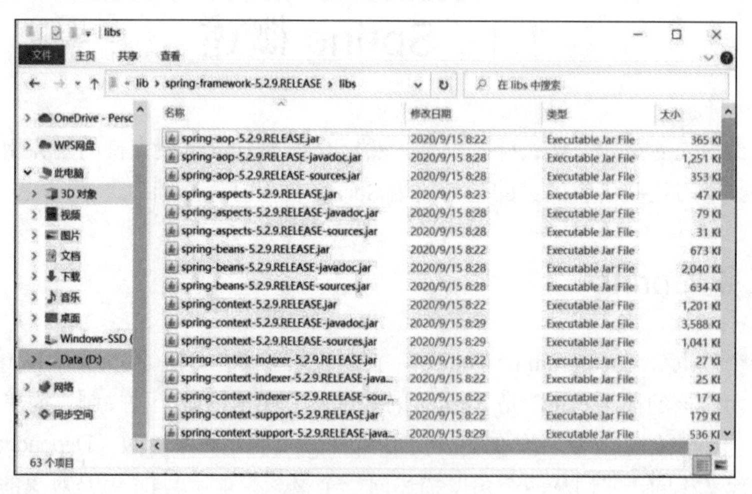

图 1.2　libs 目录

libs 目录中的 JAR 包分为如下 3 类：

- 以 RELEASE.jar 结尾的是 Spring 框架 Class 文件的压缩包。
- 以 RELEASE-javadoc.jar 结尾的是 Spring 框架 API 文档的压缩包。
- 以 RELEASE-sources.jar 结尾的是 Spring 框架源文件的压缩包。

整个 Spring 框架由 21 个模块组成，该目录下 Spring 为每个模块都提供了这 3 类压缩包。

在 libs 目录下有 4 个 Spring 的基础包，它们分别对应 Spring 核心容器的 4 个模块，具体介绍如表 1.1 所示。

表 1.1 Spring 的基础包说明

包 名	说 明
spring-core-5.2.9.RELEASE.jar	包含 Spring 框架基本的核心工具类，Spring 的其他组件都要用到这个包里的类
spring-beans-5.2.9.RELEASE.jar	所有应用都要用到的 JAR 包，包含访问配置文件、创建和管理 Bean 以及进行 IoC 或者 DI 操作相关的所有类
spring-context-5.2.9.RELEASE.jar	Spring 提供了在基础 IoC 功能上的扩展服务，还提供了许多企业级服务的支持，如任务调度、JNDI 定位、EJB 集成、远程访问、缓存、邮件服务以及各种视图层框架的封装等
spring-expression-5.2.9.RELEASE.jar	定义了 Spring 的表达式语言

2. 第三方依赖包

在使用 Spring 进行开发时，Spring 的核心容器还需要依赖 commons-logging 的 JAR 包。该 JAR 包可以通过网址 http://commons.apache.org/proper/commons-logging/download_logging.cgi 下载。下载后得到一个名为 commons-logging-1.2-bin.zip 的压缩包，将它解压后可以找到 commons-logging-1.2.jar 包。

注意：初学者学习 Spring 框架时，只需将 Spring 的 4 个基础包以及 commons-logging-1.2.jar 包复制到项目的 libs 目录下，并发布到类路径中即可。

1.2 IoC（控制反转）与 DI（依赖注入）

控制反转（Inversion of Control，IoC）和依赖注入（Dependency Inject，DI）是 Spring 的基础。在 1.1.1 节为什么会说 IoC 和 DI 描述的是同一概念呢？本节将揭晓答案。

1.2.1 什么是 IoC

“控制反转”又称为“控制反向”或者“控制倒置”。在面向对象传统编程方式中，获取对象的方式通常是用 new 关键字主动创建一个对象。Spring 中的 IoC 方式对象的生命周期由 Spring 框架提供的 IoC 容器来管理，直接从 IoC 容器中获取一个对象，控制权从应用程序交给了 IoC 容器。

IoC 理论上是借助“第三方”实现具有依赖关系的对象之间的解耦，如图 1.3 所示，即把各个对象类封装之后，通过 IoC 容器来关联这些对象类。这样对象与对象之间就通过 IoC 容器进行联系，而对象与对象之间并没有什么直接联系。

图 1.3 IoC 容器解耦

应用程序在没有引入 IoC 容器之前，对象 A 依赖对象 B，那么对象 A 在实例化或者运行到某一点的时候，必须自己主动创建对象 B 或者使用已经创建好的对象 B，其中无论是创建还是使用已创建的对象 B，控制权都在应用程序自身。应用程序引入了 IoC 容器之后，对象 A 和对象 B 之间失去了直接联系，那么当对象 A 实例化或运行时，如果需要对象 B，则 IoC 容器就会主动创建一个对象 B 注入（依赖注入）对象 A 所需要的地方。由此，对象 A 获得依赖对象 B 的过程由主动行为变成被动行为，即把创建对象交给了 IoC 容器处理，控制权颠倒了过来，这就是所谓的控制反转。

1.2.2 什么是 DI

所谓依赖注入，就是由 IoC 容器在运行期间动态地将某种依赖关系注入对象之中。例如，将对象 B 注入（赋值）对象 A 的成员变量。

事实上，依赖注入和控制反转是对同一件事情的不同描述，即它们描述的角度不同。依赖注入是从应用程序的角度来描述，即应用程序依赖容器创建并注入它所需要的外部资源；而控制反转是从容器的角度来描述，即容器控制应用程序，由容器反向地向应用程序注入所需要的外部资源。这里所说的外部资源可以是外部实例对象，也可以是外部文件对象等。

使用 IoC/DI 给软件开发带来了多方面的益处：

（1）可维护性比较好，便于单元测试、调试程序和诊断故障。代码中的每一个 Class 都可以单独测试，彼此之间互不影响，只要保证自身的功能无误即可，这就是组件之间低耦合或者无耦合带来的好处。

（2）每个开发团队的成员都只需要关注自己要实现的业务逻辑，完全不用关心其他人的工作进展，因为自己的任务跟别人的没有任何关系，只需单独测试即可。不用依赖别人的组件，也就不会扯不清责任了。所以，在一个大中型项目中，使用 IoC/DI 可以让团队成员分工明确、责任明晰，可以很容易将一个大的任务划分为细小的任务，这样开发效率和产品质量必将得到大幅度的提高。

（3）可复用性好。我们可以把具有普遍性的常用组件独立出来，反复应用到项目中的其他部分或者是其他项目，当然这也是面向对象的基本特征。IoC 更好地贯彻了这个原则，提高了模块的可复用性，符合接口标准的实现都可以插接到支持此标准的模块中。

（4）生成对象的方式转为外置方式，就是把对象生成放在配置文件中进行定义。这样，当我们需要更换一个实现子类时将会变得很简单，只需修改配置文件就可以了，完全具有热插拔的特性。

1.2.3 IoC/DI 的实现

Spring 框架的主要功能是通过其核心容器来实现的。Spring 框架提供的两种核心容器分别是BeanFactory 和 ApplicationContext。IoC/DI 通常有属性 setter()方法注入和构造方法注入两种实现方式。

注意： 如前所述，依赖注入和控制反转是对同一件事情的不同描述，所以这里讲的 IoC/DI 实现方式其实就是 DI 实现方式。

1. Spring 核心容器

Spring 框架的两个最基本和最重要的包是 org.springframework.beans（该包中的主要接口是BeanFactory）和 org.springframework.context（该包中的主要接口是 ApplicationContext）。

Spring IoC 框架的主要组件有 Beans、配置文件 applicationContext.xml、BeanFactory 接口及其相关类、ApplicationContext 接口及其相关类。

（1）Beans 是指项目中提供业务功能的 Bean，即容器要管理的 Bean。Beans 就是一个常见的JavaBean、Java 类。

（2）在 Spring 中对 Bean 的管理是在配置文件中进行的。在 Spring 容器内编辑配置文件管理Bean 又称为 Bean 的装配，实际上装配就是告诉容器需要哪些 Bean，以及容器是如何使用 IoC 将它们配合起来的。Bean 的配置文件是一个 XML 文件，可以命名为 applicationContext.xml 或其他，一般习惯使用 applicationContext.xml。

配置文件包含 Bean 的 id、类、属性及其值，包含一个<beans>元素和数个<bean>子元素。Spring IoC 框架可根据 Bean 的 id 从 Bean 配置文件中取得该 Bean 的类，并生成该类的一个实例对象，继而从配置文件中获得该对象的属性和值。常见 applicationContext.xml 配置文件的格式如下：

```
01  <?xml version="1.0" encoding="UTF-8"?>
02  <beans xmlns="http://www.springframework.org/schema/beans"
03        xmlns:xsi="http://www.w3.org/2001/XMLSchema-instance"
04        xsi:schemaLocation="http://www.springframework.org/schema/beans
05            http://www.springframework.org/schema/beans/spring-beans.xsd">
06                <!-- 将指定类配置给 Spring，让 Spring 创建其对象的实例  -->
07      <bean id= "chinese" class="com.ssm.Chinese">
08   <!--
09  <property>元素用来指定需要容器注入的属性; name 指定属性值为 language; ref 指定需要向
10  language 属性注入的 id，即注入的对象"英语"，该对象由 English 类生成。
11  -->
12          <property name="language" ref="英语"></property>
13      </bean>
14      <!--配置另外一个 Bean-->
15      <bean id="英语" class=" com.ssm..English"></bean>
16  </beans>
```

（3）BeanFactory 采用了工厂设计模式，即 Bean 容器模式，负责读取 Bean 的配置文件，管理对象的生成、加载，维护 Bean 对象与 Bean 对象之间的依赖关系，负责 Bean 的生命周期。对于简单的应用程序来说，使用 BeanFactory 就已经足够管理 Bean 了，在对象的管理上可以获得许多便利性。

org.springframework.beans.factory.BeanFactory 是一个顶级接口，包含管理 Bean 的各种方法。Spring 框架也提供了一些实现该接口的类。

org.springframework.beans.factory.xml.XmlBeanFactory 是 BeanFactory 常用的实现类，根据配置文件中的定义装载 Bean。如果要创建 XmlBeanFactory，则需要传递一个 FileInputStream 对象，该对象把 XML 文件提供给工厂。代码可以写成：

```
BeanFactory factory=new XmlBeanFactory( new FileInputStream("applicationContext.xml "));
```

BeanFactory 的常用方法如下：

- getBean(String name)：可根据 Bean 的 id 生成该 Bean 的对象。
- getBean(String name,Class requiredType)：可根据 Bean 的 id 和相应类生成该 Bean 的对象。

（4）ApplicationContext 接口是提供高级功能的容器，它的基本功能与 BeanFactory 很相似，但它还有以下功能：

- 提供访问资源文件更方便的方法。
- 支持国际化消息。
- 提供文字消息解析的方法。
- 可以发布事件，对事件感兴趣的 Bean 可以接收这些事件。

ApplicationContext 接口的常用实现类有以下 3 个：

- FileSystemXmlApplicationContext：从文件系统中的 XML 文件加载上下文中定义的信息。
- ClassPathXmlApplicationContext：从类路径中的 XML 文件加载上下文中定义的信息，把上下文定义的文件当成类路径资源。
- XmlWebApplicationContext：从 Web 系统中的 XML 文件加载上下文中定义的信息。

其中，FileSystemXmlApplicationContext 和 ClassPathXmlApplicationContext 的代码编写如下：

```
01 ApplicationContext context=new FileSystemXmlApplicationContext("d:/applicationContext.xml");
02 ApplicationContext context=new ClassPathXmlApplicationContext("applicationContext.xml ");
```

第 01 行代码使用文件系统的方式来查询配置文件，此时 applicationContext.xml 文件位于 D 盘下。第 02 行代码使用类路径来查询配置文件，此时 applicationContext.xml 文件位于项目的 src 目录下。

FileSystemXmlApplicationContext 和 ClassPathXmlApplicationContext 的区别是：FileSystemXmlApplicationContext 只能在指定的路径中查询 applicationContext.xml 配置文件，而 ClassPathXmlApplicationContext 可以在整个类路径中查询 applicationContext.xml 配置文件。

2. IoC/DI 实现方式

如前所述，依赖注入和控制反转是对同一件事情的不同描述。依赖注入的作用是在使用 Spring 框架创建对象时，动态地将它所依赖的对象注入 Bean 组件中，其实现方式通常有两种：一种是属性 setter()方法注入，另一种是构造方法注入。具体介绍如下：

- 属性 setter()方法注入：IoC 容器使用 setter()方法注入被依赖的实例。通过调用无参构造器或无参静态工厂方法实例化 Bean 后，调用该 Bean 的 setter()方法，即可实现基于 setter()方法的依赖注入。该方法简单、直观，而且容易理解，所以被大量使用。
- 构造方法注入：IoC 容器使用构造方法注入被依赖的实例。基于构造方法的依赖注入通过调用带参数的构造方法来实现，每个参数代表着一个依赖。

【示例 1-1】下面以常用的 setter()方法注入的方式为例，讲解 Spring 容器在应用中是如何实现依赖注入的。

步骤 01 在 IntelliJ IDEA 集成开发环境中创建一个名为 chapter01 的动态 Web 项目，将 Spring 的 4 个基础包以及 commons-logging 的 JAR 包复制到 lib 目录中，并发布到类路径下，如图 1.4 所示。

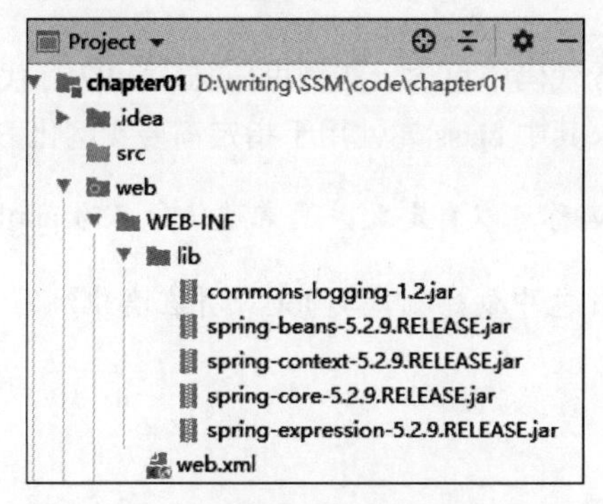

图 1.4　创建项目并导入包

步骤 02 在 src 目录下创建一个 com.ssm.ioc_di 包，并在包中创建接口 UserDao，然后在接口中定义一个 login()方法，如文件 1.1 所示。

文件 1.1　UserDao.java

```
01 package com.ssm.ioc_di;
02 public interface UserDao {
03     // 定义 login()方法
04     public void login();
05 }
```

步骤 03 在 com.ssm.ioc_di 包中创建 UserDao 接口的实现类 UserDaoImpl，该类需要实现接口中的 login()方法，并在方法中编写一条输出语句，如文件 1.2 所示。

文件 1.2　UserDaoImpl.java

```
01 package com.ssm.ioc_di;
02 public class UserDaoImpl implements UserDao {
03     // 实现 login()方法
04     public void login() {
05         System.out.println("UserDao login");
06     }
07 }
```

步骤04 在 src 目录下创建 Spring 的配置文件 applicationContext.xml，并在配置文件中创建一个 id 为 userDao 的 Bean，如文件 1.3 所示。

文件 1.3 applicationContext.xml

```
01 <?xml version="1.0" encoding="UTF-8"?>
02 <beans xmlns="http://www.springframework.org/schema/beans"
03     xmlns:xsi="http://www.w3.org/2001/XMLSchema-instance"
04     xsi:schemaLocation="http://www.springframework.org/schema/beans
05            http://www.springframework.org/schema/beans/spring-beans.xsd">
06     <!-- 将指定类配置给 Spring，让 Spring 创建其对象的实例 -->
07     <bean id="userDao" class="com.ssm.ioc_di.UserDaoImpl" />
08 </beans>
```

在文件 1.3 中，第 01~05 行代码中包含一些约束信息；第 07 行代码表示在 Spring 容器中创建一个 id 为 userDao 的 Bean 实例，其中 class 属性用于指定需要实例化 Bean 的类。

注意： Spring 配置文件的名称可以自定义，通常默认命名为 applicationContext.xml。

步骤05 在 com.ssm.ioc_di 包中创建测试类 IoC，并在类中编写 main()方法及实现 IoC 的代码，如文件 1.4 所示。

文件 1.4 IoC.java

```
01 package com.ssm.ioc_di;
02 import org.springframework.context.ApplicationContext;
03 import org.springframework.context.support.ClassPathXmlApplicationContext;
04 public class IoC {
05     public static void main(String[] args) {
06         //1.初始化 Spring 容器，加载配置文件
07         ApplicationContext applicationContext=
08                 new ClassPathXmlApplicationContext("applicationContext.xml");
09         //2.通过容器获取 userDao 实例
10         UserDao userDao=(UserDao)applicationContext.getBean("userDao");
11         //3.调用实例中的 login()方法
12         userDao.login();
13     }
14 }
```

程序执行后，控制台输出结果如图 1.5 所示。从图中可以看出，控制台成功输出了 UserDaoImpl 类中的输出语句。在文件 1.4 的 main()方法中，并没有通过 new 关键字来创建 UserDao 接口的实现类对象，而是通过 Spring 容器来获取的实现类对象，这就是 Spring IoC 的工作机制。

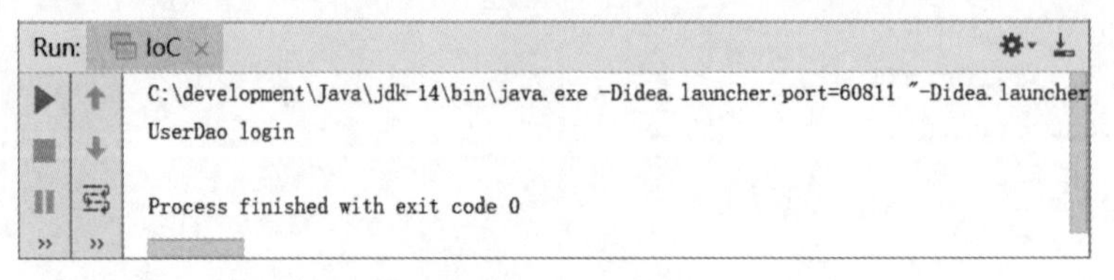

图 1.5 运行结果

步骤06 在 com.ssm.ioc_di 包中创建接口 UserService，并编写一个 login()方法，如文件 1.5 所示。

文件 1.5 UserService.java

```
01  package com.ssm.ioc_di;
02  public interface UserService {
03      public void login();
04  }
```

步骤07 在 com.ssm.ioc_di 包中创建接口 UserService 的实现类 UserServiceImpl，在类中声明 userDao 属性，并添加属性的 setter()方法；同时编写 login()方法。具体代码如文件 1.6 所示。

文件 1.6 UserServiceImpl.java

```
01  package com.ssm.ioc_di;
02  public class UserServiceImpl implements UserService {
03      //声明 userDao 属性
04      private UserDao userDao;
05      //添加 userDao 属性的 setter()方法，用于实现依赖注入
06      public void setUserDao(UserDao userDao) {
07          this.userDao = userDao;
08      }
09      //实现接口中的方法
10      public void login() {
11          //调用 userDao 属性中的 login()方法，并执行输出语句
12          this.userDao.login();
13          System.out.println("userService login");
14      }
15  }
```

步骤08 在配置文件 applicationContext.xml 中创建一个 id 为 userService 的 Bean，该 Bean 用于实例化 UserServiceImpl 类的信息，并将 name 属性为 userDao 的实例注入 userService 中，其代码如下：

```
<!-- 添加一个 id 为 userService 的 Bean -->
<bean id="userService" class="com.ssm.ioc_di.userServiceImpl">
    <!-- 将 name 为 userDao 的 Bean 实例注入 userService 实例中 -->
    <property name="userDao" ref="userDao"/>
</bean>
```

在上述代码中，<property>是<bean>元素的子元素，用于调用 Bean 实例中的 setUserDao()方法完成属性赋值，从而实现依赖注入。其 name 属性表示 Bean 实例中的相应属性名，ref 属性用于指定其属性值。

步骤09 在 com.ssm.ioc_di 包中创建测试类 DI，如文件 1.7 所示。

文件 1.7 DI.java

```
01  package com.ssm.ioc_di;
```

```
02  import org.springframework.context.ApplicationContext;
03  import org.springframework.context.support.ClassPathXmlApplicationContext;
04  public class DI {
05      public static void main(String[] args) {
06          // 1.初始化 Spring 容器，加载配置文件
07          ApplicationContext applicationContext =
08                      new ClassPathXmlApplicationContext("applicationContext.xml");
09          // 2.通过容器获取 userService 实例
10          UserService userService = (UserService) applicationContext.getBean("userService");
11          // 3.调用实例中的 login()方法
12          userService.login();
13      }
14  }
```

此时运行结果如图 1.6 所示。从图中可以看出，使用 Spring 容器通过 UserService 实现类中的 login()方法调用了 UserDao 实现类中的 login()方法，并输出了结果。这就是 Spring 容器属性 setter 注入的方式，也是实际开发中常用的一种方式。

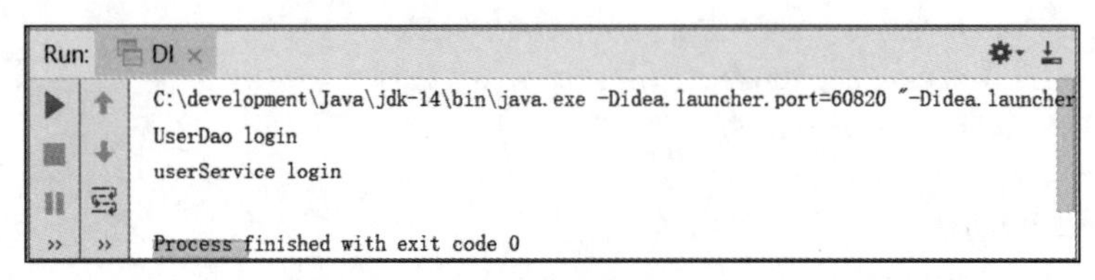

图 1.6 运行结果

为了方便读者理解，图 1.7 列出了整个项目的目录结构。

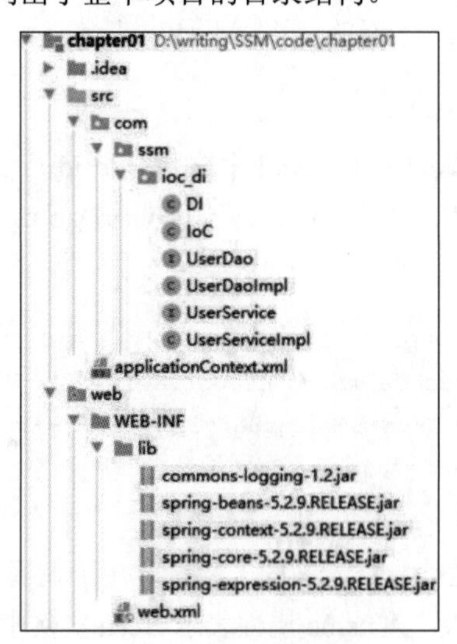

图 1.7 项目目录结构

第 2 章

Spring 中的 Bean

第 1 章讲解了 Spring 的控制反转/依赖注入并演示了它们的实现方法，本章将在第 1 章的基础上对 Spring 中的 Bean 的相关知识进行讲解。

本章主要涉及的知识点如下：

- Bean 的配置。
- Bean 的作用域。
- Bean 的装配方式。

2.1　Bean 的配置

Spring 如同一个工厂，用于生产和管理 Spring 容器中的 Bean。要使用这个工厂，需要开发者对 Spring 的配置文件进行配置。在实际开发中，最常采用 XML 格式的配置方式，即通过 XML 文件来注册并管理 Bean 之间的依赖关系。本节将使用 XML 文件的形式对 Bean 的属性和定义进行讲解。

在 Spring 中，XML 配置文件的根元素是<beans>，<beans>中可以包含多个<bean>子元素，每一个<bean>元素定义了一个 Bean，并描述了该 Bean 如何被装配到 Spring 容器中。<bean>元素中又包含多个属性和子元素，其常用的属性和子元素如表 2.1 所示。

表 2.1　<bean>元素的常用属性和子元素

属性或子元素名称	说　明
id	Bean 的唯一标识符，Spring 容器对 Bean 的配置、管理都通过该属性进行
name	Spring 容器通过此属性进行配置和管理，name 属性可以为 Bean 指定多个名称，每个名称之间用逗号或分号隔开

（续表）

属性或子元素名称	说 明
class	指定 Bean 的实现类，它必须使用类的全限定名
scope	用于设置 Bean 实例的作用域，其属性值有 singleton（单例）、prototype（原型）、request、session、global Session、application 和 websocket，默认值为 singleton
constructor-arg	<bean>元素的子元素，可以使用此元素传入构造参数进行实例化。该元素的 index 属性指定构造参数的序号（从 0 开始）；type 属性指定构造参数的类型，参数值可以通过 ref 属性或 value 属性直接指定，也可以通过 ref 或 value 子元素指定
property	<bean>元素的子元素，用于调用 Bean 实例中的 setter()方法完成属性赋值，从而完成依赖注入。该元素的 name 属性用于指定 Bean 实例中的相应属性名，ref 属性或 value 属性用于指定参数值
ref	<constructor-arg>、<property>等元素的属性或子元素，可以用于指定对 Bean 工厂中某个 Bean 实例的引用
value	<constructor-arg>、<property>等元素的属性或子元素，可以用于直接给定一个常量值
list	用于封装 List 或数组属性的依赖注入
set	用于封装 Set 类型属性的依赖注入
map	用于封装 Map 类型属性的依赖注入
entry	<map>元素的子元素，用于设置一个键-值对。其 key 属性指定字符串类型的键值，ref 属性或 value 属性直接指定其值，也可以通过 ref 或 value 子元素指定其值

表 2.1 中只介绍了<bean>元素的常用属性和子元素，实际上<bean>元素还有很多属性和子元素，读者可以到网上查阅相关资料。

在 Spring 的配置文件中，通常一个普通的 Bean 只需要定义 id（或 name）和 class 两个属性即可。定义 Bean 的方式如下：

```
<?xml version="1.0" encoding="UTF-8"?>
<beans xmlns="http://www.springframework.org/schema/beans"
      xmlns:xsi="http://www.w3.org/2001/XMLSchema-instance"
      xsi:schemaLocation="http://www.springframework.org/schema/beans
            http://www.springframework.org/schema/beans/spring-beans.xsd">
    <!-- 将指定类配置给 Spring，让 Spring 创建其对象的实例 -->
    <!—使用 id 属性定义 bean1，它对应的实现类为 com.ssm.Bean1 -->
    <bean id="bean1" class="com.ssm.Bean1" />
    <!--使用 name 属性定义 bean2，它对应的实现类为 com.ssm.Bean2 -->
    <bean name="bean2" class="com.ssm.Bean2" />
</beans>
```

在上述代码中，分别使用 id 属性和 name 属性定义了两个 Bean，并使用 class 元素指定其对应的实现类。

注意：如果在 Bean 中未指定 id 和 name，那么 Spring 会把 class 值当作 id 使用。

2.2　Bean 的作用域

通过 Spring 容器创建一个 Bean 的实例时，不仅可以完成 Bean 的实例化，还可以为 Bean 指定特定的作用域。本节将主要讲解 Bean 的作用域的相关知识。

2.2.1　作用域的种类

Spring 5.2 中为 Bean 的实例定义了 6 种作用域，如表 2.2 所示。其中，singleton 和 prototype 是常用的两种。

表 2.2　Bean 的作用域

作用域名称	说　明
singleton	使用 singleton 定义的 Bean 在 Spring 容器中将只有一个实例，也就是说，无论有多少个 Bean 引用它，都始终指向同一个对象，这也是 Spring 容器默认的作用域
prototype	每次通过 Spring 容器获取 prototype 定义的 Bean 时，容器都将创建一个新的 Bean 实例
request	在一次 HTTP 请求中，容器会返回该 Bean 的同一个实例，对不同的 HTTP 请求则会产生一个新的 Bean，而且该 Bean 仅在当前 HTTP 请求内有效
session	在一次 HTTP Session 中，容器会返回该 Bean 的同一个实例，对不同的 HTTP 请求则会产生一个新的 Bean，而且该 Bean 仅在当前 HTTP Session 内有效
application	为每个 ServletContext 对象创建一个实例，仅在 Web 相关的 ApplicationContext 中有效
websocket	为每个 websocket 对象创建一个实例，仅在 Web 相关的 ApplicationContext 中有效

2.2.2　singleton 作用域

singleton 是 Spring 容器默认的作用域，当 Bean 的作用域为 singleton 时，Spring 容器就只会存在一个共享的 Bean 实例，并且所有对 Bean 的请求，只要 id 与该 Bean 的 id 属性相匹配，就会返回同一个 Bean 的实例。singleton 作用域对于无会话状态的 Bean（如 Dao 组件、Service 组件）来说是最理想的选择。

在 Spring 配置文件中，Bean 的作用域是通过<bean>元素的 scope 属性来指定的，该属性值可以设置为 singleton、prototype、request、session、application、websocket 6 个值，分别表示表 2.2 中所示的 6 种作用域。如果要将作用域定义成 singleton，需将 scope 的属性值设置为 singleton，示例代码如下：

```
<bean id="scope"  class="com.ssm.scope.Scope"  scope="singleton" />
```

【示例 2-1】下面通过一个案例来进一步演示 singleton 作用域。

步骤 01 在 IntelliJ IDEA 中创建一个名为 chapter02 的 Web 项目，在该项目的 lib 目录中加入 Spring 支持和依赖的 JAR 包（在第 1 章相关内容基础上增加 spring-aop-5.2.9.RELEASE.jar 依赖包，并发布到类路径下）。

步骤02 在 chapter02 项目的 src 目录下创建一个 com.ssm.scope 包，在该包中创建 Scope 类，该类不需要写什么方法，如文件 2.1 所示。

文件 2.1 Scope.java

```
01  package com.ssm.scope;
02  public class Scope {
03  }
```

步骤03 在 src 目录下创建 Spring 的配置文件 applicationContext.xml，并在配置文件中创建一个 id 为 scope 的 Bean，通过 class 属性指定其对应的实现类为 Scope，如文件 2.2 所示。

文件 2.2 applicationContext.xml

```
01  <?xml version="1.0" encoding="UTF-8"?>
02  <beans xmlns="http://www.springframework.org/schema/beans"
03        xmlns:xsi="http://www.w3.org/2001/XMLSchema-instance"
04        xsi:schemaLocation="http://www.springframework.org/schema/beans
05              http://www.springframework.org/schema/beans/spring-beans.xsd">
06      <!-- 将指定类配置给 Spring，让 Spring 创建其对象的实例 -->
07      <bean id="scope" class="com.ssm.scope.Scope" />
08  </beans>
```

步骤04 在 com.ssm.scope 包中创建测试类 ScopeTest 来测试 singleton 作用域，如文件 2.3 所示。

文件 2.3 ScopeTest.java

```
01  package com.ssm.scope;
02  import org.springframework.context.ApplicationContext;
03  import org.springframework.context.support.ClassPathXmlApplicationContext;
04  public class ScopeTest {
05      public static void main(String[] args) {
06          // 1.初始化 Spring 容器，加载配置文件
07          ApplicationContext applicationContext =
08                  new ClassPathXmlApplicationContext("applicationContext.xml");
09          // 2.输出获得的实例
10          System.out.println(applicationContext.getBean("scope"));
11          System.out.println(applicationContext.getBean("scope"));
12      }
13  }
```

执行程序后，控制台的输出结果如图 2.1 所示。从图中可以看出，两次输出的结果相同，这说明 Spring 容器只创建了一个 Scope 类的实例。

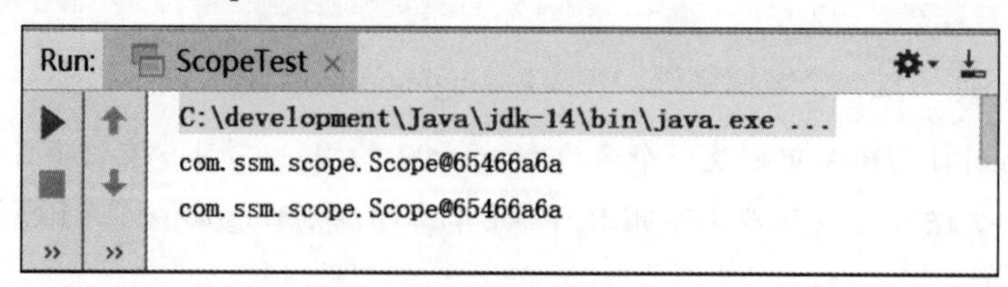

图 2.1 运行结果

注意：如果不设置 scope="singleton"，其输出结果也是一个实例，因为 Spring 容器默认的作用域就是 singleton。

2.2.3　prototype 作用域

对需要保持会话状态的 Bean 应用使用 prototype 作用域。在使用 prototype 作用域时，Spring 容器会为每个对该 Bean 的请求都创建一个新的实例。

如果要将 Bean 定义为 prototype 作用域，那么只需在配置文件中将<bean>元素的 scope 属性值设置为 prototype 即可，示例代码如下：

```
<bean id="scope" class="com.ssm.scope.Scope" scope="prototype"/>
```

将 2.2.2 节中的配置文件中的相应代码更改为上述代码形式后，再次运行测试类 ScopeTest，控制台的输出结果如图 2.2 所示。从图中可以看到，两次输出的 Bean 实例并不相同，这说明在 prototype 作用域下创建了两个不同的 Scope 实例。

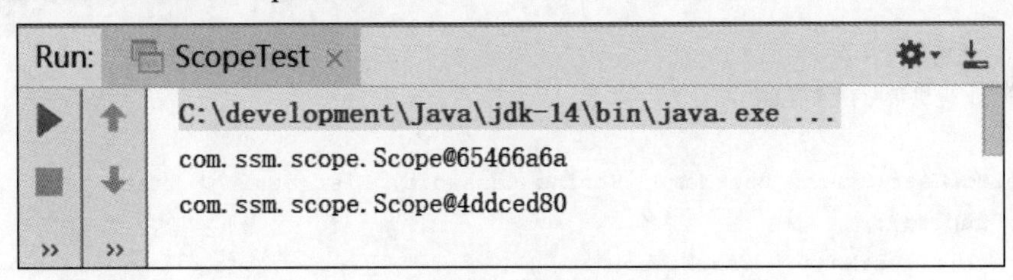

图 2.2　运行结果

2.3　Bean 的装配方式

Bean 的装配可以理解为依赖关系注入，Bean 的装配方式即 Bean 依赖注入的方式。Spring 容器支持多种形式的 Bean 装配方式，如基于 XML 的装配、基于 Annotation（注解）的装配和自动装配等。本节主要讲解这 3 种装配方式的使用。

2.3.1　基于 XML 的装配

Spring 提供了两种基于 XML 的装配方式：设值注入（Setter Injection）和构造注入（Constructor Injection）。下面讲解如何在 XML 配置文件中使用这两种注入方式来实现基于 XML 的装配。

在 Spring 实例化 Bean 的过程中，Spring 首先会调用 Bean 的默认构造方法来实例化 Bean 对象，然后通过反射的方式调用 setter()方法来注入属性值，因此设值注入要求一个 Bean 必须满足以下两点要求：

- Bean 类必须提供一个默认的无参构造方法。
- Bean 类必须为需要注入的属性提供对应的 setter()方法。

使用设值注入时，在 Spring 配置文件中需要使用<bean>元素的子元素<property>来为每个属性注入值；而使用构造注入时，在配置文件中需要使用<bean>元素的子元素<constructor-arg>来定义构造方法的参数，可以使用其 value 属性（或子元素）来设置该参数的值。

【示例 2-2】下面通过一个案例来演示基于 XML 方式的 Bean 的装配。

步骤 01 在项目 chapter02 的 src 目录下创建一个 com.ssm.assemble 包，在该包中创建 User 类，并在类中定义 userName、password 和 list 集合 3 个属性及对应的 setter()方法，如文件 2.4 所示。

文件 2.4 User.java

```
01  package com.ssm.assemble;
02  import java.util.List;
03  public class User {
04      private String userName;
05      private String password;
06      private List<String> list;
07      /**
08       * 1.使用构造注入
09       * 1.1 提供带所有参数的构造方法
10       */
11      public User(String userName, String password, List<String> list) {
12          super();
13          this.userName = userName;
14          this.password = password;
15          this.list = list;
16      }
17      @Override
18      public String toString() {
19          return "User [userName=" + userName + ",password="+password +",list="+list + "]";
20      }
21      /**
22       * 2.使用设值注入
23       * 2.1 提供默认空参构造方法
24       * 2.2 为所有属性提供 setter()方法
25       */
26      public User() {
27          super();
28      }
29      public void setUserName(String userName) {
30          this.userName = userName;
31      }
32      public void setPassword(String password) {
33          this.password = password;
34      }
35      public void setList(List<String> list) {
36          this.list = list;
```

```
37        }
38  }
```

由于要使用构造注入，因此需要编写有参（第 11~16 行）和无参（第 26~28 行）的构造方法。

步骤 02 在 Spring 的配置文件 applicationContext.xml（文件 2.2）中，增加通过构造注入和设值注入的方式装配 User 实例的两个 Bean，代码如下：

```
<bean id="user1" class="com.ssm.assemble.User">
    <constructor-arg index="0" value="zhangsan" />
    <constructor-arg index="1" value="111111" />
    <constructor-arg index="2">
        <list>
            <value>"constructorValue1"</value>
            <value>"constructorValue2"</value>
        </list>
    </constructor-arg>
</bean>
<bean id="user2" class="com.ssm.assemble.User">
    <property name="userName" value="lisi"></property>
    <property name="password" value="222222"></property>
    <property name="list">
        <list>
            <value>"listValue1"</value>
            <value>"listValue2"</value>
        </list>
    </property>
</bean>
```

在上述代码中，首先使用<constructor-arg>元素定义构造方法的参数，其中 index 属性表示索引（从 0 开始），value 属性用于设置注入的值，子元素<list>为 User 类中对应的 list 集合属性注入值。然后又使用设值注入方法装配 User 类的实例，其中<property>元素用于调用 Bean 实例中的 setter()方法完成属性赋值，从而完成依赖注入，而子元素<list>同样为 User 类中对应的 list 集合属性注入值。

步骤 03 在 com.ssm.assemble 包中创建测试类 XmlAssembleTest，在类中分别获取并输出配置文件中的 user1 和 user2 实例，如文件 2.5 所示。

文件 2.5　XmlAssembleTest.java

```
01  package com.ssm.assemble;
02  import org.springframework.context.ApplicationContext;
03  import org.springframework.context.support.ClassPathXmlApplicationContext;
04  public class XmlAssembleTest {
05      public static void main(String[] args) {
06          // 1.初始化 Spring 容器，加载配置文件
07          ApplicationContext applicationContext
08                  = new ClassPathXmlApplicationContext("applicationContext.xml");
09          // 2.输出获得的实例
```

```
10          System.out.println(applicationContext.getBean("user1"));
11          System.out.println(applicationContext.getBean("user2"));
12      }
13  }
```

执行程序后，控制台输出结果如图 2.3 所示。从图中可以看出，已经成功地使用基于 XML 装配的构造注入和设值注入两种方式装配了 User 实例。

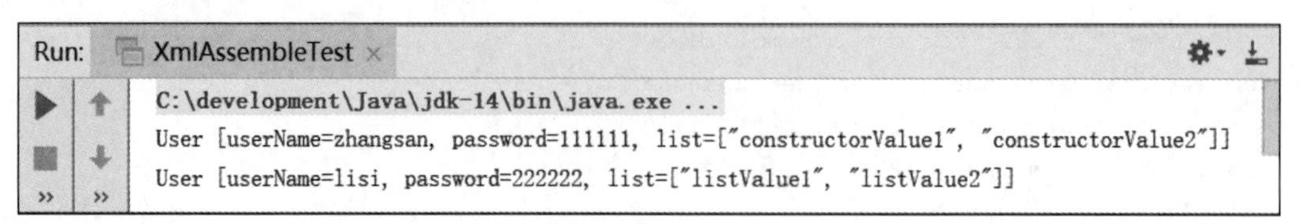

图 2.3 运行结果

2.3.2 基于 Annotation 的装配

在 Spring 中，尽管使用 XML 配置文件可以实现 Bean 的装配工作，但如果应用中有很多 Bean，就会导致 XML 配置文件过于臃肿，给以后的维护和升级工作带来一定的困难。为此，Spring 提供了对 Annotation 技术的全面支持。

Spring 中定义了一系列的注解，常用的注解如表 2.3 所示。

表 2.3 Spring 的常用注解

注解名称	说 明
@Component	可以使用此注解描述 Spring 中的 Bean，但它是一个泛化的概念，仅仅表示一个组件（Bean），并且可以作用在任何层次。使用时只需将该注解标注在相应类上即可
@Repository	用于将数据访问层（DAO 层）的类标识为 Spring 中的 Bean，其功能与@Component 相同
@Service	通常作用在业务层（Service 层），用于将业务层的类标识为 Spring 中的 Bean，其功能与@Component 相同
@Controller	通常作用在控制层（如 Spring MVC 的 Controller），用于将控制层的类标识为 Spring 中的 Bean，其功能与@Component 相同
@Autowired	用于对 Bean 的属性变量、属性的 setter()方法及构造方法进行标注，配合对应的注解处理器完成 Bean 的自动配置工作。默认按照 Bean 的类型进行装配
@Resource	其作用与@Autowired 一样，区别在于@Autowired 默认按 Bean 的类型装配，而@Resource 默认按照 Bean 的实例名称进行装配。@Resource 中有两个重要属性：name 和 type。Spring 将 name 属性解析为 Bean 实例名称，将 type 属性解析为 Bean 实例类型。若指定 name 属性，则按实例名称进行装配；若指定 type 属性，则按 Bean 的类型进行装配；若都不指定，则先按 Bean 的实例名称进行装配，不能匹配时再按照 Bean 的类型进行装配；若都无法匹配，则抛出 NoSuchBeanDefinitionException 异常
@Qualifier	与@Autowired 注解配合使用，会将默认的按 Bean 的类型装配修改为按 Bean 的实例名称装配，Bean 的实例名称由@Qualifier 注解的参数指定

注意：在表 2.3 所示的几个注解中，虽然@Repository、@Service 和@Controller 的功能与@Component 的功能相同，但为了使标注类本身用途更加清晰，建议在实际开发中使用@Repository、

@Service 和@Controller 分别对实现类进行标注。

【示例 2-3】下面通过一个案例来演示如何通过注解来装配 Bean。

步骤 01 在 chapter02 项目的 src 目录下创建一个 com.ssm.annotation 包，在该包中创建接口 UserDao，并在接口中定义一个 save()方法，如文件 2.6 所示。

文件 2.6 UserDao.java

```
01 package com.ssm.annotation;
02 public interface UserDao {
03     public void save();
04 }
```

步骤 02 在 com.ssm.annotation 包中创建 UserDao 接口的实现类 UserDaoImpl，该类需要实现接口中的 save()方法，如文件 2.7 所示。

文件 2.7 UserDaoImpl.java

```
01 package com.ssm.annotation;
02 import org.springframework.stereotype.Repository;
03 // 使用@Repository 注解将 UserDaoImpl 类标识为 Spring 中的 Bean
04 @Repository("userDao")
05 public class UserDaoImpl implements UserDao {
06     public void save() {
07         System. out. println("userDao.save()");
08     }
09 }
```

在文件 2.7 中，首先使用@Repository 注解将 UserDaoImpl 类标识为 Spring 中的 Bean，其写法相当于配置文件中<bean id="userDao" class=" com.ssm.annotation.UserDaoImpl />的写法。然后在 save()方法中输出一句话，用于验证是否成功调用了该方法。

步骤 03 在 com.ssm.annotation 包中创建接口 UserService，在接口中同样定义一个 save()方法，如文件 2.8 所示。

文件 2.8 UserService.java

```
01 package com.ssm.annotation;
02 public interface UserService {
03     public void save();
04 }
```

步骤 04 在 com.ssm.annotation 包中创建 UserService 接口的实现类 UserServiceImpl，该类需要实现接口中的 save()方法，如文件 2.9 所示。

文件 2.9 UserServiceImpl.java

```
01 package com.ssm.annotation;
02 import javax.annotation.Resource;
03 import org.springframework.stereotype.Service;
```

```
04  // 使用@Service 注解将 UserServiceImpl 类标识为 Spring 中的 Bean
05  @Service("userService")
06  public class UserServiceImpl implements UserService {
07  // 使用@Resource 注解注入
08      @Resource(name="userDao")
09      private UserDao userDao;
10      public void save() {
11          this.userDao. save();
12          System.out.println("执行 userService.save()");
13      }
14  }
```

在文件 2.9 中，首先使用@Service 注解将 UserServiceImpl 类标识为 Spring 中的 Bean，这相当于配置文件中<bean id= "userService" class="com.ssm.annotation.UserServiceImpl"/>的写法；然后将@Resource 注解标注在属性 userDao 上，这相当于配置文件中<property name="userDao" ref=" userDao"/>的写法；最后在该类的 save()方法中调用 userDao 中的 save()方法，并输出一句话。

步骤05 在 com.ssm.annotation 包中创建控制器类 UserController，如文件 2.10 所示。

文件 2.10 UserController.java

```
01  package com.ssm.annotation;
02  import javax.annotation.Resource;
03  import org.springframework.stereotype.Controller;
04  // 使用@Controller 注解将 UserController 类标识为 Spring 中的 Bean
05  @Controller("userController")
06  public class UserController {
07  // 使用@Resource 注解注入
08      @Resource(name="userService")
09      private UserService userService;
10      public void save(){
11          this.userService.save();
12          System.out.println("运行 userController.save()");
13      }
14  }
```

在文件 2.10 中，首先使用@Controller 注解标注了 UserController 类，这相当于在配置文件中编写<bean id=" userController" class=" com.ssm.annotation.UserController"/>；然后将@Resource 注解标注在 userService 属性上，这相当于在配置文件中编写<property name="userService" ref=" userService"/>；最后在该类的 save()方法中调用 userService 中的 save()方法，并输出一句话。

步骤06 在 com.ssm.annotation 包中创建配置文件 beans1.xml，在配置文件中编写基于 Annotation 装配的代码，如文件 2.11 所示。

文件 2.11 beans1.xml

```
01  <?xml version="1.0" encoding="UTF-8"?>
02  <beans xmlns="http://www.springframework.org/schema/beans"
```

```
03     xmlns:xsi="http://www.w3.org/2001/XMLSchema-instance"
04     xmlns:context="http://www.springframework.org/schema/context"
05     xsi:schemaLocation="http://www.springframework.org/schema/beans
06     http://www.springframework.org/schema/beans/spring-beans.xsd
07     http://www.springframework.org/schema/context
08     http://www.springframework.org/schema/context/spring-context.xsd">
09     <!-- 使用 context 命名空间在配置文件中开启相应的注解处理器 -->
10     <context:annotation-config />
11     <!-- 分别定义 3 个 Bean 实例 -->
12     <bean id="userDao" class="com.ssm.annotation.UserDaoImpl" />
13     <bean id="userService" class="com.ssm.annotation.UserServiceImpl" />
14     <bean id="userController" class="com.ssm.annotation.UserController" />
15  </beans>
```

从上述代码可以看出，文件 2.11 与之前的配置文件有很大不同。首先，在<beans>元素中增加的第 04、07 和 08 行代码中包含了 context 的约束信息；然后通过配置<context: annotation-config/>来开启注解处理器；最后分别定义了 3 个 Bean 对应的 3 个实例。与 XML 配置方式有所不同的是，这里不再需要配置子元素<property>。

上述 Spring 配置文件中的注解方式虽然较大程度地简化了 XML 文件中 Bean 的配置，但仍需在 Spring 配置文件中一一配置相应的 Bean，为此 Spring 注解提供了另一种高效的注解配置方式——对包路径下的所有 Bean 文件进行扫描，其配置方式如下：

```
<context:component- scan base-package="Bean 所在的包路径"/>
```

所以可以将文件 2.11 中的第 09~14 行代码进行如下替换：

```
<!--使用 context 命名空间通知 Spring 扫描指定包下所有 Bean 类，进行注解解析-->
<context:component-scan base-package="com.ssm.annotation"/>
```

注意： Spring 4.0 以上版本在使用上面的代码对指定包中的注解进行扫描前，需要先向项目中导入 Spring AOP 的 JAR 包 spring-aop-5.2.9.RELEASE.jar，否则程序在运行时会报出“java.lang.NoClassDefFoundError:org/springframework/aop/TargetSource”的错误。

步骤 07 在 com.ssm.annotation 包中创建测试类 AnnotationAssembleTest，在类中编写测试方法并定义配置文件的路径，然后通过 Spring 容器加载配置文件并获取 UserController 实例，最后调用实例中的 save()方法，如文件 2.12 所示。

文件 2.12 AnnotationAssembleTest.java

```
01  package com.ssm.annotation;
02  import org.springframework.context.ApplicationContext;
03  import org.springframework.context.support.ClassPathXmlApplicationContext;
04  public class AnnotationAssembleTest {
05      private static ApplicationContext applicationContext;
06      public static void main(String[] args) {
07          // 定义配置文件路径
08          String xmlPath = "com/ssm/annotation/beans1.xml";
```

```
09          applicationContext = new ClassPathXmlApplicationContext(xmlPath);
10          // 获取 UserController 实例
11          UserController userController =
12                  (UserController) applicationContext.getBean("userController");
13          // 调用 UserController 中的 save()方法
14          userController.save();
15      }
16  }
```

执行程序后，控制台的输出结果如图 2.4 所示。从图中可以看到，Spring 容器已成功获取了 UserController 实例，并通过调用实例中的方法输出了各层中的语句，这说明已成功实现了基于 Annotation 来装配 Bean 实例。

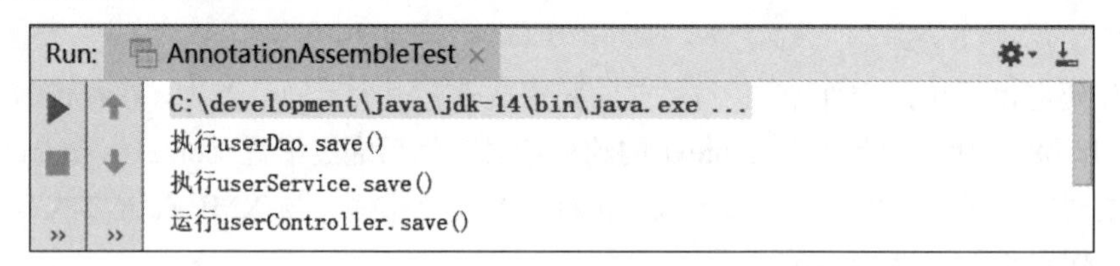

图 2.4 运行结果

注意：上述案例中，使用@Autowired 注解替换@Resource 注解也可以达到同样的效果。

2.3.3 自动装配

虽然使用注解的方式装配 Bean 在一定程度上减少了配置文件中的代码量，但是也有企业项目是没有使用注解方式开发的，那么有没有什么办法既可以减少代码量，又能够实现 Bean 的装配呢？答案肯定是有的。Spring 的<bean>元素中包含一个 autowire 属性，我们可以通过设置 autowire 的属性值来自动装配 Bean。所谓自动装配，就是将一个 Bean 自动注入其他 Bean 的 Property 中。

autowire 属性有 5 个值，其值及说明如表 2.4 所示。

表 2.4 <bean>元素的 autowire 属性值及说明

属 性 值	说 明
default（默认值）	由<bean>的上级标签<beans>的 default-autowire 属性值确定。例如 <beans default-autowire=" byName">，该<bean>元素中的 autowire 属性对应的属性值为 byName
byName	根据属性的名称自动装配。容器将根据名称查找与属性完全一致的 Bean，并将其属性自动装配
byType	根据属性的数据类型（Type）自动装配，如果一个 Bean 的数据类型兼容另一个 Bean 中属性的数据类型，则自动装配
constructor	根据构造函数参数的数据类型进行 byType 模式的自动装配
no	在默认情况下，不使用自动装配，Bean 依赖必须通过 ref 元素定义

【示例 2-4】下面通过修改【示例 2-3】来演示如何使用自动装配。

步骤 01 修改【示例 2-3】中的文件 2.9（UserServiceImpl.java）和文件 2.10（UserController.java），

分别在这两个文件中增加类属性的 setter()方法。

步骤 02 修改【示例 2-3】中的文件 2.11（beans1.xml），将它修改成自动装配形式，如文件 2.13 所示。

文件 2.13　beans2.xml

```
01  <?xml version="1.0" encoding="UTF-8"?>
02  <beans xmlns="http://www.springframework.org/schema/beans"
03     xmlns:xsi="http://www.w3.org/2001/XMLSchema-instance"
04     xmlns:context="http://www.springframework.org/schema/context"
05     xsi:schemaLocation="http://www.springframework.org/schema/beans
06     http://www.springframework.org/schema/beans/spring-beans.xsd
07     http://www.springframework.org/schema/context
08     http://www.springframework.org/schema/context/spring-context.xsd">
09     <!-- 使用 bean 元素的 autowire 属性完成自动装配 -->
10     <bean id="userDao" class="com.ssm.annotation.UserDaoImpl" />
11     <bean id="userService"
12          class="com.ssm.annotation.UserServiceImpl" autowire="byName" />
13     <bean id="userController"
14          class="com.ssm.annotation.UserController"  autowire="byName" />
15  </beans>
```

在上述配置文件中，用于配置 userService 和 userController 的<bean>元素中除了 id 和 class 属性外，还增加了 autowire 属性，并将它的属性值设置为 byName。在默认情况下，在配置文件中需要通过 ref 来装配 Bean，但设置了 autowire="byName"后，Spring 会自动寻找 userServiceBean 中的属性，并将它的属性名称与配置文件中定义的 Bean 做匹配。由于 UserServiceImpl 中定义了 userDao 属性及其 setter()方法，这与配置文件中 id 为 userDao 的 Bean 相匹配，因此 Spring 会自动地将 id 为 userDao 的 Bean 装配到 id 为 userService 的 Bean 中。

执行程序后，控制台的输出结果与图 2.4 所示的结果相同。

第 3 章

Spring AOP

Spring 的 AOP 模块是 Spring 框架体系结构中十分重要的内容，它提供了面向切面编程的实现。本章将对 Spring AOP 的相关知识进行详细讲解。

本章主要涉及的知识点如下：

- AOP 的概念和作用。
- Aspect 开发。

3.1 Spring AOP 简介

本节主要介绍 AOP 的概念和作用，以及 AOP 中的相关术语，旨在让读者熟悉另一种编程方式，为后续学习打下基础。

3.1.1 什么是 AOP

AOP 的全称是 Aspect-Oriented Programming，即面向切面编程（也称面向方面编程），是面向对象编程（OOP）的一种补充，目前已成为一种比较成熟的编程方式。

在传统的业务处理代码中，通常都会进行事务处理、日志记录等操作。虽然使用 OOP 可以通过组合或者继承的方式来达到代码的重用，但如果要实现某个功能（如日志记录），相同的代码仍然会分散到各个方法中。这样，如果想要关闭某个功能，或者对它进行修改，就必须修改所有相关的方法。这不但增加了开发人员的工作量，而且提高了代码的出错率。

为了解决这一问题，AOP 思想随之产生。AOP 采取横向抽取机制，将分散在各个方法中的重复代码提取出来，然后在程序编译或运行时，再将这些提取出来的代码应用到需要执行的地方。这种采用横向抽取机制的方式，使用传统的 OOP 编程思想显然是无法办到的，因为 OOP 只能实现父子关系的纵向重用。虽然 AOP 是一种新的编程思想，但却不是 OOP 的替代品，它是 OOP 的延伸和补充。

在 AOP 编程思想中，通过 Aspect（切面）可以分别在不同类的方法中加入事务、日志、权限和异常等功能。

AOP 的使用使开发人员在编写业务逻辑时可以专心于核心业务，而不用过多地关注其他业务逻辑的实现，这不但提高了开发效率，而且增强了代码的可维护性。

目前流行的 AOP 框架有两个，分别为 Spring AOP 和 AspectJ。Spring AOP 使用纯 Java 实现，不需要专门的编译过程和类加载器，在运行期间通过代理方式向目标类植入增强的代码。AspectJ 是一个基于 Java 语言的 AOP 框架，从 Spring 2.0 开始，Spring AOP 引入了对 AspectJ 的支持，AspectJ 扩展了 Java 语言，提供了一个专门的编译器，在编译时提供横向代码的植入。

3.1.2 AOP 术语

在学习使用 AOP 之前，首先要了解一下 AOP 的专业术语。这些术语包括 Aspect、Joinpoint、Pointcut、Advice、Target Object、Proxy 和 Weaving，对于这些专业术语的具体解释如下：

- Aspect（切面）：在实际应用中，切面通常是指封装的用于横向插入系统功能（如事务、日志等）的类，该类要被 Spring 容器识别为切面，需要在配置文件中通过<bean>元素指定。
- Joinpoint（连接点）：在程序执行过程中的某个阶段点，它实际上是对象的一个操作，例如方法的调用或异常的抛出。在 Spring AOP 中，连接点就是指方法的调用。
- Pointcut（切入点）：是指切面与程序流程的交叉点，即那些需要处理的连接点。通常在程序中，切入点指的是类或者方法名，如某个通知要应用到所有以 add 开头的方法中，那么所有满足这一规则的方法都是切入点。
- Advice（通知增强处理）：AOP 框架在特定的切入点执行增强处理，即在定义好的切入点处所要执行的程序代码。可以将它理解为切面类中的方法，它是切面的具体实现。
- Target Object（目标对象）：是指所有被通知的对象，也称为被增强对象。如果 AOP 框架采用的是动态的 AOP 实现，那么该对象就是一个被代理对象。
- Proxy（代理）：将通知应用到目标对象之后，被动态创建的对象。
- Weaving（织入）：将切面代码插入目标对象上，从而生成代理对象的过程。

3.2 AspectJ 开发

AspectJ 是一个基于 Java 语言的 AOP 框架，它提供了强大的 AOP 功能。Spring 2.0 以后，Spring AOP 引入了对 AspectJ 的支持，并允许直接使用 AspectJ 进行编程，而 Spring 自身的 AOP API 也尽量与 AspectJ 保持一致。新版本的 Spring 框架建议使用 AspectJ 来开发 AOP。

使用 AspectJ 实现 AOP 有两种方式：一种是基于 XML 的声明式 AspectJ，另一种是基于注解的声明式 AspectJ。本节将对这两种 AspectJ 的开发方式进行讲解。

3.2.1 基于 XML 的声明式 AspectJ

基于 XML 的声明式 AspectJ 是指通过 XML 文件来定义切面、切入点及通知，所有的切面、切入点和通知都必须定义在<aop:config>元素内。Spring 配置文件中的<bean>元素下可以包含多个<aop:config>元素，一个<aop:config>元素中又可以包含属性和子元素，其子元素包括<aop:pointcut>、<aop:advisor>和<aop:aspect>。在配置时，这 3 个子元素必须按照此顺序来定义。在<aop:aspect>元素下，同样包含属性和多个子元素，通过使用<aop:aspect>元素及其子元素就可以在 XML 文件中配置切面、切入点和通知。常用元素的配置代码如下：

```
<!-- 定义切面 Bean -->
<bean id="myAspect" class="com.smm. aspectj.xmI.MyAspect />
<aop:config>
    <!-- 1.配置切面 -->
    <aop:aspect id="aspect" ref="myAspect">
        <!-- 2.配置切入点 -->
        <aop:pointcut expression="execution(* com.ssm.aspectj.*.*(..))" id="myPointCut"/>
        <!-- 3.配置通知 -->
            <!-- 前置通知 -->
            <aop:before method="myBefore" pointcut-ref="myPointCut" />
            <!--后置通知-->
            <aop:after-returning method="myAfterReturning"
                    pointcut-ref="myPointCut" returning="returnVal" />
            <!--环绕通知 -->
            <aop:around method="myAround" pointcut-ref="myPointCut" />
            <!--异常通知 -->
            <aop:after-throwing method="myAfterThrowing"
                    pointcut-ref="myPointCut" throwing="e" />
            <!--最终通知 -->
            <aop:after method="myAfter" pointcut-ref="myPointCut" />
    </aop:aspect>
</aop:config>
```

为了让读者能够快速掌握上述代码中的配置信息，下面对上述代码的配置内容进行详细讲解。

1. 配置切面

在 Spring 的配置文件中，配置切面使用的是<aop:aspect>元素，该元素会将一个已定义好的 Spring Bean 转换成切面 Bean，所以要在配置文件中先定义一个普通的 Spring Bean（如上述代码中定义的 myAspect）。定义完成后，通过<aop:aspect>元素的 ref 属性即可引用该 Bean。

配置<aop:aspect>元素时，通常会指定 id 和 ref 两个属性，这两个属性及其描述如表 3.1 所示。

表 3.1　<aop:aspect>元素的属性及其描述

属性名称	描　述
id	用于定义该切面的唯一标识名称
ref	用于引用普通的 Spring Bean

2. 配置切入点

在 Spring 的配置文件中，切入点是通过<aop:pointcut>元素来定义的。当<aop:pointcut>元素作为<aop:config>元素的子元素时，表示该切入点是全局切入点，可以被多个切面共享；当<aop:pointcut>元素作为<aop:aspect>元素的子元素时，表示该切入点只对当前切面有效。在定义<aop:pointcut>元素时，通常会指定 id 和 expression 两个属性，这两个属性及其描述如表 3.2 所示。

表 3.2　<aop:pointcut>元素的属性及其描述

属性名称	描　述
id	用于定义切入点的唯一标识名称
expression	用于指定切入点关联的切入点表达式

在上述配置代码片段中，execution(* com.ssm.jdk.*.*(..))就是定义的切入点表达式，该切入点表达式的意思是匹配 com.ssm.jdk 包中任意类的任意方法的执行。其中 execution 是表达式的主体；第 1 个“*”表示的是返回类型，使用*代表所有类型；com.ssm.jdk 表示的是需要拦截的包名，后面第 2 个“*”表示的是类名，使用*代表所有的类；第 3 个“*”表示的是方法名，使用*表示所有方法；后面的()表示方法的参数，其中的“..”表示任意参数。需要注意的是，第 1 个“*”与包名之间有一个空格。

上面示例中定义的切入点表达式只是开发中常用的配置方式，而 Spring AOP 中切入点表达式的基本格式如下:

```
execution(modifiers-pattern? ret-type-pattern declaring-type-pattern?name-pattern(param-
pattern) throws-pattern?)
```

各部分说明如下：

- modifiers-pattern：表示定义的目标方法的访问修饰符，如 public、private 等。
- ret-type-pattern：表示定义的目标方法的返回值类型，如 void、String 等。
- declaring-type-pattern：表示定义的目标方法的类路径，如 com.ssm.jdk.UserDaoImpl。
- name-pattern：表示具体需要被代理的目标方法，如 add()方法。
- param-pattern：表示需要被代理的目标方法包含的参数，本章示例中目标方法参数都为空。
- throws-pattern：表示需要被代理的目标方法抛出的异常类型。

提示：带有问号（?）的部分（如 modifiers-pattern、declaring-type-pattern 和 throws-pattern）表示可选配置项，其他部分属于必须配置项。

如果想要了解更多切入点表达式的配置信息，读者可以参考 Spring 官方文档的切入点声明部分（Declaring a pointcut）。

3. 配置通知

在配置代码中，分别使用<aop:aspect>的子元素配置了 5 种常用通知，这些子元素不支持再使用子元素，但可以指定一些属性，如表 3.3 所示。

表 3.3 通知的常用属性及其描述

属性名称	描述
pointcut	用于指定一个切入点表达式，Spring 将在匹配该表达式的连接点时植入通知
pointcut-ref	指定一个已经存在的切入点名称，如配置代码中的 myPointcut。通常 pointcut 和 pointcut--ref 两个属性只需要使用其中之一即可
method	指定一个方法名，指定将切面 Bean 中的该方法转换为增强处理
throwing	只对<after-throwing>元素有效，用于指定一个形参名，异常通知方法可以通过该形参访问目标方法所抛出的异常
returning	只对<after-returning>元素有效，用于指定一个形参名，后置通知方法可以通过该形参访问目标方法的返回值

【示例 3-1】在了解了如何在 XML 中配置切面、切入点和通知后，接下来通过一个案例来演示如何在 Spring 中使用基于 XML 的声明式 AspectJ，具体实现步骤如下：

步骤 01 创建一个名为 chapter03 的动态 Web 项目，导入 Spring 构架所需的 JAR 包到项目的 lib 目录中，并发布到类路径下。同时，导入 AspectJ 框架相关的 JAR 包，具体说明如下：

- spring- aspects-5.2.9.RELEASE.jar：Spring 为 AspectJ 提供的实现，Spring 的包中已经提供。
- aspectjweaver-1.9.9.1.jar：是 AspectJ 框架所提供的规范，读者可以通过网址 http://mvnrepository.com/artifact/org.aspectj/aspectjweaver/1.9.9.1 下载。

步骤 02 在 chapter03 项目的 src 目录下创建一个 com.ssm.aspectj 包，在该包中创建接口 UserDao，并在接口中编写添加和删除用户的方法，如文件 3.1 所示。

文件 3.1 UserDao.java

```
01  package com.ssm.aspectj;
02  public interface UserDao {
03      //添加用户方法
04      public void addUser();
05      //删除用户方法
06      public void deleteUser();
07  }
```

步骤 03 在 com.ssm.aspectj 包中创建 UserDao 接口的实现类 UserDaoImpl，该类需要实现接口中的方法，如文件 3.2 所示。

文件 3.2 UserDaoImpl.java

```
01  package com.ssm.aspectj;
02  public class UserDaoImpl implements UserDao {
03      public void addUser() {
04          System. out. println("添加用户");
```

```
05      }
06      public void deleteUser() {
07          System.out.println("删除用户");
08      }
09  }
```

本案例中将实现类 UserDaoImpl 作为目标类，对其中的方法进行增强处理。

步骤04 在 chapter03 项目的 src 目录下创建一个 com.ssm.aspectj.xml 包，在该包中创建切面类 MyAspect，并在类中分别定义不同类型的通知，如文件 3.3 所示。

文件 3.3 MyAspect.java

```
01  package com.ssm.aspectj.xml;
02  import org.aspectj.lang.JoinPoint;
03  import org.aspectj.lang.ProceedingJoinPoint;
04  /**
05   * 切面类，在此类中编写通知
06   */
07  public class MyAspect {
08      //前置通知
09      public void myBefore(JoinPoint joinPoint){
10          System.out.print("前置通知：模拟执行权限检查...，");
11          System.out.print("目标类是："+joinPoint.getTarget());
12          System.out.println("，被植入增强处理的目标方法为："+
13                      joinPoint.getSignature().getName());
14      }
15      //后置通知
16      public void myAfterReturning(JoinPoint joinPoint) {
17          System.out.print("后置通知：模拟记录日志...，");
18          System.out.println("被植入增强处理的目标方法为：" +
19                      joinPoint.getSignature().getName());
20      }
21      /**
22       * 环绕通知
23       * ProceedingJoinPoint 是 JoinPoint 的子接口，表示可执行目标方法
24       * 1.必须是 Object 类型的返回值
25       * 2.必须接收一个参数，类型为 ProceedingJoinPoint
26       * 3.必须是 throws Throwable
27       */
28      public Object myAround(ProceedingJoinPoint proceedingJoinPoint) throws Throwable{
29          //开始
30          System.out.println("环绕开始：执行目标方法之前，模拟开启事务...，");
31          //执行当前目标方法
32          Object obj=proceedingJoinPoint.proceed();
33          //结束
34          System.out.println("环绕结束：执行目标方法之后，模拟关闭事务...，");
35          return obj;
36      }
37      //异常通知
38      public void myAfterThrowing(JoinPoint joinPoint,Throwable e){
```

```
39            System.out.println("异常通知：出错了"+e.getMessage());
40        }
41        //最终通知
42        public void myAfter(){
43            System.out.println("最终通知：模拟方法结束后释放资源...");
44        }
45    }
```

在文件 3.3 中，分别自定义了 5 种不同类型的通知，在通知中使用了 JoinPoint 接口及其子接口 ProceedingJoinPoint 作为参数，来获得目标对象的类名、目标方法名和目标方法参数等。

注意：环绕通知必须接收一个类型为 ProceedingJoinPoint 的参数，返回值也必须是 Object 类型，且必须抛出异常。异常通知中可以传入 Throwable 类型的参数来输出异常信息。

步骤 05 在 com.ssm.aspectj.xml 包中创建配置文件 applicationContext.xml，并编写相关配置，如文件 3.4 所示。

文件 3.4 applicationContext.xml

```
01 <?xml version="1.0" encoding="UTF-8"?>
02 <beans xmlns="http://www.springframework.org/schema/beans"
03     xmlns:xsi="http://www.w3.org/2001/XMLSchema-instance"
04     xmlns:aop="http://www.springframework.org/schema/aop"
05     xsi:schemaLocation="http://www.springframework.org/schema/beans
06       http://www.springframework.org/schema/beans/spring-beans.xsd
07       http://www.springframework.org/schema/aop
08       http://www.springframework.org/schema/aop/spring-aop.xsd">
09     <!-- 1. 目标类 -->
10     <bean id="userDao" class="com.ssm.aspectj.UserDaoImpl" />
11     <!-- 2. 切面 -->
12     <bean id="myAspect" class="com.ssm.aspectj.xml.MyAspect" />
13     <!-- 3. aop 编程 -->
14     <aop:config>
15         <!-- 3.1 配置切面 -->
16         <aop:aspect id="aspect" ref="myAspect">
17             <!-- 3.2 配置切入点 -->
18             <aop:pointcut expression="execution(* com.ssm.aspectj.*.*(..))"id="myPointCut"/>
19             <!-- 3.3 配置通知 -->
20             <!-- 前置通知 -->
21             <aop:before method="myBefore" pointcut-ref="myPointCut" />
22             <!--后置通知-->
23             <aop:after-returning method="myAfterReturning" pointcut-ref="myPointCut"
24                returning="joinPoint"/>
25             <!--环绕通知 -->
26             <aop:around method="myAround" pointcut-ref="myPointCut" />
27             <!--异常通知 -->
28             <aop:after-throwing method="myAfterThrowing" pointcut-ref="myPointCut"
29                throwing="e" />
30             <!--最终通知 -->
31             <aop:after method="myAfter" pointcut-ref="myPointCut" />
32         </aop:aspect>
33     </aop:config>
34 </beans>
```

注意：在 AOP 的配置信息中，使用<aop:after-returning>配置的后置通知和使用<aop:after>配置的最终通知，虽然都是在目标方法执行之后执行，但它们是有区别的：后置通知只有在目标方法成功执行后才会被植入；而最终通知不论目标方法如何结束（包括成功执行和异常中止两种情况），它都会被植入。另外，如果程序没有异常，异常通知将不会执行。

步骤 06 在 com.ssm.aspectj.xml 包下创建测试类 TestXmlAspectJ，为了在类中更加清晰地演示几种通知的执行情况，这里只对 addUser()方法进行增强测试，如文件 3.5 所示。

文件 3.5 TestXmlAspectJ.java

```
01  package com.ssm.aspectj.xml;
02  import org.springframework.context.ApplicationContext;
03  import org.springframework.context.support.ClassPathXmlApplicationContext;
04  import com.ssm.aspectj.UserDao;
05  public class TestXmlAspectJ {
06      public static void main(String[] args) {
07          // 定义配置文件路径
08          String xmlPath="com/ssm/aspectj/xml/applicationContext.xml";
09          // 初始化 Spring 容器，加载配置文件
10          ApplicationContext applicationContext=new ClassPathXmlApplicationContext(xmlPath);
11          // 从容器中获得 userDao 实例
12          UserDao userDao=(UserDao)applicationContext.getBean("userDao");
13          // 执行添加用户方法
14          userDao.addUser();
15      }
16  }
```

执行程序后，控制台的输出结果如图 3.1 所示。

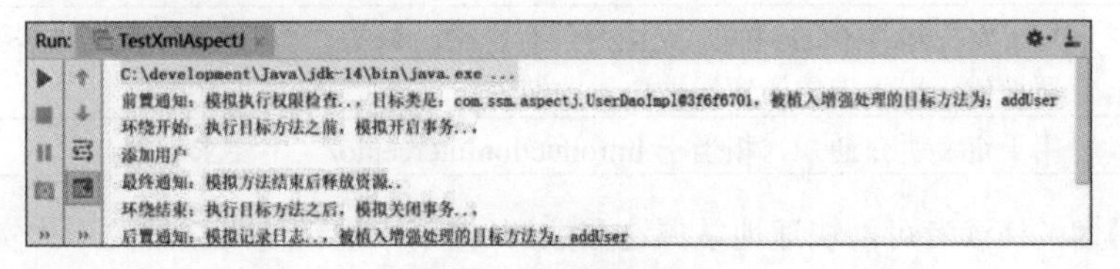

图 3.1 运行结果 1

要查看异常通知的执行效果，可以在 UserDaoImpl 类的 addUser()方法中添加出错代码，如“int i=10/0;”。重新运行测试类，可以看到异常通知的执行，此时控制台的输出结果如图 3.2 所示。

```
Run: TestXmlAspectJ
C:\development\Java\jdk-14\bin\java.exe ...
前置通知：模拟执行权限检查..，目标类是：com.ssm.aspectj.UserDaoImpl@3f6f6701，被植入增强处理的目标方法为：addUser
环绕开始：执行目标方法之前，模拟开启事务...
添加用户
最终通知：模拟方法结束后释放资源..
异常通知：出错了/ by zero
Exception in thread "main" java.lang.ArithmeticException: / by zero
    at com.ssm.aspectj.UserDaoImpl.addUser(UserDaoImpl.java:8)
```

图 3.2 运行结果 2

从图 3.1 和图 3.2 可以看出，使用基于 XML 的声明式 AspectJ 已经实现了 AOP 开发。

3.2.2 基于注解的声明式 AspectJ

基于 XML 的声明式 AspectJ 实现 AOP 编程虽然便捷，但是存在一些缺点，那就是要在 Spring 文件中配置大量的代码信息。为了解决这个问题，AspectJ 框架为 AOP 的实现提供了一套注解，用以取代 Spring 配置文件中为实现 AOP 功能所配置的臃肿代码。

关于 AspectJ 注解的介绍如表 3.4 所示。

表 3.4 AspectJ 的注解及其描述

注解名称	描 述
@Aspect	用于定义一个切面
@Pointcut	用于定义切入点表达式。在使用时还需定义一个包含名字和任意参数的方法签名来表示切入点名称。实际上，这个方法签名就是一个返回值为 void 且方法体为空的普通方法
@Before	用于定义前置通知，相当于 BeforeAdvice。在使用时，通常需要指定一个 value 属性值，该属性值用于指定一个切入点表达式（可以是已有的切入点，也可以直接定义切入点表达式）
@AfterReturning	用于定义后置通知，相当于 AfterReturningAdvice。在使用时可以指定 pointcut/value 和 returning 属性，其中 pointcut、value 两个属性的作用一样，都用于指定切入点表达式；returning 属性值用于表示 Advice()方法中可定义与此同名的形参，该形参可用于访问目标方法的返回值
@Around	用于定义环绕通知，相当于 MethodInterceptor。在使用时需要指定一个 value 属性值，该属性值用于指定通知被植入的切入点
@AfterThrowing	用于定义异常通知来处理程序中未处理的异常，相当于 ThrowAdvice。在使用时可指定 pointcut/value 和 throwing 属性。其中，pointcut、value 用于指定切入点表达式，而 throwing 属性值用于指定一个形参名，来表示 Advice()方法中可定义与此同名的形参，该形参可用于访问目标方法抛出的异常
@After	用于定义最终 final 通知，无论是否有异常，该通知都会执行。使用时需要指定一个 value 属性值，用于指定该通知被植入的切入点
@DeclareParents	用于定义引介通知，相当于 IntroductionInterceptor（不要求掌握）

【示例 3-2】为了使读者可以快速地掌握这些注解，接下来重新使用注解的形式实现【示例 3-1】，具体步骤如下：

步骤 01 在 chapter03 项目的 src 目录下创建 com.ssm.aspectj.annotation 包，将文件 3.3 的切面类 MyAspect 复制到该包下，并对该文件进行修改，如文件 3.6 所示。

文件 3.6 MyAspect.java

```
01  package com.ssm.aspectj.annotation;
02  import org.aspectj.lang.JoinPoint;
03  import org.aspectj.lang.ProceedingJoinPoint;
04  import org.aspectj.lang.annotation.After;
05  import org.aspectj.lang.annotation.AfterReturning;
06  import org.aspectj.lang.annotation.AfterThrowing;
07  import org.aspectj.lang.annotation.Around;
08  import org.aspectj.lang.annotation.Aspect;
```

```
09 import org.aspectj.lang.annotation.Before;
10 import org.aspectj.lang.annotation.Pointcut;
11 import org.springframework.stereotype.Component;
12 /**
13  * 切面类，在此类中编写通知
14  */
15 @Aspect
16 @Component
17 public class MyAspect {
18     //定义切入点表达式
19     @Pointcut("execution(* com.ssm.aspectj.*.*(..))")
20     //使用一个返回值为void、方法体为空的方法来命名切入点
21     public void myPointCut(){}
22     //前置通知
23     @Before("myPointCut()")
24     public void myBefore(JoinPoint joinPoint){
25         System.out.print("前置通知：模拟执行权限检查..，");
26         System.out.print("目标类是："+joinPoint.getTarget());
27         System.out.println("，被植入增强处理的目标方法为："+
28             joinPoint.getSignature().getName());
29     }
30     //后置通知
31     @AfterReturning(value="myPointCut()")
32     public void myAfterReturning(JoinPoint joinPoint) {
33         System.out.print("后置通知：模拟记录日志..，");
34         System.out.println("被植入增强处理的目标方法为：" +
35             joinPoint.getSignature().getName());
36     }
37     /**
38      * 环绕通知
39      * ProceedingJoinPoint 是 JoinPoint 的子接口，表示可执行目标方法
40      * 1.必须是 Object 类型的返回值
41      * 2.必须接收一个参数，类型为 ProceedingJoinPoint
42      * 3.必须 throws Throwable
43      */
44     @Around("myPointCut()")
45     public Object myAround(ProceedingJoinPoint proceedingJoinPoint) throws Throwable{
46         //开始
47         System.out.println("环绕开始：执行目标方法之前，模拟开启事务..，");
48         //执行当前目标方法
49         Object obj=proceedingJoinPoint.proceed();
50         //结束
51         System.out.println("环绕结束：执行目标方法之后，模拟关闭事务..，");
52         return obj;
53     }
54     //异常通知
```

```
55        @AfterThrowing(value="myPointCut()",throwing="e")
56        public void myAfterThrowing(JoinPoint joinPoint,Throwable e){
57            System.out.println("异常通知：出错了"+e.getMessage());
58        }
59        //最终通知
60        @After("myPointCut()")
61        public void myAfter(){
62            System.out.println("最终通知：模拟方法结束后释放资源..");
63        }
64    }
```

在文件 3.6 中，首先使用@Aspect 注解定义了切面类，由于该类在 Spring 中是作为组件使用的，因此还需要添加@Component 注解才能生效。然后使用@Pointcut 注解来配置切入表达式，并通过定义方法来表示切入点名称。接下来在每个通知相应的方法上添加相应的注解，并将切入点名称"myPointcut"作为参数传递给需要执行增强的通知方法。如果需要其他参数（如异常通知的异常参数），则可以根据代码提示传递相应的属性值。

步骤02 在目标类 com.ssm.aspectj.UserDaoImpl 中添加注解@Repository("userDao")。

步骤03 在 com.ssm.aspectj.annotation 包下创建配置文件 applicationContext.xml，并对该文件进行编辑，如文件 3.7 所示。

文件 3.7 applicationContext.xml

```
01  <?xml version="1.0" encoding="UTF-8"?>
02  <beans xmlns="http://www.springframework.org/schema/beans"
03      xmlns:xsi="http://www.w3.org/2001/XMLSchema-instance"
04      xmlns:aop="http://www.springframework.org/schema/aop"
05      xmlns:context="http://www.springframework.org/schema/context"
06      xsi:schemaLocation="http://www.springframework.org/schema/beans
07         http://www.springframework.org/schema/beans/spring-beans.xsd
08         http://www.springframework.org/schema/aop
09         http://www.springframework.org/schema/aop/spring-aop.xsd
10         http://www.springframework.org/schema/context
11         http://www.springframework.org/schema/context/spring-context.xsd">
12      <!-- 指定需要扫描的包，使注解生效 -->
13      <context:component-scan base-package="com.ssm" />
14      <!-- 启动基于注解的声明式 AspectJ 支持 -->
15      <aop:aspectj-autoproxy />
16  </beans>
```

在文件 3.7 中，首先引入 context 约束信息，然后使用<context>元素设置需要扫描的包，使注解生效。由于此案例中的目标类位于 com.ssm.aspectj 包中，因此这里设置 base-package 的值为"com.ssm"。最后，使用<aop. aspectj-autoproxy />来启动 Spring 对基于注解的声明式 AspectJ 的支持。

步骤04 在 com.ssm.aspectj.annotation 包中创建测试类 TestAnnotation，该类与文件 3.5 基本一致，只是配置文件的路径有所不同，如文件 3.8 所示。

文件 3.8 TestAnnotation.java

```
01 package com.ssm.aspectj.annotation;
02 import org.springframework.context.ApplicationContext;
03 import org.springframework.context.support.ClassPathXmlApplicationContext;
04 import com.ssm.aspectj.UserDao;
05 public class TestAnnotation {
06     public static void main(String[] args) {
07         String xmlPath="com/ssm/aspectj/annotation/applicationContext.xml";
08         ApplicationContext applicationContext=new ClassPathXmlApplicationContext(xmlPath);
09         //从容器中获得 userDao 实例
10         UserDao userDao=(UserDao)applicationContext.getBean("userDao");
11         //执行添加用户方法
12         userDao.addUser();
13     }
14 }
```

执行程序后，控制台的输出结果如图 3.3 所示。

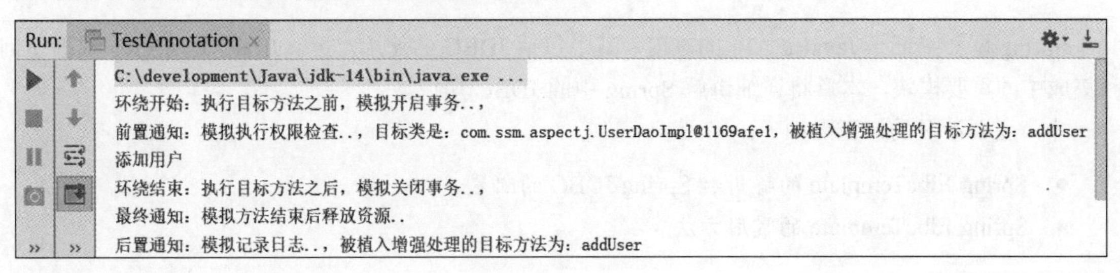

图 3.3 运行结果

在 UserDaoImpl 类的 addUser()方法中加上出错代码来演示异常通知的执行，控制台的输出结果如图 3.2 所示。

从图 3.1 和图 3.3 可以看出，基于注解的方式与基于 XML 的方式的执行结果相同，只是在目标方法前后通知的执行顺序发生了变化。相对来说，使用注解的方式更加简单、方便，所以在实际开发中推荐使用注解的方式进行 AOP 开发。

注意：如果在同一个连接点有多个通知需要执行，那么在同一切面中，目标方法之前的前置通知和环绕通知的执行顺序是未知的，目标方法之后的后置通知和环绕通知的执行顺序也是未知的。

第 4 章

Spring 的数据库开发

Spring 框架降低了 JavaEE API 的难度，其中包括 JDBC 的使用难度。JDBC 是 Spring 数据访问和集成中的重要模块，本章将详细讲解 Spring 中的 JDBC 知识。

本章主要涉及的知识点如下：

- Spring JdbcTemplate 的解析和 Spring JDBC 的配置。
- Spring JdbcTemplate 的常用方法。

4.1　Spring JDBC

Spring 的 JDBC 模块负责数据库资源管理和错误处理，大大简化了开发人员对数据库的操作，使得开发人员可以从烦琐的数据库操作中解脱出来，从而将更多的精力投入编写业务逻辑中。

4.1.1　Spring JdbcTemplate 的解析

针对数据库的操作，Spring 框架提供了 JdbcTemplate 类，该类是 Spring 框架数据抽象层的基础，其他更高层次的抽象类是构建于 JdbcTemplate 类之上的。可以说，JdbcTemplate 类是 Spring JDBC 的核心类。

JdbcTemplate 类的继承关系十分简单。它继承自抽象类 JdbcAccessor，同时实现了 JdbcOperations 接口。

（1）JdbcOperations 接口定义了在 JdbcTemplate 类中可以使用的操作集合，包括添加、修改、查询和删除等操作。

（2）JdbcTemplate 类的直接父类是 JdbcAccessor，该类为子类提供了一些访问数据库时使用的公共属性，具体如下：

- DataSource：其主要功能是获取数据库连接，具体实现时还可以引入对数据库连接的缓冲池和分布事务的支持，它可以作为访问数据库资源的标准接口。
- SQLExceptionTranslator：org.springframework.jdbc.support.SQLExceptionTranslator 接口负责对 SQLException 进行转译工作。通过必要的设置或者获取 SQLExceptionTranslator 中的方法，可以使 JdbcTemplate 在需要处理 SQLException 时委托 SQLExceptionTranslator 的实现类来完成相关的转译工作。

4.1.2　Spring JDBC 的配置

Spring JDBC 模块主要由 4 个包组成，分别是 core（核心包）、dataSource（数据包）、object（对象包）和 support（支持包）。关于这 4 个包的具体说明如表 4.1 所示。

表 4.1　Spring JDBC 中的主要包及说明

包　名	说　明
core	包含 JDBC 的核心功能，包括 JdbcTemplate 类、SimpleJdbcInsert 类、SimpleJdbcCall 类以及 NamedParameterJdbcTemplate 类
dataSource	访问数据源的实用工具类，它有多种数据源的实现，可以在 Java EE 容器外部测试 JDBC 代码
object	以面向对象的方式访问数据库，它允许执行查询并将返回结果作为业务对象，可以在数据表的列和业务对象的属性之间映射查询结果
support	包含 core 和 object 包的支持类，例如提供异常转换功能的 SQLException 类

从表 4.1 可以看出，Spring 对数据库的操作都封装在了这几个包中，如果想要使用 Spring JDBC，就需要对它进行配置。在 Spring 中，JDBC 的配置是在配置文件 applicationContext.xml 中完成的，其配置模板如下：

```
<?xml version="1.0" encoding="UTF-8"?>
<beans xmlns="http://www.springframework.org/schema/beans"
    xmlns:xsi="http://www.w3.org/2001/XMLSchema-instance"
    xsi:schemaLocation="http://www.springframework.org/schema/beans
      http://www.springframework.org/schema/beans/spring-beans.xsd">
    <!--1.配置数据源 -->
    <bean id="dataSource"
        class="org.springframework.jdbc.datasource.DriverManagerDataSource">
        <!--数据库驱动 -->
        <property name="driverClassName" value="com.mysql.cj.jdbc.Driver" />
        <!--连接数据库的 url -->
        <property name="url"
value="jdbc:mysql://localhost:3306/db_spring?serverTimezone=UTC" />
        <!--连接数据库的用户名 -->
        <property name="username" value="root" />
        <!--连接数据库的密码 -->
        <property name="password" value="root" />
    </bean>
```

```
    <!--2.配置 JDBC 模板 -->
    <bean id="jdbcTemplate" class="org.springframework.jdbc.core.JdbcTemplate">
        <!--默认必须使用数据源 -->
        <property name="dataSource" ref="dataSource" />
    </bean>
    <!--3.配置注入类 -->
    <bean id="xxx" class="Xxx">
        <property name="jdbcTemplate" ref="jdbcTemplate" />
    </bean>
</beans>
```

在上述代码中定义了 3 个 Bean，分别是 dataSource、jdbcTemplate 和需要注入类的 Bean。其中 dataSource 对应的 org.springframework.jdbc.datasource.DriverManagerDataSource 类用于对数据源进行配置，jdbcTemplate 对应的 org.springframework.jdbc.core.JdbcTemplate 类中定义了 JdbcTemplate 的相关配置。上述代码中 dataSource 的配置就是 JDBC 连接数据库时所需的 4 个属性，如表 4.2 所示。

表 4.2 dataSource 的 4 个属性

属 性 名	含 义
driverClassName	所使用的驱动名称，对应驱动 JAR 包中的 Driver 类
url	数据源所在地址
username	访问数据库的用户名
password	访问数据库的密码

表 4.2 中的 4 个属性需要根据数据库类型或者机器配置的不同设置相应的属性值。例如，如果数据库类型不同，则需要更改驱动名称；如果数据库不在本地，则需要将地址中的 localhost 替换成相应的主机 IP；如果修改过 MySQL 数据库的端口号（默认为 3306），则需要加上修改后的端口号，如果未修改，那么端口号可以省略；同时连接数据库的用户名和密码需要与数据库创建时设置的用户名和密码保持一致。本示例中 Spring 数据库的用户名和密码都是 root。

定义 jdbcTemplate 时，需要将 dataSource 注入 jdbcTemplate 中，而其他需要使用 jdbcTemplate 的 Bean 也需要将 jdbcTemplate 注入该 Bean 中（通常注入 Dao 类中，在 Dao 类中进行数据库的相关操作）。

4.2 Spring JdbcTemplate 的常用方法

在 JdbcTemplate 类中提供了大量更新和查询数据库的方法，我们可以使用这些方法来操作数据库。本节将介绍这些方法的使用。

4.2.1 execute()——执行 SQL 语句

execute(String sql)方法能够完成执行 SQL 语句的功能。

【示例 4-1】下面以创建数据表的 SQL 语句为例演示 execute()方法的使用，具体步骤如下：

步骤01 在 MySQL 中创建一个名为 db_spring 的数据库，如图 4.1 所示。

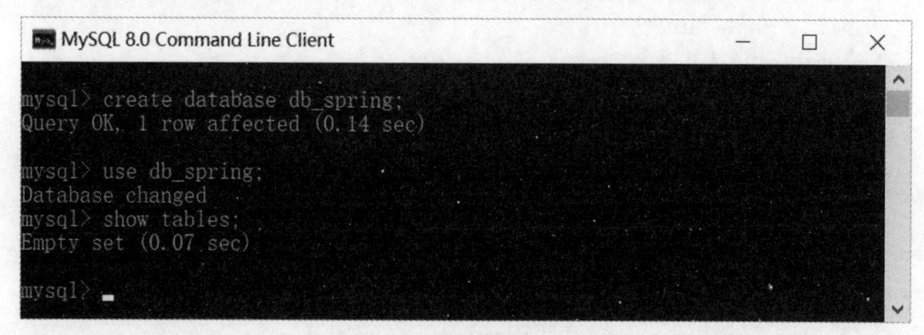

图 4.1 创建数据库

在图 4.1 中，首先使用 SQL 语句创建了数据库 db_spring，然后选择使用 db_spring。为了便于后续验证数据表是通过 execute(String sql)方法创建的，这里使用了 show tables 语句查看数据库中的表，其结果显示为空。

步骤02 在 IntelliJ IDEA 中创建一个名为 chapter04 的 Web 项目，将运行 Spring 框架所需的 5 个基础 JAR 包、MySQL 数据库的驱动 JAR 包、Spring JDBC 的 JAR 包以及 Spring 事务处理的 JAR 包复制到项目的 lib 目录，并发布到类路径中。项目中所添加的 JAR 包如图 4.2 所示。

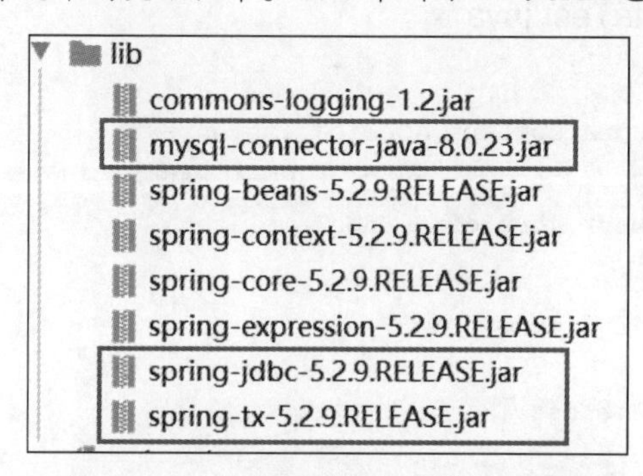

图 4.2 Spring JDBC 操作相关的 JAR 包

步骤03 在 src 目录下创建配置文件 applicationContext.xml，在该文件中配置 id 为 dataSource 的数据源 Bean 和 id 为 jdbcTemplate 的 JDBC 模板 Bean，并将数据源注入 JDBC 模板中，如文件 4.1 所示。

文件 4.1 applicationContext.xml

```
01 <?xml version="1.0" encoding="UTF-8"?>
02 <beans xmlns="http://www.springframework.org/schema/beans"
03     xmlns:xsi="http://www.w3.org/2001/XMLSchema-instance"
04     xsi:schemaLocation="http://www.springframework.org/schema/beans
05       http://www.springframework.org/schema/beans/spring-beans.xsd">
06     <!--1.配置数据源 -->
07     <bean id="dataSource"
08         class="org.springframework.jdbc.datasource.DriverManagerDataSource">
```

```
09          <!--数据库驱动 -->
10          <property name="driverClassName" value="com.mysql.cj.jdbc.Driver" />
11          <!--连接数据库的 url -->
12          <property name="url"
13          value="jdbc:mysql://localhost:3306/db_spring?serverTimezone=UTC" />
14          <!--连接数据库的用户名 -->
15          <property name="username" value="root" />
16          <!--连接数据库的密码 -->
17          <property name="password" value="root" />
18      </bean>
19      <!--2.配置 JDBC 模板 -->
20      <bean id="jdbcTemplate" class="org.springframework.jdbc.core.JdbcTemplate">
21          <!--默认必须使用数据源 -->
22          <property name="dataSource" ref="dataSource" />
23      </bean>
24  </beans>
```

步骤 04 在 src 目录下创建一个 com.ssm.jdbc 包，在该包中创建测试类 JdbcTemplateTest。在该类的 main()方法中，通过 Spring 容器获取在配置文件中定义的 JdbcTemplate 实例，然后使用实例的 execute(String s)方法执行创建数据表的 SQL 语句，如文件 4.2 所示。

文件 4.2 JdbcTemplateTest.java

```
01  package com.ssm.jdbc;
02  import org.springframework.context.ApplicationContext;
03  import org.springframework.context.support.ClassPathXmlApplicationContext;
04  import org.springframework.jdbc.core.JdbcTemplate;
05  /**
06   *使用 execute()方法创建表
07   */
08  public class JdbcTemplateTest {
09      public static void main(String[] args) {
10          //加载配置文件
11          ApplicationContext applicationContext =
12                  new ClassPathXmlApplicationContext("applicationContext.xml");
13          //获取 JdbcTemplate 实例
14          JdbcTemplate jdbcTemplate =
15                  (JdbcTemplate) applicationContext.getBean("jdbcTemplate");
16          //使用 execute()方法执行 SQL 语句，创建用户表 user
17          jdbcTemplate.execute("create table user(" +
18                      "id int primary key auto_increment," +
19                      "username varchar(40)," +
20                      "password varchar(40))");
21      }
22  }
```

成功运行程序后，再次查询 db_spring 数据库，其结果如图 4.3 所示。从图中可以看出，程序使用 execute(String sql)方法执行 SQL 语句成功创建了数据表 user。

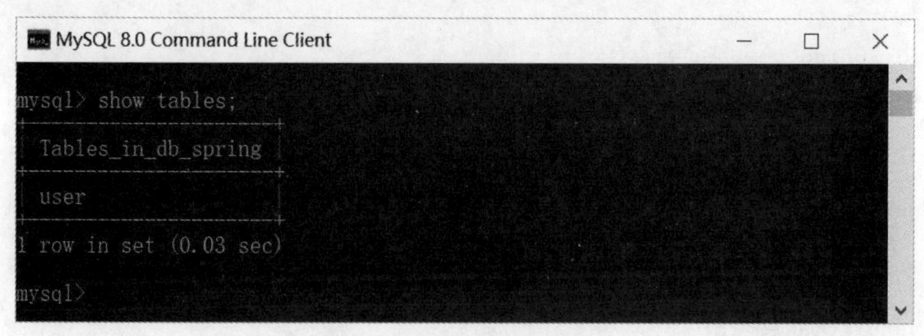

图 4.3 db_spring 数据库中的表

4.2.2 update()——更新数据

update()方法可以完成插入、更新和删除数据的操作。在 JdbcTemplate 类中提供了一系列 update()方法，其常用格式如表 4.3 所示。

表 4.3 JdbcTemplate 类中常用的 update()方法

方 法	说 明
int update(String sql)	该方法是最简单的 update 方法重载形式，直接执行传入的 SQL 语句，并返回受影响的行数
int update(PreparedStatementCreator psc)	该方法执行从 PreparedStatementCreator 返回的语句，然后返回受影响的行数
int update(String sql, PreparedStatementSetter pss)	该方法通过 PreparedStatementsetter 设置 SQL 语句中的参数，并返回受影响的行数
int update(String sql, Object... args)	该方法使用 Object...设置 SQL 语句中的参数，要求参数不能为 NULL，并返回受影响的行数

【示例 4-2】通过一个用户管理的案例来演示 update()方法的使用，具体步骤如下：

步骤 01 在 chapter04 项目的 com.ssm.jdbc 包中创建 User 类，在该类中定义 id、username 和 password 属性，以及对应的 getter()方法和 setter()方法，如文件 4.3 所示。

文件 4.3 User.java

```
01  package com.ssm.jdbc;
02  // User 类
03  public class User {
04      private Integer id;  // 用户 id
05      private String username; // 用户名
06      private String password; // 密码
07      public Integer getId() {
08          return id;
09      }
10      public void setId(Integer id) {
```

```
11          this.id = id;
12      }
13      public String getUsername() {
14          return username;
15      }
16      public void setUsername(String username) {
17          this.username = username;
18      }
19      public String getPassword() {
20          return password;
21      }
22      public void setPassword(String password) {
23          this.password = password;
24      }
25      public String toString() {
26          return "User [id=" + id + ", username=" + username + ", password="+password + "]";
27      }
28  }
```

步骤02 在 com.ssm.jdbc 包中创建接口 UserDao，并在接口中定义添加、更新和删除用户的方法，如文件 4.4 所示。

文件 4.4 UserDao.java

```
01  package com.ssm.jdbc;
02  public interface UserDao {
03   // 添加用户方法
04      public int addUser(User user);
05   // 更新用户方法
06      public int updateUser(User user);
07   // 删除用户方法
08      public int deleteUser(int id);
09  }
```

步骤03 在 com.ssm.jdbc 包中创建 UserDao 接口的实现类 UserDaoImpl，并在类中实现添加、更新和删除用户的方法，如文件 4.5 所示。

文件 4.5 UserDaoImpl.java

```
01  package com.ssm.jdbc;
02  import org.springframework.jdbc.core.JdbcTemplate;
03  public class UserDaoImpl implements UserDao {
04      private JdbcTemplate jdbcTemplate;
05      public void setJdbcTemplate(JdbcTemplate jdbcTemplate) {
06          this.jdbcTemplate = jdbcTemplate;
07      }
08   // 添加用户方法
09      public int addUser(User user) {
```

```
10          String sql="insert into user(username,password) value(?,?)";
11          Object[] obj=new Object[]{
12              user.getUsername(),
13              user.getPassword()
14          };
15          int num=this.jdbcTemplate.update(sql,obj);
16          return num;
17      }
18      // 更新用户方法
19      public int updateUser(User user) {
20          String sql="update user set username=?,password=? where id=?";
21          Object[] params=new Object[]{
22              user.getUsername(),
23              user.getPassword(),
24              user.getId()
25          };
26          int num=this.jdbcTemplate.update(sql,params);
27          return num;
28      }
29      // 删除用户方法
30      public int deleteUser(int id) {
31          String sql="delete from user where id=?";
32          int num=this.jdbcTemplate.update(sql,id);
33          return num;
34      }
35  }
```

从上述 3 种操作的代码可以看出，添加、更新和删除操作的实现步骤类似，只是定义的 SQL 语句有所不同。

步骤 04 在 applicationContext.xml 中定义一个 id 为 userDao 的 Bean，该 Bean 用于将 jdbcTemplate 注入 userDao 实例中，代码如下：

```
<!-- 定义 id 为 userDao 的 Bean -->
<bean id="userDao" class="com.ssm.jdbc.UserDaoImpl">
    <!--将 jdbcTemplate 注入 userDao 实例中 -->
    <property name="jdbcTemplate" ref="jdbcTemplate" />
</bean>
```

步骤 05 在测试类 JdbcTemplateTest 中添加一个测试方法 addUserTest()，该方法主要用于添加用户信息，代码如下：

```
// 测试添加用户方法
@Test
public void addUserTest(){
//加载配置文件
    ApplicationContext applicationContext=
```

```
                new ClassPathXmlApplicationContext("applicationContext.xml");
    //获取 userDao 实例
    UserDao userDao=(UserDao)applicationContext.getBean("userDao");
    //创建 user 实例
    User user = new User();
    //设置 user 实例属性值
    user.setUsername("zhangsan");
    user.setPassword("123456");
    //添加用户
    int num=userDao.addUser(user);
    if(num>0){
        System.out.println("成功插入了"+num+"条数据。");
    }else{
        System.out.println("插入操作执行失败。");
    }
}
```

在上述代码中，获取 UserDao 的实例后又创建了 User 对象，并向 User 对象中添加了属性值。然后调用 UserDao 对象的 addUser()方法向数据表中添加一条数据。最后，通过返回的受影响的行数来判断数据是否插入成功。

使用 JUnit4 测试运行后，控制台的输出结果如图 4.4 所示。

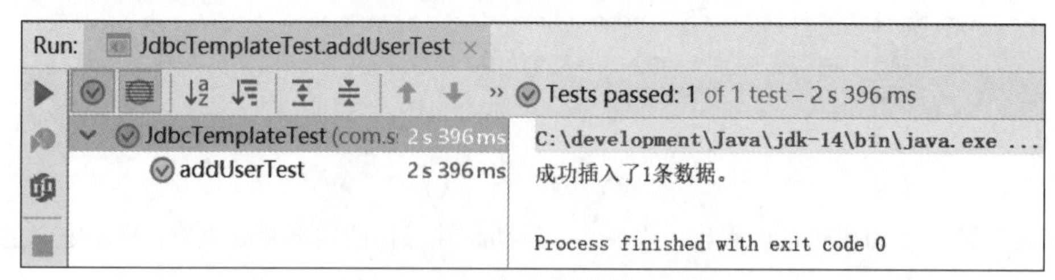

图 4.4　运行结果

此时再次查询数据库中的 user 表，结果如图 4.5 所示。从图中可以看出，使用 JdbcTemplate 的 update()方法已成功地向数据表中插入了一条数据。

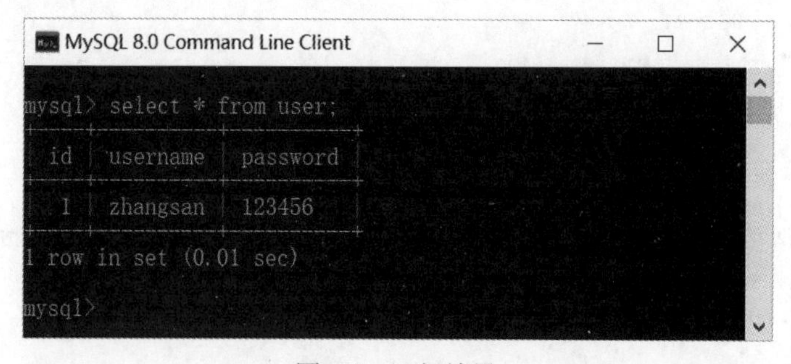

图 4.5　运行结果

步骤 06 执行完插入操作后，接下来使用 JdbcTemplate 类的 update()方法执行更新操作。在测试类 JdbcTemplateTest 中添加一个测试方法 updateUser Test()，代码如下：

```
@Test
public void updateUserTest(){
    ApplicationContext applicationContext=
            new ClassPathXmlApplicationContext("applicationContext.xml");
    UserDao userDao=(UserDao)applicationContext.getBean("userDao");
    User user = new User();
    user.setId(1);
    user.setUsername("tom");
    user.setPassword("111111");
    //更新用户
    int num=userDao.updateUser(user);
    if(num>0){
        System.out.println("成功更新了"+num+"条数据。");
    }else{
        System.out.println("更新操作执行失败。");
    }
}
```

与 addUserTest()方法相比，更新操作的代码增加了 id 属性值的设置，并在修改用户名和密码后调用了 UserDao 对象中的 updateUser()方法，执行对数据表的更新操作。使用 JUnit4 运行方法后，再次查询数据库中的 user 表，运行结果如图 4.6 所示。从图中可以看出，使用 update()方法已成功更新了 user 表中 id 为 1 的用户的用户名和密码。

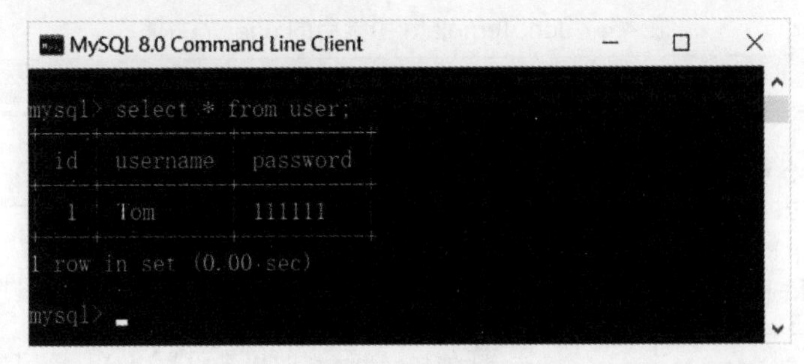

图 4.6　运行结果

步骤 07 在测试类 JdbcTemplateTest 中添加一个测试方法 deleteUserTest()来执行删除操作，代码如下：

```
@Test
public void deleteUserTest(){
    ApplicationContext applicationContext=
                new ClassPathXmlApplicationContext("applicationContext.xml");
    UserDao userDao=(UserDao)applicationContext.getBean("userDao");
    //删除用户
    int num=userDao.deleteUser(1);
    if(num>0){
        System.out.println("成功删除了"+num+"条数据。");
    }else{
```

```
        System.out.println("删除操作执行失败。");
    }
}
```

在上述代码中，获取了 UserDao 的实例后，执行实例中的 deleteUser()方法来删除 id 为 1 的数据。

使用 JUnit4 测试运行方法后，查询 user 表中的数据，运行结果如图 4.7 所示。从图中可以看出，已成功通过 deleteUser()方法删除了 id 为 1 的数据。由于 user 表中只有一条数据，因此删除后表中数据为空。

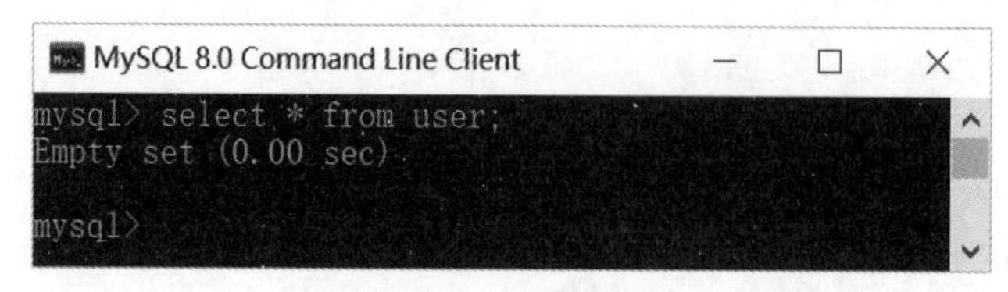

图 4.7　运行结果

4.2.3 query()——查询数据

JdbcTemplate 类中还提供了大量的 query()方法来处理各种对数据库表的查询操作，其中常用的 query()方法格式如表 4.4 所示。

表 4.4　JdbcTemplate 中常用的 query()方法

方　法	说　明
List query(String sql,RowMapper rowMapper)	执行 String 类型参数提供的 SQL 语句，并通过 RowMapper 返回一个 List 类型的结果
List query(String sql,PreparedStatementSetter pss,RowMapper rowMapper)	根据 String 类型参数提供的 SQL 语句创建 PreparedStatement 对象，通过 RowMapper 将结果返回到 List 中
List query(String sql,RowMapper rowMapper)	使用 Object 的值来设置 SQL 语句中的参数值，采用 RowMapper 回调方法可以直接返回 List 类型的数据
queryForObject(String sql, RowMapper rowMapper,Object...args)	将 args 参数绑定到 SQL 语句中，并通过 RowMapper 返回一个 Object 类型的单行记录
queryForList(string sql, Object[] args, class<T> elementType)	该方法可以返回多行数据的结果，但必须是列表形式，elementType 参数返回的是 List 元素类型

【示例 4-3】通过一个具体的案例演示 query()方法的使用，其实现步骤如下：

步骤 01 向数据表 user 中插入几条数据，插入后 user 表中的数据如图 4.8 所示。

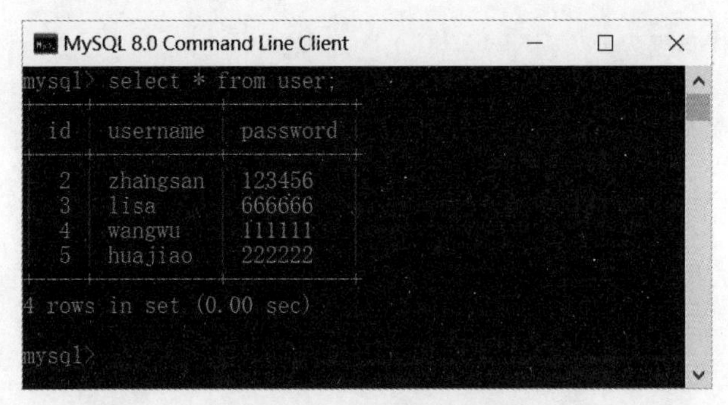

图 4.8　插入数据后的结果

步骤 02 在 UserDao 中分别创建一个通过 id 查询单个用户和查询所有用户的方法，代码如下：

```
//通过 id 查询单个用户
public User findUserById(int id);
//查询所有用户
public List<User> findAllUser();
```

步骤 03 在 UserDao 接口的实现类 UserDaoImpl 中实现接口中的方法，并使用 query()方法分别进行查询，代码如下：

```
//通过 id 查询单个用户数据信息
public User findUserById(int id) {
    String sql="select * from user where id=?";
    RowMapper<User> rowMapper=new BeanPropertyRowMapper<User>(User.class);
    return this.jdbcTemplate.queryForObject(sql,rowMapper,id);
}
//查询所有用户数据信息
public List<User> findAllUser() {
    String sql="select * from user";
    RowMapper<User> rowMapper=new BeanPropertyRowMapper<User>(User.class);
    return this.jdbcTemplate.query(sql,rowMapper);
}
```

在上面两个方法代码中，BeanPropertyRowMapper 是 RowMapper 接口的实现类，可以自动地将数据表中的数据映射到用户自定义的类中（前提是用户自定义类中的字段与数据表中的字段相对应）。创建完 BeanPropertyRowMapper 对象后，在 findUserById()方法中通过 queryForObject()方法返回了一个 Object 类型的单行记录，而在 findAllUser()方法中通过 query()方法返回了一个结果集合。

步骤 04 在测试类 JdbcTemplateTest 中添加一个测试方法 findUserByIdTest()来测试条件查询，代码如下：

```
@Test
public void findUserByIdTest(){
    ApplicationContext applicationContext=
            new ClassPathXmlApplicationContext("applicationContext.xml");
```

```
    //通过 id 查询单个用户数据信息
    UserDao userDao=(UserDao)applicationContext.getBean("userDao");
    User user=userDao.findUserById(2);
    System.out.println(user);
}
```

上述代码通过执行 findUserById()方法获取了 id 为 1 的对象信息，并通过输出语句输出查询结果。使用 JUnit4 测试运行后，控制台的输出结果如图 4.9 所示。

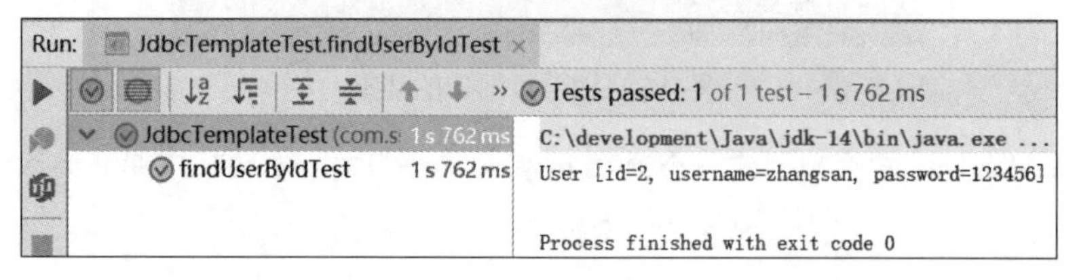

图 4.9 运行结果

步骤 05 在测试类 JdbcTemplateTest 中添加一个测试方法 findAllUserTest()来测试所有用户信息，代码如下：

```
@Test
public void findAllUserTest(){
    ApplicationContext applicationContext=
                new ClassPathXmlApplicationContext("applicationContext.xml");
    UserDao userDao=(UserDao)applicationContext.getBean("userDao");
    //查询所有用户数据信息
    List<User> list=userDao.findAllUser();
    //循环输出用户数据信息
    for(User user:list){
        System.out.println(user);
    }
}
```

在上述代码中，调用了 UserDao 对象的 findAllUser()方法查询所有用户的账户信息，并通过 for 循环输出查询结果。

使用 JUnit4 成功运行 findAllUser()方法后，控制台的显示信息如图 4.10 所示。从图中可以看出，数据表 user 中的 4 条记录都被查询出来了。

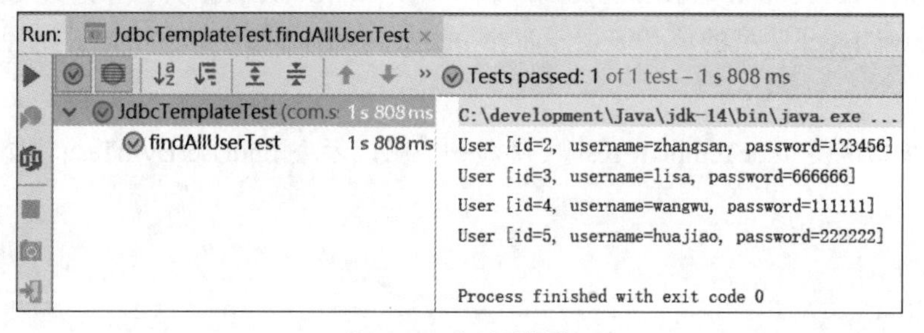

图 4.10 运行结果

第 5 章

Spring 的事务管理

第 4 章介绍了如何使用 Spring 对数据库进行基本操作，操作数据库时还涉及事务管理问题，为此 Spring 提供了专门用于事务处理的 APl。本章将详细讲解 Spring 的事务管理功能。

本章主要涉及的知识点如下：

- 事务管理的核心接口和事务管理的方式。
- 声明式事务管理。

5.1 Spring 事务管理概述

Spring 的事务管理简化了传统的事务管理流程，并且在一定程度上减少了开发者的工作量。

5.1.1 事务管理的核心接口

在 Spring 的所有 JAR 包中包含一个名为 Spring-tx-5.2.9.RELEASE 的 JAR 包，该包就是 Spring 提供的用于事务管理的依赖包。在该 JAR 包的 org.springframework.transaction 包中有 3 个接口文件，分别是 PlatformTransactionManager、TransactionDefinition 和 TransactionStatus。

1. PlatformTransactionManager

PlatformTransactionManager 接口是 Spring 提供的平台事务管理器，主要用于管理事务。该接口中提供了 3 个事务操作的方法，具体如下：

- TransactionStatus getTransaction(TransactionDefinition definition)：用于获取事务状态信息。该方法会根据 TransactionDefinition 参数返回一个 TransactionStatus 对象。TransactionStatus 对象表示一个事务，被关联在当前执行的线程上。

- void commit(TransactionStatus status)：用于提交事务。
- void rollback(TransactionStatus status)：用于回滚事务。

PlatformTransactionManager 接口只是代表事务管理的接口，并不知道底层是如何管理事务的，它只需要事务管理器提供上面的 3 个操作方法，但具体如何管理事务则由它的实现类来完成。

PlatformTransactionManager 接口有许多不同的实现类，常见的 3 个实现类如下：

- org.springframework.jdbc.datasource.DataSourceTransactionManager：用于配置 JDBC 数据源的事务管理器。
- org.springframework.orm.Hibernate5.HibernateTransactionManager：用于配置 Hibernate 的事务管理器。
- org.springframework.transaction.jta.JtaTransactionManager：用于配置全局事务管理器。

当底层采用不同的持久层技术时，系统只需使用不同的 PlatformTransactionManager 实现类即可。

2. TransactionDefinition

TransactionDefinition 接口是事务定义（描述）的对象，该对象中定义了事务规则，并提供了获取事务相关信息的方法，具体如下：

- string getName()：获取事务对象名称。
- int getlsolationLeve()：获取事务的隔离级别。
- int getPropagationBehavior()：获取事务的传播行为。
- int setTimeout()：获取事务的超时时间。
- boolean isReadOnly()：获取事务是否只读。

上述方法中，事务的传播行为是指在同一个方法中不同操作前后所使用的事务。传播行为有很多种，具体如表 5.1 所示。

表 5.1 传播行为的种类

属性名称	值	描 述
PROPAGATION_REQUIRED	REQUIRED	表示当前方法必须运行在一个事务环境中。如果当前方法已处于事务环境中，那么可以直接使用该方法，否则会在开启一个新事务后执行该方法
PROPAGATION_SUPPORTS	SUPPORTS	如果当前方法处于事务环境中，那就使用当前事务，否则不使用事务
PROPAGATION_MANDATORY	MANDATORY	表示调用该方法的线程必须处于当前事务环境中，否则将抛出异常
PROPAGATION_REQUIRES_NEW	REQUIRES_NEW	要求方法在新的事务环境中执行。如果当前方法已在事务环境中，那就先暂停当前事务，在启动新的事务后执行该方法；如果当前方法不在事务环境中，那就在启动一个新的事务后执行该方法

（续表）

属性名称	值	描　述
PROPAGATION_NOT_SUPPORTED	NOT_SUPPORTED	不支持当前事务，总是以非事务状态执行。如果调用该方法的线程处于事务环境中，那就先暂停当前事务，然后执行该方法
PROPAGATION_NEVER	NEVER	不支持当前事务。如果调用该方法的线程处于事务环境中，将抛出异常
PROPAGATION_NESTED	NESTED	即使当前执行的方法处于事务环境中，依然会启动一个新的事务，并且方法在嵌套的事务里执行；即使当前执行的方法不在事务环境中，也会启动一个新事务，然后执行该方法

在事务管理过程中，传播行为可以控制是否需要创建事务以及如何创建事务。通常情况下，数据的查询不会影响原数据的改变，所以不需要进行事务管理；而对于数据的插入、更新和删除操作，则必须进行事务管理。如果没有指定事务的传播行为，则 Spring 默认传播行为是 REQUIRED。

3. TransactionStatus

TransactionStatus 接口描述了某一时间点上事务的状态信息。该接口中包含 6 个方法，具体如下：

- void flush()：刷新事务。
- boolean hasSavepoint()：获取是否存在保存点。
- boolean isCompleted()：获取事务是否完成。
- boolean isNewTransaction()：获取是否为新事务。
- boolean isRollbackOnly()：获取是否回滚。
- void setRollbackOnly()：设置事务回滚。

5.1.2 事务管理的方式

Spring 中的事务管理分为两种方式：一种是传统的编程序事务管理，另一种是声明式事务管理。

- 编程序事务管理：通过编写代码实现的事务管理，包括定义事务的开始、正常执行后的事务提交和异常时的事务回滚。
- 声明式事务管理：通过 AOP 技术实现的事务管理，主要思想是将事务管理作为一个“切面”代码单独编写，然后通过 AOP 技术将事务管理的“切面”代码植入业务目标类中。

声明式事务管理最大的优点在于，开发者无须通过编程的方式来管理事务，只需在配置文件中进行相关的事务规则声明就可以将事务规则应用到业务逻辑中。这使得开发人员可以更加专注于核心业务逻辑代码的编写，在一定程度上减少了工作量，提高了开发效率。在实际开发中，通常都推荐使用声明式事务管理。

5.2 声明式事务管理

Spring 的声明式事务管理可以通过两种方式来实现：一种是基于 XML 的方式，另一种是基于 Annotation 的方式。本节将对这两种声明式事务管理方式进行详细讲解。

5.2.1 基于 XML 方式的声明式事务管理

基于 XML 方式的声明式事务管理是通过在配置文件中配置事务规则的相关声明来实现的。

Spring 2.0 以后提供了 tx 命名空间来配置事务，tx 命名空间下提供了<tx:advice>元素来配置事务的通知（增强处理）。当使用<tx:advice>元素配置了事务的增强处理后，就可以通过编写的 AOP 配置让 Spring 自动对目标生成代理。

配置<tx:advice>元素时，通常需要指定 id 和 transaction-manager 属性，其中 id 属性是配置文件中的唯一标识，transaction-manager 属性用于指定事务管理器。除此之外，还需要配置一个<tx:attributes>子元素，该子元素可通过配置多个<tx:method>子元素来配置执行事务的细节。

关于<tx:method>元素的属性描述如表 5.2 所示。

表 5.2 <tx:method>元素的属性

属性名称	描 述
name	必选属性，指定了与事务属性相关的方法名。其属性值支持使用通配符，如'*'、'get*'、'handle*'、'* Order'等
propagation	用于指定事务的传播行为，其属性值就是表 5.1 中的值，默认值为 REQUIRED
isolation	用于指定事务的隔离级别，其属性值可以为 DEFAULT、READ_UNCO MMITTED、READ_COMMITTED、REPEATABLE_READ 和 SERIALIZABLE，其默认值为 DEFAULT
read-only	用于指定事务是否只读，其默认值为 false
timeout	用于指定事务超时的时间，其默认值为-1，即永不超时
rollback-for	用于指定触发事务回滚的异常类，在指定多个异常类时，异常类之间以英文逗号分隔
no-rollback-for	用于指定不触发事务回滚的异常类，在指定多个异常类时，异常类之间以英文逗号分隔

了解了如何在 XML 文件中配置事务后，接下来通过一个案例来演示如何通过 XML 方式实现 Spring 的声明式事务管理。

【示例 5-1】本案例以第 4 章的项目代码和数据表为基础，模拟一个会员赠送积分的功能，要求在赠送积分时通过 Spring 对事务进行控制，具体实现步骤如下：

步骤 01 在 IntelliJ IDEA 中创建一个名为 chapter05 的 Web 项目，在项目的 lib 目录中导入 chapter05 项目中的所有 JAR 包，并将 AOP 所需 JAR 包也导入 lib 目录中。导入后的目录如图 5.1 所示。

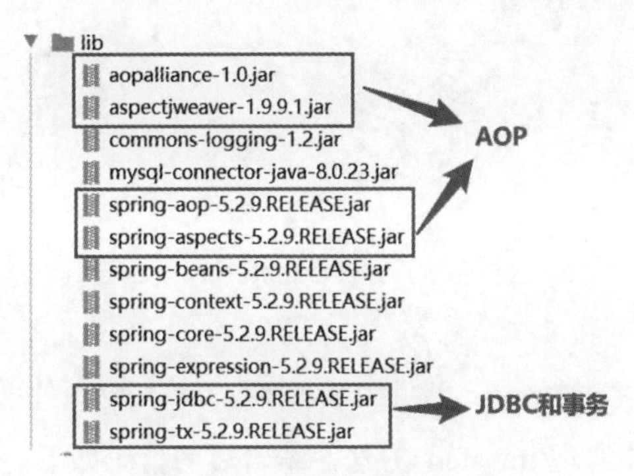

图 5.1　项目所需的 JAR 包

步骤 02 在 MySQL 中，修改数据库 db_spring 中的数据表 user，增加字段 jf（积分），如图 5.2 所示。同时，为了便于后续操作数据表，在数据表 user 中设置所有用户的初始积分为 1000，如图 5.3 所示。

```
MySQL 8.0 Command Line Client
mysql> alter table user add jf int;
Query OK, 0 rows affected (0.21 sec)
Records: 0  Duplicates: 0  Warnings: 0

mysql> desc user;
+----------+-------------+------+-----+---------+----------------+
| Field    | Type        | Null | Key | Default | Extra          |
+----------+-------------+------+-----+---------+----------------+
| id       | int(11)     | NO   | PRI | NULL    | auto_increment |
| username | varchar(40) | YES  |     | NULL    |                |
| password | varchar(40) | YES  |     | NULL    |                |
| jf       | int(11)     | YES  |     | NULL    |                |
+----------+-------------+------+-----+---------+----------------+
4 rows in set (0.04 sec)

mysql>
```

图 5.2　修改数据表

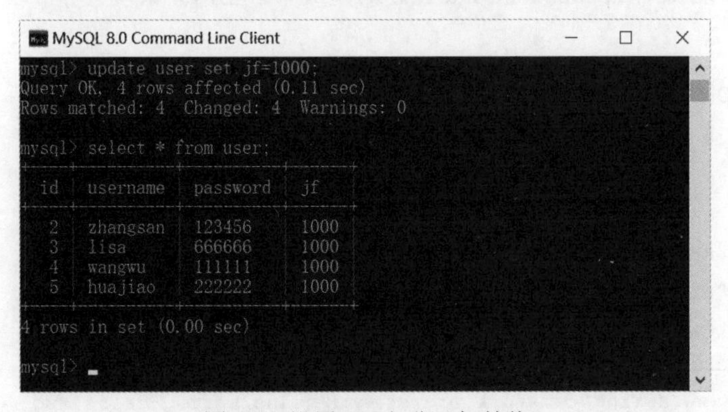

图 5.3　设置 jf（积分）初始值

步骤 03 将 chapter04 项目中的代码和配置文件复制到 chapter05 项目的 src 目录下。接下来，在 User 类中增加 jf（积分）成员和对应的 getter()和 setter()方法，代码如下：

```
private Integer jf;
public Integer getJf() {
    return jf;
```

```
}
public void setJf(Integer jf) {
    this.jf = jf;
}
```

在 UserDao 接口中创建一个赠送积分的方法 transfer()，代码如下：

```
//赠送积分
public void transfer(String outUser, String inUser, Integer jf);
```

在实现类 UserDaoImpl 中实现 transfer()方法，编辑后的代码如下：

```
//赠送积分
public void transfer(String outUser, String inUser, Integer jf) {
    //接收积分
    this.jdbcTemplate.update("update user set jf=jf+? where username=?",jf,inUser);
    //模拟系统运行时的突发性问题
    int i=1/0;
    //赠送（送出）积分
    this.jdbcTemplate.update("update user set jf =jf-? where username=? ", jf, outUser);
}
```

在上述代码中，使用了两个 update()方法对 user 表中的数据执行接收积分和赠送积分的更新操作。在两个操作之间添加了一行代码"int i=1/0;"来模拟系统运行时的突发性问题。如果没有事务控制，那么在 transfer()方法执行后，接收积分用户的积分会增加，而赠送积分用户的积分会因为系统出现问题而不变，这显然是有问题的；如果增加了事务控制，那么在 transfer()方法执行后，接收积分用户的积分和赠送积分用户的积分在问题出现前后都应该保持不变。

步骤 04 修改配置文件 applicationContext.xml，添加命名空间并编写事务管理的相关配置代码，如文件 5.1 所示。

文件 5.1 applicationContext.xml

```
01  <?xml version="1.0" encoding="UTF-8"?>
02  <beans xmlns="http://www.springframework.org/schema/beans"
03      xmlns:xsi="http://www.w3.org/2001/XMLSchema-instance"
04      xmlns:aop="http://www.springframework.org/schema/aop"
05      xmlns:tx="http://www.springframework.org/schema/tx"
06      xmlns:context="http://www.springframework.org/schema/context"
07      xsi:schemaLocation="http://www.springframework.org/schema/beans
08        http://www.springframework.org/schema/beans/spring-beans.xsd
09        http://www.springframework.org/schema/tx
10        http://www.springframework.org/schema/tx/spring-tx.xsd
11        http://www.springframework.org/schema/context
12        http://www.springframework.org/schema/context/spring-context.xsd
13        http://www.springframework.org/schema/aop
14        http://www.springframework.org/schema/aop/spring-aop.xsd">
15      <!--1.配置数据源 -->
16      <bean id="dataSource"
```

```
17            class="org.springframework.jdbc.datasource.DriverManagerDataSource">
18            <!--数据库驱动 -->
19            <property name="driverClassName" value="com.mysql.cj.jdbc.Driver" />
20            <!--连接数据库的 url -->
21            <property name="url"
22    value="jdbc:mysql://localhost:3306/db_spring?serverTimezone=UTC" />
23            <!--连接数据库的用户名 -->
24            <property name="username" value="root" />
25            <!--连接数据库的密码 -->
26            <property name="password" value="root" />
27        </bean>
28        <!--2.配置 JDBC 模板 -->
29        <bean id="jdbcTemplate" class="org.springframework.jdbc.core.JdbcTemplate">
30            <!--默认必须使用数据源 -->
31            <property name="dataSource" ref="dataSource" />
32        </bean>
33        <!--3.定义 id 为 userDao 的 Bean -->
34        <bean id="userDao" class="com.ssm.jdbc.UserDaoImpl">
35            <!--将 jdbcTemplate 注入 userDao 实例中 -->
36            <property name="jdbcTemplate" ref="jdbcTemplate" />
37        </bean>
38        <!--4.事务管理器，依赖于数据源 -->
39        <bean id="transactionManager"
40          class="org.springframework.jdbc.datasource.DataSourceTransactionManager">
41            <property name="dataSource" ref="dataSource"/>
42        </bean>
43        <!--5.编写通知：对事务进行增强（通知），需要编写切入点和具体执行事务细节 -->
44        <tx:advice id="txAdvice" transaction-manager="transactionManager">
45            <tx:attributes>
46                <tx:method name="*" propagation="REQUIRED"
47              isolation="DEFAULT" read-only="false"/>
48            </tx:attributes>
49        </tx:advice>
50        <!--6.编写 AOP，让 Spring 自动对目标生成代理，需要使用 AspectJ 的表达式 -->
51        <aop:config>
52            <!--切入点 -->
53            <aop:pointcut expression="execution(* com.ssm.jdbc.*.*(..))" id="txPointCut" />
54            <!--切面 -->
55            <aop:advisor advice-ref="txAdvice" pointcut-ref="txPointCut"/>
56        </aop:config>
57    </beans>
```

在文件 5.1 中定义了 id 为 transactionManager 的事务管理器，接下来通过编写的通知来声明事务，最后通过声明 AOP 的方式让 Spring 自动生成代理。

步骤 05 在 com.ssm.jdbc 包中创建测试类 TransactionTest，并在类中编写测试方法 xmlTest()，

如文件 5.2 所示。

文件 5.2 TransactionTest.java

```
01 package com.ssm.jdbc;
02 import org.junit.Test;
03 import org.springframework.context.ApplicationContext;
04 import org.springframework.context.support.ClassPathXmlApplicationContext;
05 public class TransactionTest {
06     @Test
07     public void xmlTest(){
08         ApplicationContext applicationContext =
09                 new ClassPathXmlApplicationContext("applicationContext.xml");
10         UserDao userDao=(UserDao)applicationContext.getBean("userDao");
11         //赠送积分
12         userDao.transfer("zhangsan","lisi", 100);
13         System.out.println("赠送积分成功！");
14     }
15 }
```

在文件 5.2 中，获取 UserDao 实例后调用了实例中的赠送积分方法，由 zhangsan 向 lisi 转入 100 积分。如果事务代码起作用，那么在整个赠送积分方法执行完毕后，zhangsan 和 lisi 的积分应该都是原来的值。

执行完文件 5.2 中的测试方法后，JUnit 控制台的显示结果如图 5.4 所示。从图中可以看到，JUnit 控制台中报出了"/ by zero"的算术异常信息。

在执行赠送积分操作后，查看 user 表中的数据，结果如图 5.3 所示，即 zhangsan 和 lisi 的积分没有发生变化，这说明 Spring 中的事务管理配置已经生效。

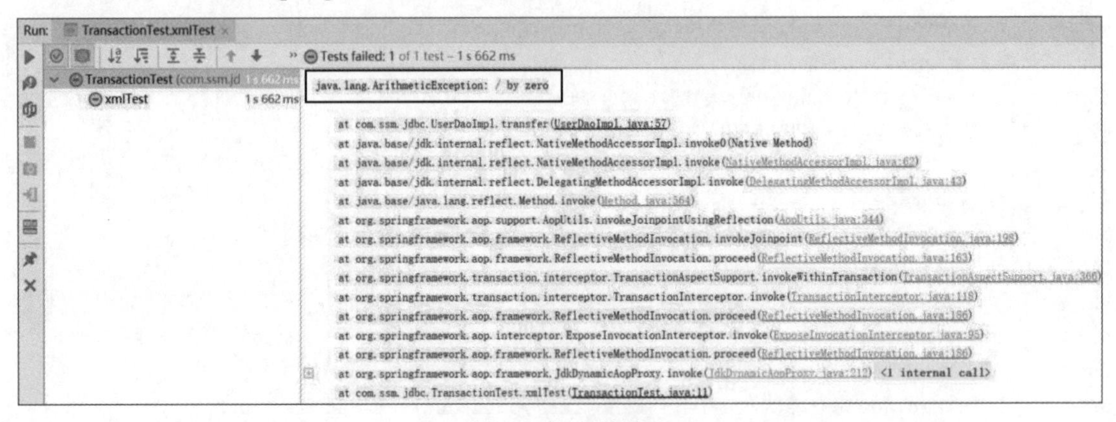

图 5.4 运行结果

5.2.2 基于 Annotation 方式的声明式事务管理

Spring 的声明式事务管理还可以通过 Annotation 的方式来实现。这种方式的使用非常简单，开发者只需做如下两件事情：

（1）在 Spring 容器中注册事务注解驱动，代码如下：

```
<tx:annotation-driven transaction-managers transactionManager"/>
```

（2）在需要使用事务的 Spring Bean 类或者 Bean 类的方法上添加@Transactional 注解。如果将注解添加在 Bean 类上，就表示事务的设置对整个 Bean 类的所有方法都起作用；如果将注解添加在 Bean 类中的某个方法上，就表示事务的设置只对该方法有效。

使用@Transactional 注解时，可以通过其参数配置事务详情。@Transactional 注解可配置的参数信息如表 5.3 所示。

表 5.3　@Transactional 注解的参数及其描述

参数名称	描　述
value	用于指定需要使用的事务管理器，默认为""，其别名为 transactionManager
transactionManager	指定事务的限定符值，可用于确定目标事务管理器，匹配特定的限定值（或者 Bean 的 name 值），默认为""，其别名为 value
isolation	用于指定事务的隔离级别，默认为 Isolation.DEFAULT（底层事务的隔离级别）
noRollbackFor	用于指定遇到特定异常时强制不回滚事务
noRolbackFor ClassName	用于指定遇到特定的多个异常时强制不回滚事务。其属性值可以指定多个异常类名
propagation	用于指定事务的传播行为，默认为 Propagation REQUIRED
read-only	用于指定事务是否只读，默认为 false
rollbackFor	用于指定遇到特定异常时强制回滚事务
rollbackForClassName	用于指定遇到特定的多个异常时强制回滚事务。其属性值可以指定多个异常类名
timeout	用于指定事务的超时时长，默认为 TransactionDefinition.TIMEOUT _DEFAULT（底层事务系统的默认时间）

从表 5.3 可以看出，@Transactional 注解与<tx:method>元素中的事务属性基本是对应的，并且其含义也基本相似。

【示例 5-2】为了让读者更加清楚地掌握@Transactiona 注解的使用，接下来对【示例 5-1】进行修改，以 Annotation 方式实现项目中的事务管理，具体实现步骤如下：

步骤 01 在 src 目录下创建一个 Spring 配置文件 applicationContext-annotation.xml，在该文件中声明事务管理器等配置信息，如文件 5.3 所示。

文件 5.3　applicationContext-annotation.xml

```
01  <?xml version="1.0" encoding="UTF-8"?>
02  <beans xmlns="http://www.springframework.org/schema/beans"
03      xmlns:xsi="http://www.w3.org/2001/XMLSchema-instance"
04      xmlns:aop="http://www.springframework.org/schema/aop"
05      xmlns:tx="http://www.springframework.org/schema/tx"
06      xmlns:context="http://www.springframework.org/schema/context"
07      xsi:schemaLocation="http://www.springframework.org/schema/beans
08        http://www.springframework.org/schema/beans/spring-beans.xsd
```

```
09         http://www.springframework.org/schema/tx
10         http://www.springframework.org/schema/tx/spring-tx.xsd
11         http://www.springframework.org/schema/context
12         http://www.springframework.org/schema/context/spring-context.xsd
13         http://www.springframework.org/schema/aop
14         http://www.springframework.org/schema/aop/spring-aop.xsd">
15      <!--1.配置数据源 -->
16      <bean id="dataSource"
17          class="org.springframework.jdbc.datasource.DriverManagerDataSource">
18          <!--数据库驱动 -->
19          <property name="driverClassName" value="com.mysql.cj.jdbc.Driver" />
20          <!--连接数据库的url -->
21          <property name="url"
22  value="jdbc:mysql://localhost:3306/db_spring?serverTimezone=UTC" />
23          <!--连接数据库的用户名 -->
24          <property name="username" value="root" />
25          <!--连接数据库的密码 -->
26          <property name="password" value="root" />
27      </bean>
28      <!--2.配置JDBC模板 -->
29      <bean id="jdbcTemplate" class="org.springframework.jdbc.core.JdbcTemplate">
30          <!--默认必须使用数据源 -->
31          <property name="dataSource" ref="dataSource" />
32      </bean>
33      <!--3.定义id为userDao的Bean -->
34      <bean id="userDao" class="com.ssm.jdbc.UserDaoImpl">
35          <!--将 jdbcTemplate注入 userDao实例中 -->
36          <property name="jdbcTemplate" ref="jdbcTemplate" />
37      </bean>
38      <!--4.事务管理器，依赖于数据源 -->
39      <bean id="transactionManager"
40        class="org.springframework.jdbc.datasource.DataSourceTransactionManager">
41          <property name="dataSource" ref="dataSource"/>
42      </bean>
43      <!--5.注册事务管理器驱动 -->
44      <tx:annotation-driven transaction-manager="transactionManager"/>
45  </beans>
```

与基于XML方式的配置文件相比，文件5.3通过注册事务管理器驱动替换了文件5.1中的编写通知和编写AOP的配置，大大减少了配置文件中的代码量。

注意：如果案例中使用了注解式开发，那么就需要在配置文件中开启注解处理器，指定扫描哪些包下的注解。这里没有开启注解处理器，是因为在配置文件中已经配置了UserDaoImpl类的Bean，而@Transactional注解就配置在该Bean类中，所以可以直接生效。

步骤02 在UserDaoImpl类的transfer()方法上添加事务注解，添加后的代码如下：

```
@Transactional(propagation=Propagation.REQUIRED,isolation=Isolation.DEFAULT,
readOnly=false)
public void transfer(String outUser, String inUser, Integer jf) {
```

```
    //接收积分
    this.jdbcTemplate.update("update user set jf=jf+? where username=?",jf,inUser);
    //模拟系统运行时的突发性问题
    int i=1/0;
    //赠送（送出）积分
    this.jdbcTemplate.update("update user set jf =jf-? where username=? ", jf, outUser);
}
```

上述方法已经添加了@Transactional 注解，并且使用注解的参数配置了事务详情，各个参数之间要用英文逗号（,）进行分隔。

注意：在实际开发中，事务的配置信息通常是在 Spring 的配置文件中完成的，而在业务层类上只需使用@Transactional 注解即可，不需要配置@Transactional 注解的属性。

步骤 03 在 TransactionTest 类中创建测试方法 annotationTest()，编辑后的代码如下：

```
@Test
public void annotationTest(){
    ApplicationContext applicationContext = new
ClassPathXmlApplicationContext("applicationContext-annotation.xml");
    UserDao userDao=(UserDao)applicationContext.getBean("userDao");
    //赠送积分
    userDao.transfer("zhangsan","lisi", 200);
    System.out.println("赠送积分成功！");
}
```

从上述代码可以看出，与 XML 方式的测试方法相比，该方法只是对配置文件的名称进行了修改。程序执行后，会出现与 XML 方式同样的执行结果。

第6章

初识 MyBatis

MyBatis是当前主流的Java持久层框架之一，与Hibernate一样，也是一种ORM（Object Relational Mapping，对象关系映射）框架。MyBatis因其性能优异，且具有高度的灵活性、可优化性和易于维护等特点，所以受到了广大互联网企业的青睐，是目前大型互联网项目首选的持久层框架。接下来的几章将详细讲解MyBatis框架的相关知识。

本章主要涉及的知识点如下：

- MyBatis概述。
- MyBatis入门程序。

6.1 MyBatis概述

MyBatis是一个支持普通SQL查询、存储过程以及高级映射的持久层框架，它消除了几乎所有的JDBC代码和参数的手动设置以及对结果集的检索，并使用简单的XML或注解进行配置和原始映射，用以将接口和Java的POJO（Plain Old Java Object，普通Java对象）映射成数据库中的记录，使得Java开发人员可以使用面向对象的编程思想来操作数据库。

6.1.1 什么是MyBatis

MyBatis框架也被称为ORM框架。所谓ORM，就是一种解决面向对象与关系数据库中数据类型不匹配问题的技术，通过描述Java对象与数据库表之间的映射关系，自动将Java应用程序中的对象持久化到关系数据库的表中。

使用ORM框架后，应用程序不再直接访问底层数据库，而是以面向对象的方式来操作持久化对象（Persistent Object，PO），而ORM框架则会通过映射关系将这些面向对象的操作转换成底层的SQL操作。

当前的ORM框架产品有很多，常见的ORM框架有Hibernate和MyBatis。

（1）Hibernate：一个全表映射的框架。通常开发者只需定义好持久化对象到数据库表的映射关系，就可以通过 Hibernate 提供的方法完成持久层操作。开发者并不需要熟练地掌握 SQL 语句的编写，Hibernate 会根据制定的存储逻辑自动生成对应的 SQL，并调用 JDBC 接口来执行，所以其开发效率会高于 MyBatis。然而 Hibernate 自身存在着一些缺点，例如它在多表关联时，对 SQL 查询的支持较差；更新数据时，需要发送所有字段；不支持存储过程；不能通过优化 SQL 来优化性能；等等这些问题导致 Hibernate 只适合在场景不太复杂且对性能要求不高的项目中使用。

（2）MyBatis：一个半自动映射的框架。这里所谓的“半自动”是相对于 Hibernate 全表映射而言的，MyBatis 需要手动匹配提供 POJO、SQL 和映射关系，而 Hibernate 只需提供 POJO 和映射关系即可。与 Hibernate 相比，虽然使用 MyBatis 手动编写 SQL 要比使用 Hibernate 的工作量大，但 MyBatis 可以配置动态 SQL 并优化 SQL，可以通过配置决定 SQL 的映射规则，它还支持存储过程等。对于一些复杂的和需要优化性能的项目来说，显然使用 MyBatis 更加合适。

6.1.2 MyBatis 的下载和使用

MyBatis 可以到官网上下载。本书使用的版本是 mybatis-3.5.10。下载后解压 mybatis-3.5.10.zip 压缩包，会得到一个名为 mybatis-3.5.4 的文件夹，里面有核心包 mybatis-3.5.10.jar 和一个 lib 文件夹（里面包含 MyBatis 的依赖包）。

使用 MyBatis 框架非常简单，只需在应用程序中引入 MyBatis 的核心包 mybatis-3.5.10.jar 和 lib 目录中的依赖包即可。

注意：如果底层采用的是 MySQL 数据库，还需要将 MySQL 数据库的驱动 JAR 包添加到应用程序的类路径中；如果采用其他类型的数据库，同样需要将对应类型的数据库驱动包添加到应用程序的类路径中。

6.2 MyBatis 入门程序

上一节对 MyBatis 框架进行了初步的介绍，本节将通过一个用户信息管理的入门案例来讲解 MyBatis 框架的基本使用。

6.2.1 查询用户

在实际开发中，查询操作通常都会涉及单条数据的精确查询以及多条数据的模糊查询。那么使用 MyBatis 框架是如何进行这两种查询的呢？接下来，本小节将讲解如何使用 MyBatis 框架根据用户编号查询客户信息，以及根据用户名模糊查询用户信息。

1. 根据用户编号查询用户信息

【示例 6-1】根据用户编号查询用户信息主要是通过查询用户表中的主键（这里表示唯一的用户编号）来实现的，具体实现步骤如下：

步骤01 在 MySQL 数据库中创建一个名为 db_mybatis 的数据库,在此数据库中创建一个 t_user 表，同时预先添加几条数据，对应的 SQL 语句如下：

```
#创建数据库 db_mybatis
CREATE DATABASE db_mybatis;
//使用数据库 db_mybatis
USE db_mybatis;
#创建数据表 t_user
CREATE TABLE t_user(
    id int(32) PRIMARY KEY AUTO_INCREMENT,
    username varchar(50),
    jobs varchar(50),
    phone varchar (16)
);
#向数据表 t_user 中插入数据记录
INSERT INTO t_user VALUES(1,'zhangsan','teacher','13907998372');
INSERT INTO t_user VALUES(2,'lisi','worker','13907396542');
INSERT INTO t_user VALUES(3,'wangwu','doctor','13817348729');
```

执行上述代码后，t_user 表中的数据如图 6.1 所示。

```
MySQL 8.0 Command Line Client
mysql> select * from t_user;
+----+----------+---------+-------------+
| id | username | jobs    | phone       |
+----+----------+---------+-------------+
|  1 | zhangsan | teacher | 13907998372 |
|  2 | lisi     | worker  | 13907396542 |
|  3 | wangwu   | doctor  | 13817348729 |
+----+----------+---------+-------------+
3 rows in set (0.00 sec)
```

图 6.1　t_user 表中的数据

步骤02 在 IntelliJ IDEA 中创建一个名为 chapter06 的 Web 项目，将 MyBatis 的核心 JAR 包、lib 目录中的依赖 JAR 包，以及 MySQL 数据库的驱动 JAR 包一同添加到项目的 lib 目录下，并发布到类路径中。添加后的 lib 目录如图 6.2 所示。

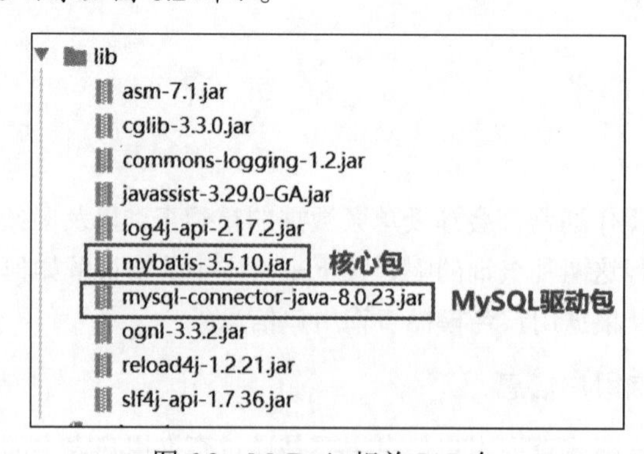

图 6.2　MyBatis 相关 JAR 包

步骤03 由于 MyBatis 默认使用 log4j 输出日志信息,因此如果要查看控制台的输出 SQL 语句,就需要在 classpath 路径下配置日志文件。在项目的 src 目录下创建 log4j.properties 文件，内容如文件 6.1 所示。

文件 6.1 log4j.properties

```
01 # 全局的日志配置
02 log4j.rootLogger=ERROR, stdout
03 # MyBatis 的日志配置
04 log4j.logger.com.ssm=DEBUG
05 Console output...
06 log4j.appender.stdout=org.apache.log4j.ConsoleAppender
07 log4j.appender.stdout.layout=org.apache.log4j.PatternLayout
08 log4j.appender.stdout.layout.ConversionPattern=%5p [%t] - %m%n
```

在文件 6.1 中包含全局的日志配置、MyBatis 的日志配置和控制台输出，其中 MyBatis 的日志配置用于将 com.ssm 包下所有类的日志记录级别设置为 DEBUG。

步骤 04 在 src 目录下创建一个 com.ssm.po 包，在该包下创建持久化类 User，并在类中声明 id、username、jobs 和 phone 属性，及其对应的 getter()和 setter()方法，如文件 6.2 所示。

文件 6.2 User.java

```
01 package com.ssm.po;
02 //用户类User
03 public class User {
04     private Integer id; //用户 id
05     private String username; //用户姓名
06     private String jobs;  //用户职业
07     private String phone; //用户电话号码
08     public Integer getId() {
09         return id;
10     }
11     public void setId(Integer id) {
12         this.id = id;
13     }
14     public String getUsername() {
15         return username;
16     }
17     public void setUsername(String username) {
18         this.username = username;
19     }
20     public String getJobs() {
21         return jobs;
22     }
23     public void setJobs(String jobs) {
24         this.jobs = jobs;
25     }
26     public String getPhone() {
```

```
27          return phone;
28      }
29      public void setPhone(String phone) {
30          this.phone = phone;
31      }
32      public String toString() {
33          return "User [id=" + id + ", username=" + username + ", jobs=" + jobs +
34              ", phone=" + phone + "]";
35      }
36  }
```

从上述代码可以看出，持久化类 User 与普通的 JavaBean 并没有什么区别，只是它的属性字段与数据库中的表字段相对应。实际上，User 就是一个 POJO（普通 Java 对象）。MyBatis 就是采用 POJO 作为持久化类来完成数据库操作的。

步骤 05 在 src 目录下创建一个 com.ssm.mapper 包，并在包中创建映射文件 UserMapper.xml，内容如文件 6.3 所示。

文件 6.3 UserMapper.xml

```
01  <?xml version="1.0" encoding="UTF-8"?>
02  <!DOCTYPE mapper PUBLIC "-//mybatis.org//DTD Mapper 3.0//EN"
03      "http://mybatis.org/dtd/mybatis-3-mapper.dtd">
04  <mapper namespace="com.ssm.mapper.UserMapper">
05      <!--根据用户编号获取用户信息 -->
06      <select id="findUserById" parameterType="Integer" resultType="com.ssm.po.User">
07          select * from t_user where id=#{id}
08      </select>
09  </mapper>
```

在文件 6.3 中，第 02、03 行代码是 MyBatis 的约束配置，第 04~09 行代码是需要程序员编写的映射信息。其中，<mapper>元素是配置文件的根元素，包含一个 namespace 属性，该属性为<mapper>元素指定了唯一的命名空间，通常会设置成“包名+SQL 映射文件名”的形式。子元素<select>中的信息是用于执行查询操作的配置，其 id 属性是<select>元素在映射文件中的唯一标识；parameterType 属性用于指定传入参数的类型，这里表示传递给执行 SQL 的是一个 Integer 类型的参数；resultType 属性用于指定返回结果的类型，这里表示返回的数据是 User 类型。在定义的查询 SQL 语句中，“#{}”用于表示一个占位符，相当于“?”，而“#{id}”表示该占位符待接收参数的名称为 id。

步骤 06 在 src 目录下创建 MyBatis 的核心配置文件 mybatis-config.xml，内容如文件 6.4 所示。

文件 6.4 mybatis-config.xml

```
01  <?xml version="1.0" encoding="UTF-8"?>
02  <!DOCTYPE configuration PUBLIC "-//mybatis.org//DTD Config 3.0//EN"
03      "http://mybatis.org/dtd/mybatis-3-config.dtd">
04  <configuration>
05  <!--1.配置环境，默认的环境 id 为 mysql -->
06      <environments default="mysql">
```

```
07        <!--1.2 配置 id 为 mysql 的数据库环境 -->
08            <environment id="mysql">
09                <!-- 使用 JDBC 的事务管理 -->
10                <transactionManager type="JDBC" />
11                <!-- 数据库连接池 -->
12                <dataSource type="POOLED">
13                    <property name="driver" value="com.mysql.cj.jdbc.Driver" />
14                    <property name="url"
15    value="jdbc:mysql://localhost:3306/db_mybatis?serverTimezone=UTC" />
16                    <property name="username" value="root" />
17                    <property name="password" value="root" />
18                </dataSource>
19            </environment>
20        </environments>
21        <!—2.配置 Mapper 的位置 -->
22        <mappers>
23            <mapper resource="com/ssm/mapper/UserMapper.xml" />
24        </mappers>
25    </configuration>
```

在文件 6.4 中，第 02、03 行代码是 MyBatis 的配置文件的约束信息。下面<configuration>元素中的内容是开发人员需要编写的配置信息。这里按照<configuration>子元素的功能将配置分为了两步：第 1 步，配置环境；第 2 步，配置 mapper 的位置。

步骤 07 在 src 目录下创建一个 com.ssm.test 包，在该包下创建测试类 MybatisTest，并在类中编写测试方法 findUserByIdTest()，如文件 6.5 所示。

文件 6.5　MybatisTest.java

```
01    package com.ssm.test;
02    import java.io.InputStream;
03    import org.apache.ibatis.io.Resources;
04    import org.apache.ibatis.session.SqlSession;
05    import org.apache.ibatis.session.SqlSessionFactory;
06    import org.apache.ibatis.session.SqlSessionFactoryBuilder;
07    import org.junit.Test;
08    import com.ssm.po.User;
09    // 入门程序测试类
10    public class MybatisTest {
11    // 根据用户 id 查询用户信息
12        @Test
13        public void findUserByIdTest() throws Exception{
14            //1.读取配置文件
15            String resource="mybatis-config.xml";
16            InputStream inputStream=Resources.getResourceAsStream(resource);
17            //2.根据配置文件构建 SqlSessionFactory 实例
18            SqlSessionFactory sqlSessionFactory=new SqlSessionFactoryBuilder().build(inputStream);
```

```
19          //3.通过 SqlSessionFactory 创建 SqlSession 实例
20          SqlSession sqlSession=sqlSessionFactory.openSession();
21          //4.SqlSession 执行映射文件中定义的 SQL，并返回映射结果
22          User user=sqlSession.selectOne("com.ssm.mapper.UserMapper.findUserById",1);
23          //5.打印输出结果
24          System.out.println(user.toString());
25          //6.关闭 SqlSession
26          sqlSession.close();
27      }
28  }
```

在文件 6.5 的 findUserByIdTest()方法中，首先通过输入流读取了配置文件，然后根据配置文件构建了 SqlSessionFactory 对象，接下来通过 SqlSessionFactory 对象又创建了 SqlSession 对象，并通过 SqlSession 对象的 selectOne()方法执行查询操作。selectOne()方法的第 1 个参数表示映射 SQL 的标识字符串，由 UserMapper.xml 中<mapper>元素的 namespace 属性值+<select>元素的 id 属性值组成；第 2 个参数表示查询所需要的参数，这里查询的是用户表中 id 为 1 的用户。为了查看查询结果，这里使用了输出语句输出查询结果信息。最后，程序执行完毕时，关闭 SqlSession。

使用 JUnit4 测试执行 findUserByIdTest()方法后，控制台的输出结果如图 6.3 所示。

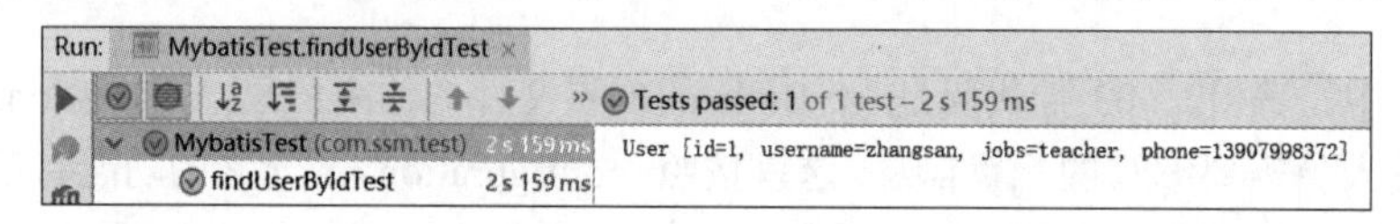

图 6.3 运行结果

2. 根据用户名模糊查询用户信息

接下来讲解如何根据用户名来模糊查询相关的用户信息。

【示例 6-2】模糊查询的实现只需要在映射文件中通过<select>元素编写相应的 SQL 语句，并通过 SqlSession 的查询方法执行该 SQL 语句即可。具体实现步骤如下：

步骤 01 在映射文件 UserMapper.xml 中添加根据用户名模糊查询用户信息列表的 SQL 语句，具体实现代码如下：

```
<!--根据用户名模糊查询用户信息 -->
<select id="findUserByName" parameterType="String" resultType="com.ssm.po.User">
select * from t_user where username like '%${value}%'
</select>
```

与根据用户编号查询相比，上述配置代码中的属性 id、parameterType 和 SQL 语句都发生了相应变化。其中，SQL 语句中的“${}”用来表示拼接 SQL 的字符串，即不加解释地原样输出。“${value}”表示要拼接的是简单类型参数。

注意：在使用“${}”进行 SQL 字符串拼接时，无法防止 SQL 注入问题。如果想要既能实现模糊查询，又能防止 SQL 注入，可以对上述映射文件 UserMapper.xml 中模糊查询的 select 语句进行修改，使用 MySQL 中的 concat 函数进行字符串拼接。具体修改示例如下：

```
select * from t_user where username like concat('%', '#{value}','%')
```

步骤02 在测试类MybatisTest中添加一个测试方法findUserByNameTest()，代码如下：

```
@Test
public void findUserByNameTest() throws Exception{
    String resource = "mybatis-config.xml";
    InputStream inputStream = Resources.getResourceAsStream(resource);
    SqlSessionFactory sqlSessionFactory =
                     new SqlSessionFactoryBuilder().build(inputStream);
    SqlSession sqlSession = sqlSessionFactory.openSession();
    List<User> users =
           sqlSession.selectList("com.ssm.mapper.UserMapper.findUserByName", "g");
    //循环输出结果集
    for (User user : users) {
        System.out.println(user.toString());
    }
    sqlSession.close();
}
```

在上述代码中，由于可能查询出多条数据，因此调用SqlSession的selectList()方法来查询返回结果的集合对象，并使用for循环输出结果集对象。执行findUserByNameTest()方法后，控制台的输出结果如图6.4所示。

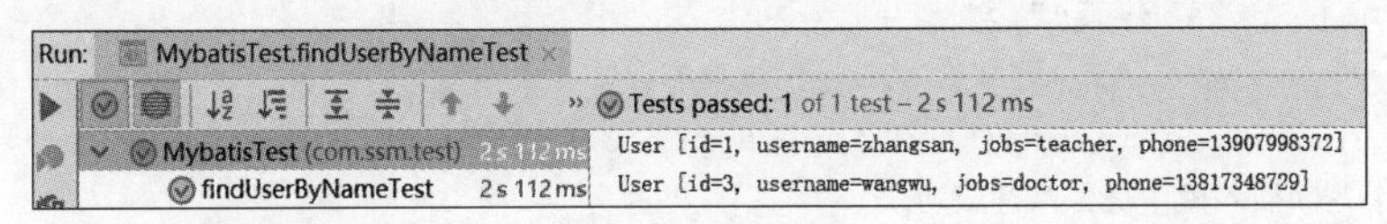

图6.4 运行结果

从图6.4可以看出，使用MyBatis框架已成功查询出了用户表中用户名带有“g”的两条客户信息。

至此，MyBatis入门程序的查询功能已经讲解完毕。从上面两个查询方法中可以发现，MyBatis的操作大致可分为以下几个步骤：

步骤01 读取配置文件。

步骤02 根据配置文件构建SqlSessionFactory。

步骤03 通过SqlSessionFactory创建SqlSession。

步骤04 使用SqlSession对象操作数据库（包括查询、添加、修改、删除以及提交事务等）。

步骤05 关闭SqlSession。

6.2.2 添加用户

在MyBatis的映射文件中，添加操作是通过<insert>元素来实现的。例如，向数据库中的t_user表中插入一条数据，可以通过如下配置来实现。

```
<!--添加用户信息 -->
<insert id="addUser" parameterType="com.ssm.po.User">
```

```
    insert into  t_user(username,jobs,phone)values(#{username},#{jobs},#{phone})
</insert>
```

在上述配置代码中，传入的参数是一个 User 类型，该类型的参数对象被传递到语句中时，#{username}会查找参数对象 User 的 username 属性，#{jobs}和#{phone}也是一样的，并将其属性值传入 SQL 语句中。

【示例 6-3】为了验证上述配置是否正确，下面编写一个测试方法来执行添加操作。

在测试类 MybatisTest 中，添加测试方法 addUserTest()，其代码如下：

```
@Test
public void addUserTest() throws Exception {
    String resource = "mybatis-config.xml";
    InputStream inputStream = Resources.getResourceAsStream(resource);
    SqlSessionFactory sqlSessionFactory =
                 new SqlSessionFactoryBuilder().build(inputStream);
    SqlSession sqlSession = sqlSessionFactory.openSession();
    //创建 User 对象，并向对象中添加数据
    User user = new User();
    user.setUsername("tom");
    user.setJobs("worker");
    user.setPhone("13624589654");
    //执行 SqlSession 的插入方法，返回受 SQL 语句影响的行数
    int rows = sqlSession.insert("com.ssm.mapper.UserMapper.addUser", user);
    //通过返回结果判断插入操作是否执行成功
    if (rows > 0) {
        System.out.println("成功添加" + rows + "条数据！");
    } else {
        System.out.println("添加数据失败!");
    }
    //提交事务
    sqlSession.commit();
    //关闭 SqlSession
    sqlSession.close();
}
```

上述代码中，创建了 User 对象，并向 User 对象中添加了属性值；然后通过 SqlSession 对象的 insert()方法执行插入操作，并通过该操作返回的数据来判断插入操作是否执行成功；最后通过 SqlSession 的 commit()方法提交事务，并通过 sqlSession.close()方法关闭了 SqlSession。

6.2.3 更新用户

MyBatis 的更新操作在映射文件中是通过配置<update>元素来实现的。如果需要更新用户数据，可以通过如下配置来实现。

```
<!--更新用户信息 -->
```

```
<update id="updateUser" parameterType="com.ssm.po.User">
    update t_user set username=#{username},jobs=#{jobs},phone=#{phone} where id=#{id}
</update>
```

与插入数据的配置相比，更新操作配置中的元素与 SQL 语句都发生了相应变化，但其属性名却没有变。

【示例 6-4】为了验证配置是否正确，下面以第 6.2.2 节中新插入的数据为例进行更新用户测试。

在测试类 MybatisTest 中添加测试方法 updateUserTest()，将 id 为 4 的用户的 jobs 属性值修改为 teacher，代码如下：

```
@Test
public void updateUserTest() throws Exception {
    String resource = "mybatis-config.xml";
    InputStream inputStream = Resources.getResourceAsStream(resource);
    SqlSessionFactory sqlSessionFactory = new SqlSessionFactoryBuilder().build(inputStream);
    SqlSession sqlSession = sqlSessionFactory.openSession();
    //创建 User 对象，并对对象中的数据进行模拟更新
    User user = new User();
    user.setId(4);
    user.setUsername("tom");
    user.setJobs("teacher");
    user.setPhone("13624589654");
    //执行 SqlSession 的更新方法，返回受 SQL 语句影响的行数
    int rows = sqlSession.update("com.ssm.mapper.UserMapper.updateUser", user);
    if (rows > 0) {
        System.out.println("成功修改了" + rows + "条数据！");
    } else {
        System.out.println("修改数据失败！");
    }
    sqlSession.commit();
    sqlSession.close();
}
```

与添加用户的方法相比，更新操作的代码增加了 id 属性值的设置，并调用 SqlSession 的 update() 方法对 id 为 4 的用户的 jobs 属性值进行了修改。

6.2.4 删除用户

在映射文件中，MyBatis 的删除操作是通过配置<delete>元素来实现的。在映射文件 UserMapper.xml 中添加删除客户信息的 SQL 语句，示例代码如下：

```
<!--删除用户信息-->
<delete id="deleteUser" parameterType="Integer">
    delete from t_user where id=#{id}
</delete>
```

从上述配置的 SQL 语句中可以看出，我们只需要传递一个 id 值就可以将数据表中相应的数据删除。要测试删除操作的配置十分简单，只需使用 SqlSession 对象的 delete()方法传入需要删除数据的 id 值即可。

【示例 6-5】在测试类 MybatisTest 中添加测试方法 deleteUserTest()，该方法用于将 id 为 4 的用户信息删除，代码如下：

```
@Test
public void deleteUserTest() throws Exception {
    String resource = "mybatis-config.xml";
    InputStream inputStream = Resources.getResourceAsStream(resource);
    SqlSessionFactory sqlSessionFactory = new
SqlSessionFactoryBuilder().build(inputStream);
    SqlSession sqlSession = sqlSessionFactory.openSession();
    //执行 SqlSession 的删除方法，返回受 SQL 语句影响的行数
    int rows = sqlSession.delete("com.ssm.mapper.UserMapper.deleteUser", 4);
    if (rows > 0) {
        System.out.println("成功删除了" + rows + "条数据！");
    } else {
        System.out.println("删除数据失败！");
    }
    sqlSession.commit();
    sqlSession.close();
}
```

至此，MyBatis 入门程序的增、删、改、查操作已经讲解完毕。关于程序中的映射文件和配置文件中的元素信息将在第 7 章详细讲解，本章入门程序只需要了解所使用的元素即可。

第 7 章

MyBatis 的核心配置

如果想要熟练地使用 MyBatis 框架进行实际开发，那么需要对框架中的核心对象以及映射文件和配置文件有深入的了解。本章将对这些内容进行详细讲解。

本章主要涉及的知识点如下：

- MyBatis 核心对象。
- MyBatis 配置文件的元素。
- MyBatis 映射文件。

7.1 MyBatis 核心对象

MyBatis 框架主要涉及两个核心对象：SqlSessionFactory 和 SqlSession。本节将详细介绍这两个对象。

7.1.1 SqlSessionFactory

SqlSessionFactory 是单个数据库映射关系经过编译后的内存镜像，用于创建 SqlSession。SqlSessionFactory 对象的实例通过 SqlSessionFactoryBuilder 对象来构建，通过 XML 配置文件或一个预先定义好的 Configuration 实例构建出 SqlSessionFactory 的实例。通过 XML 配置文件构建出 SqlSessionFactory 实例的实现代码如下：

```
//读取配置文件
InputStream inputStream = Resources.getResourceAsStream("配置文件位置");
//根据配置文件构建 SqlSessionFactory
```

```
SqlSessionFactory sqlSessionFactory = new SqlSessionFactoryBuilder().build(inputStream);
```

SqlSessionFactory 对象是线程安全的，一旦被创建，在整个应用执行期间都会存在。如果多次创建同一个数据库的 SqlSessionFactory，那么此数据库的资源将很容易被耗尽，所以在构建 SqlSessionFactory 实例时，建议使用单例模式。

7.1.2 SqlSession

SqlSession 是应用程序与持久层之间执行交互操作的一个单线程对象，其主要作用是执行持久化操作。SqlSession 对象包含数据库中所有执行 SQL 操作的方法，底层封装了 JDBC 连接，所以可以直接使用它的实例来执行已映射的 SQL 语句。SqlSession 实例是不能被共享的，也是线程不安全的，因此它的使用范围最好限定在一次请求或一个方法中，绝不能将它放在一个类的静态字段、实例字段或任何类型的管理范围中使用。使用完 SqlSession 对象之后，要及时将它关闭，通常可以放在 finally 块中关闭。

SqlSession 对象常用方法如下：

（1）<T> T selectOne(String statement);

查询方法。参数 statement 是在配置文件中定义的<select>元素的 id。该方法返回 SQL 语句查询结果的一个泛型对象。

（2）<T>I selectOne(String statement, Object parameter);

查询方法。参数 statement 是在配置文件中定义的<select>元素的 id，parameter 是查询所需的参数。该方法返回 SQL 语句查询结果的一个泛型对象。

（3）<E> List<E> selectList(String statement);

查询方法。参数 statement 是在配置文件中定义的<select>元素的 id。该方法返回 SQL 语句查询结果的泛型对象的集合。

（4）<E> List<E> selectList(String statement, Object parameter);

查询方法。参数 statement 是在配置文件中定义的<select>元素的 id，parameter 是查询所需的参数。该方法返回 SQL 语句查询结果的泛型对象的集合。

（5）<E> List<E> selectList(String statement, Object parameter, RowBounds rowBounds);

查询方法。参数 statement 是在配置文件中定义的<select>元素的 id，parameter 是查询所需的参数，rowBounds 是用于分页的参数对象。该方法返回 SQL 语句查询结果的泛型对象的集合。

（6）<K, V> Map<K, V> selectMap(String statement, String mapKey);

查询方法。参数 statement 是在配置文件中定义的<select>元素的 id，mapKey 是查询所需的 key 值。该方法返回 SQL 语句查询结果的泛型对象的键-值对。

（7）<K, V> Map<K, V> selectMap(String statement, Object parameter, String mapKey);

查询方法。参数 statement 是在配置文件中定义的<select>元素的 id，parameter 是查询所需的参数，mapKey 是查询所需的 key 值。该方法返回 SQL 语句查询结果的泛型对象的键-值对。

（8）<K, V> Map<K, V> selectMap(String statement, Object parameter, String mapKey, RowBounds rowBounds);

查询方法。参数 statement 是在配置文件中定义的<select>元素的 id，parameter 是查询所需的参数，mapKey 是查询所需的 key 值，rowBounds 是用于分页的参数对象。该方法返回 SQL 语句查询结果的泛型对象的键-值对。

（9）void select(String statement, Object parameter, ResultHandler handler);

查询方法。参数 statement 是在配置文件中定义的<select>元素的 id，parameter 是查询所需的参数，ResultHandler 对象用于处理查询返回的复杂结果集，通常用于多表查询。

（10）int insert(String statement);

插入方法。参数 statement 是在配置文件中定义的<insert>元素的 id。该方法返回执行 SQL 语句所影响的行数。

（11）int insert(String statement, Object parameter);

插入方法。参数 statement 是在配置文件中定义的<insert>元素的 id，parameter 是插入所需的参数。该方法返回执行 SQL 语句所影响的行数。

（12）int update(String statement);

更新方法。参数 statement 是在配置文件中定义的<update>元素的 id。该方法返回执行 SQL 语句所影响的行数。

（13）int update(String statement, Object parameter);

更新方法。参数 statement 是在配置文件中定义的<update>元素的 id，parameter 是更新所需的参数。该方法返回执行 SQL 语句所影响的行数。

（14）int delete(String statement);

删除方法。参数 statement 是在配置文件中定义的<delete>元素的 id。该方法返回执行 SQL 语句所影响的行数。

（15）int delete(String statement, Object parameter);

删除方法。参数 statement 是在配置文件中定义的<delete>元素的 id，parameter 是删除所需的参数。该方法返回执行 SQL 语句所影响的行数。

（16）void commit();

提交事务的方法。

（17）void rollback();

回滚事务的方法。

（18）List<BatchResult> flushStatements();

获得被操作的数据记录。

（19）void close();

关闭 SqlSession 对象。

（20）<T>T getMapper(Class<T> type)

返回 Mapper 接口的代理对象，该对象关联了 SqlSession 对象，开发人员可以使用该对象直接调用方法操作数据库。参数 type 是 Mapper 的接口类型。

（21）Connection getConnection();

获取 JDBC 数据库连接对象的方法。

注意：为了简化开发，可以将构建 SqlSessionFactory 对象、SqlSession 对象等重复性代码封装到一个工具类中，然后通过工具类来创建 SqlSession。

7.2 MyBatis 配置文件的元素

使用 MyBatis 框架进行开发，需要创建 MyBatis 的核心配置文件，该配置文件包含重要的元素，熟悉配置文件中各个元素的功能十分重要。本节将对 MyBatis 配置文件中的主要元素进行详细讲解。

在 MyBatis 框架的核心配置文件中，<configuration>元素是配置文件的根元素，其他元素都要在<contiguration>元素内配置。

MyBatis 配置文件中的主要元素如下：

```
<configuration>
    <!-- 属性 -->
    <properties/>
    <!-- 设置 -->
    <settings/>
    <!-- 类型命名 -->
    <typeAliases/>
    <!-- 类型处理器 -->
    <typeHandlers/>
    <!-- 对象工厂 -->
    <objectFactory/>
    <!-- 插件 -->
    <plugins/>
    <!-- 配置环境 -->
    <environments>
        <!-- 环境变量 -->
        <environment>
            <!-- 事务管理器 -->
            <transactionManager/>
            <!-- 数据源-->
            <dataSource/>
        </environment>
    </environments>
    <!-- 数据库厂商标识 -->
    <databaseIdProvider/>
```

```
    <!-- 映射器 -->
    <mappers/>
</configuration>
```

在 MyBatis 的配置文件中包含多个元素，这些元素在配置文件中分别发挥着不同的作用。开发人员需要熟悉的是<configuration>元素中的各个子元素的配置。

注意：<configuration>的子元素必须按照上述代码中的顺序进行配置，否则 MyBatis 在解析 XML 配置文件的时候会报错。

7.2.1 <properties>元素

<properties>是一个配置属性的元素，通过外部配置来动态替换内部定义的属性。

【示例 7-1】配置数据库的连接等属性，具体方式如下：

步骤 01 在项目的 src 目录下创建一个名称为 db.properties 的配置文件，代码如下：

```
jdbc.driver=com.mysql.cj.jdbc.Driver
jdbc.url=jdbc:mysql: //localhost: 3306/db_mybatis?serverTimezone=UTC
jdbc.username=root
jdbc.password=root
```

步骤 02 在 MyBatis 配置文件 mybatis-config.xml 中配置<properties/>属性，具体如下：

```
<properties resource="db.properties"/>
```

步骤 03 修改配置文件中数据库连接的信息，具体如下：

```
<dataSource type="POOLED">
    <!--数据库驱动 -->
    <property name="driver" value="${jdbc.driver}" />
    <!--连接数据库的 url -->
    <property name="url" value="${jdbc.url}" />
    <!--连接数据库的用户名 -->
    <property name="username" value="${jdbc.username}" />
    <!--连接数据库的密码-->
    <property name="password" value="${jdbc.password}" />
</dataSource>
```

完成上述配置，dataSource 中连接数据库的 4 个属性（driver、url、username 和 password）值将会由 db.properties 文件中对应的值来动态替换。这样就为配置提供了灵活性。

另外，还可以通过配置<properties>元素的子元素<property>以及通过方法参数传递的方式来获取属性值。

由于使用 properties 配置文件来配置属性值，可以方便地在多个配置文件中使用这些属性值，并且方便维护和修改，因此它在实际开发中最常用。

7.2.2 <settings>元素

<settings>元素主要用于改变 MyBatis 运行时的行为，例如开启二级缓存、开启延迟加载等。<settings>元素中的常见配置如表 7.1 所示。

表 7.1 <settings>元素中的常见配置

配置参数	描 述	有 效 值	默 认 值
cacheEnabled	该配置影响所有映射器中的缓存全局开关	true/false	false
lazyLoadingEnabled	延迟加载的全局开关。开启时，所有关联对象都会延迟加载。在特定关联关系中可以通过设置 fetchType 属性来覆盖该项的开关状态	true/false	false
aggressiveLazyLoading	关联对象属性的延迟加载开关。当启用时，对任意延迟属性的调用会使带有延迟加载属性的对象完整加载；反之，每种属性都会按需加载	true/false	true
multipleResultSetsEnabled	是否允许单一语句返回多结果集（需要兼容驱动）	true/false	true
useColumnLabel	使用列标签代替列名。不同的驱动在这方面有不同的表现。具体可参考驱动文档或通过测试两种模式来观察所用驱动的行为	true/false	true
useGeneratedKeys	允许 JDBC 支持自动生成主键，需要驱动兼容。如果设置为 true，那么这个设置强制使用自动生成主键，尽管一些驱动不兼容，但仍可正常工作	true/false	false
autoMappingBehavior	指定 MyBatis 应如何自动映射列到字段或属性。NONE 表示取消自动映射，PARTIAL 只会自动映射没有定义嵌套结果集映射的结果集，FULL 会自动映射任意复杂的结果集（无论是否嵌套）	NONE、PARTIAL、FULL	PARTIAL
defaultExecutorType	配置默认的执行器。SIMPLE 就是普通的执行器，REUSE 执行器会重用预处理语句（Prepared Statement），BATCH 执行器将重用语句并执行批量更新	SIMPLE、REUSE、BATCH	SIMPLE
defaultStatementTimeout	设置超时时间，决定驱动等待数据库响应的秒数。在没有设置值的时候，它取的是驱动默认的时间	任何正整数	没有设置
mapUnderscoreTocamelcase	是否开启自动驼峰命名规则（Camel-Case）映射	true/false	false
jdbcTypeForNull	当没有为参数提供特定的 JDBC 类型时，为空值指定 JDBC 类型。某些驱动需要指定列的 JDBC 类型，多数情况下直接用一般类型即可，比如 NULL、VARCHAR 或 OTHER	NULL、VARCHAR、OTHER	OTHER
returnInstanceForEmptyRow	当返回行的所有列都是空时，MyBatis 默认返回 null。当开启这个设置时，MyBatis 会返回一个空实例。同样适用于嵌套的结果集。	true/false	false

（续表）

配置参数	描　述	有效值	默认值
configurationFactory	指定一个提供 Configuration 实例的类。这个被返回的 Configuration 实例用来加载被反序列化对象的延迟加载属性值。这个类必须包含一个签名为 static Configuration getConfiguration()的方法	一个类别名或完全限定类名	未设置

表 7.1 中介绍了<settings>元素中的常见配置，这些配置在配置文件中的使用方式如下：

```
<!--设置-->
<settings>
    <setting name="cacheEnabled" value="true" />
    <setting name="lazyLoadingEnabled" value="true" />
    <setting name="multipleResultsetsEnabled" value="true" />
    <setting name="useColumnLabel" value="true" />
    <setting name="useGeneratedKeys" value="false"/>
    <setting name="autoMappingBehavior" value="PARTIAL"/>
    …
</settings>
```

上面介绍的配置内容通常在需要时只配置少数几项即可。

7.2.3 <typeAliases>元素

<typeAliases>元素用于为配置文件中的 Java 类型设置一个简短的名字，即设置别名。别名的设置与 XML 配置相关，其使用的意义在于减少全限定类名的冗余。

使用<typeAliases>元素配置别名的方法如下：

```
<!--定义别名-->
<typeAliases>
    <typeAlias alias="user" type="com.ssm.po.User"/>
</typeAliases>
```

在上述示例中，<typeAliases>元素的子元素<typeAlias>中的 type 属性用于指定需要被定义别名的类的全限定名；alias 属性的属性值“user”就是自定义的别名，可以代替“com.ssm.po.User”使用在 MyBatis 文件的任何位置。如果省略 alias 属性，MyBatis 会默认将类名首字母小写后的名称作为别名。

当 POJO 类过多时，还可以通过自动扫描包的形式自定义别名，具体示例如下：

```
<!-使用自动扫描包来定义别名-->
<typeAliases>
    <package name="com.ssm.po"/>
</typeAliases>
```

在上述示例中，<typeAliases>元素的子元素<package>中的 name 属性用于指定要被定义别名的包，MyBatis 会将所有 com.ssm.po 包中的 POJO 类以首字母小写的非限定类名作为它的别名。

注意： 上述方式的别名只适用于没有使用注解的情况。如果在程序中使用了注解，那么别名为其注解的值，具体如下：

```
@Alias(value ="user")
public class User {
    //user 的属性和方法
    ...
}
```

除了可以使用<typeAliases>元素自定义别名外，MyBatis 框架还默认为许多常见的 Java 类型（如数值、字符串、日期和集合等）提供相应的类型别名，如表 7.2 所示。

表 7.2　MyBatis 默认别名

别　名	映射的类型	别　名	映射的类型
_byte	byte	double	Double
_short	short	float	Float
_int	int	boolean	Boolean
_integer	int	date	Date
_long	long	decimal	BigDecimal
float	float	bigdecimal	BigDecimal
_double	double	object	Object
_boolean	boolean	map	Map
string	String	hashmap	HashMap
byte	Byte	list	List
long	Long	arraylist	ArrayList
short	Short	collection	Collection
int	Integer	iterator	Iterator
integer	Integer		

提示： 表 7.2 所列举的别名可以在 MyBatis 中直接使用，但由于别名不区分大小写，因此在使用时要注意重复定义的覆盖问题。

7.2.4 <typeHandler>元素

MyBatis 在预处理语句中设置一个参数或者从结果集中取出一个值时，都会使用其框架内部注册了的 typeHandler（类型处理器）进行相关处理。typeHandler 的作用就是将预处理语句中传入的参数从 javaType（Java 类型）转换为 jdbcType（JDBC 类型），或者从数据库取出结果时将 jdbcType 转换为 javaType。

为了方便转换，MyBatis 框架提供了一些默认的类型处理器，如表 7.3 所示。

表 7.3　常用的类型处理器

类型处理器	Java 类型	JDBC 类型
BooleanTypeHandler	java.Lang.Boolean,boolean	数据库兼容的 BOOLEAN
ByteTypeHandler	java.Jang.Byte,byte	数据库兼容的 NUMERIC 或 BYTE
ShortTypeHandler	java.lang.Short,short	数据库兼容的 NUMERIC 或 SHORT INTEGER
IntegerTypeHandler	java.lang.Integer,int	数据库兼容的 NUMERIC 或 INTEGER
LongTypeHandler	java.lang.Long,long	数据库兼容的 NUMERIC 或 LONG INTEGER
FloatTypeHandler	java.lang.Float,float	数据库兼容的 NUMERIC 或 FLOAT
DoubleTypeHandler	java.lang.Double,double	数据库兼容的 NUMERIC 或 DOUBLE
BigDecimalTypeHandler	java.math.BigDecimal	数据库兼容的 NUMERIC 或 DECIMAL
StringTypeHandler	java.langString	CHAR、VARCHAR
ClobTypeHandler	java.lang.String	CLOB、LONGVARCHAR
ByteArrayTypeHandler	byte[]	数据库兼容的字节流类型
BlobTypeHandler	byte[]	BLOB、LONGVARBINARY
DateTypeHandler	java.util.Date	TIMESTAMP
SqlTimestampTypeHandler	Java.sql.Timestamp	TIMESTAMP
SalDateTypeHandlel	java.sql.Date	DATE
sqlTimeTypeHandler	java.sql.Time	TIME

当 MyBatis 框架所提供的这些类型处理器不能够满足需求时，还可以通过自定义的方式对类型处理器进行扩展。自定义类型处理器可以通过实现 TypeHandler 接口或者继承 BaseTypeHandle 类来定义。<typeHandler>元素就是用于在配置文件中注册自定义的类型处理器的。它的使用方式有如下两种：

（1）注册一个类的类型处理器

```
<typeHandlers>
    <!--以单个类的形式配置-->
    <typeHandler handler="com.ssm.type.UsertypeHandlerl" />
</typeHandlers>
```

上述代码中，子元素<typeHandler>的 handler 属性用于指定在程序中自定义的类型处理器。

（2）注册一个包中所有的类型处理器

```
<typeHandlers>
    <!-注册一个包中所有的 typeHandler，系统在启动时会自动扫描包下的所有文件-->
    <package name="com.ssm.type" />
</typeHandlers>
```

在上述代码中，子元素<package>的 name 属性用于指定类型处理器所在的包名，使用这种方式后，系统会在启动时自动扫描 com.ssm.type 包下所有的文件，并把它们作为类型处理器。

7.2.5 <objectFactory>元素

MyBatis 框架每次创建结果对象的新实例时，都会使用一个对象工厂（ObjectFactory）的实例来完成。MyBatis 中默认的 ObjectFactory 的作用就是实例化目标类，我们既可以通过默认构造方法来实例化，也可以在参数映射存在的时候通过参数构造方法来实例化。

在通常情况下，我们使用默认的 ObjectFactory 即可。MyBatis 中默认的 ObjectFactory 是由 org.apache.ibatis.reflection.factory.DefaultObjectFactory 来提供服务的，大部分场景下都不用配置和修改。如果想覆盖 ObjectFactory 的默认行为，那么可以通过自定义 ObjectFactory 来实现。

7.2.6 <plugins>元素

MyBatis 允许在已映射语句执行过程中的某一点进行拦截调用（通过插件来实现）。<plugins>元素的作用就是配置用户所开发的插件。关于插件的使用，本书不做详细讲解，有兴趣的读者请查找官方文档等资料自行学习。

7.2.7 <environments>元素

<environments>元素用于在配置文件中对环境进行配置。MyBatis 的环境配置实际上就是数据源的配置，可以通过<environments>元素配置多种数据源，即配置多种数据库。

【示例 7-2】使用<environments>元素进行环境配置的示例如下：

```
<environments default="development">
    <environment id="development">
        <!--使用 JDBC 事务管理-->
        <transactionManager type="JDBC"/>
        <!--配置数据源-->
        <dataSource type="POOLED">
            <property name="driver" value="${jdbc.driver}" />
            <property name="url" value="${jdbc.url}" />
            <property name="username" value="${jdbc. username }" />
            <property name=" password " value="${jdbc. password }" />
        </datasource>
    </environment>
    …
</environments>
```

在上述示例代码中，<environments>元素是环境配置的根元素，它包含一个 default 属性，该属性用于指定默认的环境 id。<environment>是<environments>元素的子元素，它可以被定义多个，其 id 属性用于表示所定义环境的 id 值。在<environment>元素内，包含事务管理和数据源的配置信息，其中<transactionManager>元素用于配置事务管理，它的 type 属性用于指定事务管理的方式，即使用哪种事务管理器；<dataSource>元素用于配置数据源，它的 type 属性用于指定使用哪种数据源。

在 MyBatis 中，可以配置两种类型的事务管理器，分别是 JDBC 和 MANAGED。

- JDBC：此配置直接使用 JDBC 的提交和回滚设置，依赖从数据源得到的连接来管理事务的作用域。
- MANAGED：此配置从来不提交或回滚一个连接，而是让容器来管理事务的整个生命周期。在默认情况下，它会关闭连接，但一些容器并不希望这样，为此可以将 closeConnection 属性设置为 false 来阻止它默认的关闭行为。

注意：如果项目中使用 Spring+MyBatis，那就没有必要在 MyBatis 中配置事务管理器，因为实际开发中会使用 Spring 自带的管理器来实现事务管理。

对于数据源的配置，MyBatis 框架提供了 UNPOOLED、POOLED 和 JNDI 三种数据源类型。

- UNPOOLED：配置此数据源类型后，在每次被请求时会打开和关闭连接。它对没有性能要求的简单应用程序来说是一个很好的选择。
- POOLED：此数据源利用"池"的概念将 JDBC 连接对象组织起来，避免在创建新的连接实例时花费时间进行初始化和认证。这种方式使得并发 Web 应用可以快速地响应请求，是当前流行的处理方式。
- JNDI：此数据源可以在 EJB 或应用服务器等容器中使用。容器可以集中或在外部配置数据源，然后放置一个 JNDI 上下文的引用。

7.2.8　<mappers>元素

在配置文件中，<mappers>元素用于指定 MyBatis 映射文件的位置，一般可以使用以下 4 种方法来引入映射文件：

（1）使用类路径引入

```
<mappers>
    <mapper resource="com/ssm/mapper/UserMapper.xmI"/>
</mappers>
```

（2）使用本地文件路径引入

```
<mappers>
    <mapper url=file: ///D:/com/ssm/mapper/UserMapper.xml"/>
</mappers>
```

（3）使用接口类引入

```
<mappers>
    <mapper class="com.ssm.mapper.UserMapper"/>
</mappers>
```

（4）使用包名引入

```
<mappers>
    <package name=""com.ssm.mapper"/>
</mappers>
```

7.3 映射文件

映射文件是 MyBatis 框架中十分重要的文件。在映射文件中，<mapper>元素是映射文件的根元素，其他元素都是它的子元素。映射文件中的主要元素如下：

```
<mapper>
   <!-- 映射查询语句，可自定义参数、返回结果集等 -->
   <select/>
   <!-- 映射插入语句，执行后返回一个整数，代表插入的条数 -->
   <insert/>
   <!-- 映射更新语句，执行后返回一个整数，代表更新的条数 -->
   <update/>
   <!-- 映射删除语句，执行后返回一个整数，代表删除的条数 -->
   <delete/>
   <!-- 用于定义一部分 SQL，然后其他语句可引用此 SQL -->
   <sql/>
   <!-- 给定命名空间的缓存配置 -->
   <cache/>
   <!-- 其他命名空间缓存配置的引用 -->
   <cache-ref/>
   <!-- 用于描述如何从数据库结果集中加载对象 -->
   <resultMap/>
</mapper>
```

7.3.1 <select>元素

<select>元素用于映射查询语句，从数据库中读取数据，并组装数据给业务开发人员。示例如下：

```
<select id="findUserById" parameterType="Integer" resultType="com.ssm.po.User">
   select * from t_user where id=#{id}
</select>
```

上述语句中的唯一标识为 findUserById，它接收一个 Integer 类型的参数，并返回一个 User 类型的对象。

<select>元素中的常用属性如表 7.4 所示。

表 7.4 <select>元素的常用属性

属 性	说 明
id	表示命名空间中的唯一标识符，常与命名空间组合起来使用。如果组合后不唯一，那么 MyBatis 就会抛出异常
parameterType	该属性表示传入 SQL 语句的参数类的全限定名或者别名。它是一个可选属性，因为 MyBatis 可以通过 TypeHandler 推断出具体传入语句的参数。其默认值是 unset（依赖于驱动）
resultType	从 SQL 语句中返回的类型的类的全限定名或者别名。如果是集合类型，那么返回的应该是集合可以包含的类型，而不是集合本身。返回时可以使用 resultType 和 resultMap 中的一个
resultMap	表示外部 resultMap 的命名引用。返回时可以使用 resultType 和 resultMap 中的一个
flushCache	表示在调用 SQL 语句之后，是否需要 MyBatis 清空之前查询本地缓存和二级缓存。其值为布尔类型（true/false），默认值为 false。如果设置为 true，那么任何时候只要 SQL 语句被调用，都会清空本地缓存和二级缓存
useCache	用于控制二级缓存的开启和关闭。其值为布尔类型（true/false），默认值为 true，表示将查询结果存入二级缓存中
timeout	用于设置超时时间，单位为秒。超时时将抛出异常
fetchSize	获取记录的总条数设定，默认值是 unset（依赖于驱动）
statementType	用于设置 MyBatis 使用哪个 JDBC 的 Statement 工作，其值为 STATEMENT、PREPARED（默认值）或 CALLABLE，分别对应 JDBC 中的 Statement、PreparedStatement 和 CallableStatement
resultSetType	表示结果集的类型，其值可设置为 FORWARD_ONLY、SCROLL_SENSITIVE 或 SCROLL_INSENSITIVE，默认值是 unset（依赖于驱动）

7.3.2 <insert>元素

<insert>元素用于映射插入语句，在执行完元素中定义的 SQL 语句后，会返回一个表示插入记录数的整数。<insert>元素的配置示例如下：

```
<insert id="addUser" parameterType="com.ssm.po.User" flushCache="true"
   statementType="PREPARED"  keyProperty="id" keyColumn=""
   useGeneratedKeys="" timeout="20">
   insert into t_user(username,jobs,phone)values(#{username},#{jobs},#{phone})
</insert>
```

从上述示例代码中可以看出，<insert>元素的属性与<select>元素的属性大部分相同，只是多了 3 个属性（仅对<insert>和<update>元素有用），如表 7.5 所示。

表 7.5 <insert>元素的 3 个属性

属 性	说 明
keyProperty	此属性的作用是将插入或更新操作时的返回值赋给 PO 类的某个属性，通常会设置为主键对应的属性。如果需要设置联合主键，那么可以在多个值之间用逗号隔开

（续表）

属　性	说　明
keyColumn	此属性用于设置第几列是主键，当主键列不是表中的第一列时，需要设置。在需要主键联合时，值可以用逗号隔开
useGeneratedKeys	此属性会使 MyBatis 用 JDBC 的 getGeneratedKeys()方法来获取由数据库内部产生的主键，如 MySQL 和 SQL Server 等自动递增的字段，默认值为 false

执行插入操作后，很多时候我们需要返回插入成功的数据生成的主键值，此时就可以通过上面所讲解的 3 个属性来实现。

【示例 7-3】如果使用的数据库支持主键自动增长（如 MySQL），那么可以通过 keyProperty 属性指定 PO 类的某个属性接收主键返回值（通常会设置到 id 属性上），然后将 useGeneratedKeys 的属性值设置为 true。使用上述配置执行插入后，会返回插入成功的行数以及插入行的主键值。测试代码如下：

```
@Test
public void addUserTest() throws Exception {
    String resource = "mybatis-config.xml";
    InputStream inputStream = Resources.getResourceAsStream(resource);
    SqlSessionFactory sqlSessionFactory =
new SqlSessionFactoryBuilder().build(inputStream);
    SqlSession sqlSession = sqlSessionFactory.openSession();
    User user = new User();
    user.setUsername("jack");
    user.setJobs("worker");
    user.setPhone("13324585254");
    int rows = sqlSession.insert("com.ssm.mapper.UserMapper.addUser", user);
    //输出插入数据的主键 id 值
    System.out.println(user.getId());
    if (rows > 0) {
        System.out.println("成功添加" + rows + "条数据！");
    } else {
        System.out.println("添加数据失败！");
    }
    sqlSession.commit();
    sqlSession.close();
}
```

如果使用的数据库不支持主键自动增长（如 Oracle），或者支持增长的数据库取消了主键自增的规则，那就可以使用 MyBatis 提供的另一种方式来自定义生成主键，具体配置示例如下：

```
<insert id="insertUser" parameterType="com.ssm.po.User">
    <selectKey keyProperty="id" resultType="Integer" order="BEFORE">
        select if(max(id) is null, 1, max(id)+1) as newId from t_user
    </select>
    insert into t_user(id,username,jobs,phone) values(#{id},#{username},#{jobs},#{phone})
```

```
</insert>
```

在执行上述示例代码时，首先运行<selectKey>元素，它会通过自定义的语句来设置数据表中的主键（如果 t_uesr 表中没有记录，那就将 id 设置为 1，否则将 id 的最大值加 1 作为新的主键），然后调用插入语句。

<selectKey>元素在使用时可以设置以下几种属性：

```
<selectKey
   keyProperty="id"
   resultType="Integer"
   order="BEFORE"
   statement="PREPARED">
</selectKey>
```

在上述<selectKey>元素的几个属性中，keyProperty、resultType 和 statement 的作用与前面讲解的相同，order 属性可以被设置为 BEFORE 或 AFTER。如果设置为 BEFORE，那么它会先执行<selectKey>元素中的配置内容来设置主键，再执行插入语句；如果设置为 AFTER，那么它会先执行插入语句，再执行<selectKey>元素中的配置内容。

7.3.3　<update>元素和<delete>元素

<update>元素和<delete>元素的使用比较简单，它们的属性配置也基本相同（<delete>元素中不包含表 7.5 所示的 3 个属性），其常用属性配置如下：

```
<update
    id="updateUser"
    parameterType="com.ssm.po.User"
    flushCache="true"
    statementType="PREPARED"
    timeout="20">
</update>
<delete
    id="deleteUser"
    parameterType="com.ssm.po.User"
    flushCache="true"
    statementType="PREPARED"
    timeout="20">
</delete>
```

从上述配置代码中可以看出，<update>元素和<delete>元素的属性基本与<select>元素中的属性一致。与<insert>元素一样，<update>元素和<delete>元素在执行完之后也会返回一个表示影响记录条数的整数，使用示例如下：

```
<!--更新用户信息 -->
<update id="updateUser" parameterType="com.ssm.po.User">
    update t_user set username=#{username},jobs=#{jobs},phone=#{phone} where id=#{id}
```

```
</update>
<!--删除用户信息-->
<delete id="deleteUser" parameterType="Integer">
    delete from t_user where id=#{id}
</delete>
```

7.3.4 <sql>元素

在一个映射文件中，通常需要定义多条 SQL 语句，这些 SQL 语句的组成可能有一部分是相同的（如多条 select 语句中都查询相同的 id、username、jobs 字段），如果每一个 SQL 语句都重写一遍相同的部分，势必会增加代码量，导致映射文件过于臃肿。那么有没有什么办法可以将这些 SQL 语句中相同的组成部分抽取出来，然后在需要的地方引用呢？答案肯定是有的，我们可以在映射文件中使用 MyBatis 提供的<sql>元素来解决上述问题。

<sql>元素的作用是定义可重用的 SQL 代码片段，然后在其他语句中引用这一代码片段。

例如，定义一个包含 id、username、jobs 和 phone 字段的代码片段：

```
<sql id="user Columns">id,username,jobs, phone</sql>
```

这一代码片段可以在其他语句中使用，具体如下：

```
<select id="findUserById" parameterType="Integer" resultType="com.ssm.po.User">
    select <include refid="user Columns">
    from t_user
    where id=#{id}
</select>
```

在上述代码中，使用<include>元素的 refid 属性引用了自定义的代码片段，refid 属性值为自定义代码片段的 id。

上面的示例只是一个简单的引用查询。在实际开发中，可以更加灵活地定义 SQL 片段，读者可以查找相关资料来进一步了解。

7.3.5 <resultMap>元素

<resultMap>元素表示结果映射集，主要作用是定义映射规则、级联更新以及定义类型转换器等。

< resultMap>元素中包含一些子元素，元素结构如下：

```
<!-- resultMap 的元素结构-->
<resultMap type="" id="">
    <constructor>                       <!--类在实例化时，用来注入结果到构造方法中-->
        <idArg />                       <!--id 参数，标记结果作为 id-->
        <arg />                         <!--注入构造方法的一个普通结果-->
    </constructor>
    <id />                              <!--用于表示哪个列是主键-->
    <result />                          <!--注入字段或 Javabean 属性的普通结果-->
```

```
    <association property="" />  <!--用于一对一关联-->
    <collection property="" />  <!--用于一对多关联-->
    <discriminator javaType="">  <!--使用结果值来决定使用哪个结果映射-->
        <case value="" />         <!--基于某些值的结果映射-->
    </discriminator>
</resultMap>
```

<resultMap>元素的 type 属性表示需要映射的 POJO，id 属性是这个 resultMap 的唯一标识。它的子元素<constructor>用于配置构造方法（当一个 POJO 中未定义无参的构造方法时，就可以使用<constructor>元素进行配置）。子元素<id>用于表示哪个列是主键，而<result>用于表示 POJO 和数据表中普通列的映射关系。<association>和<collection>用于处理多表时的关联关系，而<discriminator>元素主要用于处理一个单独的数据库查询返回包含很多不同数据类型的结果集的情况。

在默认情况下，MyBatis 程序在运行时会自动地将查询到的数据与需要返回的对象的属性进行匹配赋值（需要表中的列名与对象的属性名称完全一致）。然而在实际开发时，数据表中的列和需要返回的对象的属性可能不会完全一致，这种情况下 MyBatis 是不会自动赋值的。此时，就可以使用<resultMap>元素进行处理，UserMapper.xml 中的代码如下：

```
<?xml version="1. 0" encoding="UTE-8"?>
<!-- DOCTYPE mapper
        PUBLIC "-//mybatis.org//DTD Mapper 3.0//EN"
        "http://mybatis.org/dtd/mybatis-3-mapper.dtd">
<mapper namespace="com.ssm.mapper.UserMapper">
<resultMap type="com.ssn.po.User" id="resultMap">
    <id property="id" column="t_id"/>
    <result property="name" column="t_username"/>
    <result property="age" column="t_age"/>
</resultMap>
<select id="findAllUser" resultMap="resultMap">
    select * from t_user
</select>
```

在上述代码中，<resultMap>的子元素<id>和<result>的 property 属性表示 User 类的属性名，column 属性表示数据表 t_user 的列名。<select>元素的 resultMap 属性表示引用上面定义的 resultMap。接下来可以在配置文件 mybatis-config.xml 中引入 UserMapper.xml。

除此之外，还可以通过<resultMap>元素中的<association>和<collection>处理多表的关联关系，这将在后续章节中进行详细讲解。

第 8 章

动态 SQL

MyBatis 提供对 SQL 语句进行动态组装的功能，能很好地解决开发人员在使用 JDBC 或框架进行数据库开发时手动拼装 SQL 这一非常烦琐的工作。动态 SQL 是 MyBatis 的强大特性之一，MyBatis 3 采用了功能强大的基于 OGNL 的表达式来完成动态 SQL。

本章主要涉及的知识点如下：

- <if>元素。
- <choose>、<when>、<otherwise>元素。
- <where>、<trim>元素。
- <set>元素。
- <foreach>元素。
- <bind>元素。

8.1 <if>元素

在 MyBatis 中，<if>元素是常用的判断语句，主要用于实现某些简单的条件选择。在实际应用中，我们可能会通过多个条件来查询某个数据。例如，要查找某个用户信息，可以通过姓名和职业来查找；也可以不填写职业，直接通过姓名来查找；还可以都不填写而查询出所有用户，此时姓名和职业就是非必需条件。类似这种情况，在 MyBatis 中就可以通过<if>元素来实现。

【示例 8-1】下面通过一个具体的案例来演示<if>元素的使用。

步骤 01 在 IntelliJ IDEA 中创建一个名为 chapter08 的 Web 项目。

步骤02 将第6章中chapter06项目中的JAR包和src目录下的文件复制到chapter08中。

步骤03 将配置文件中的数据库信息修改为外部引用的形式，即在项目的src目录下创建一个名称为db.properties的配置文件，其内容如下：

```
jdbc.driver=com.mysql.cj.jdbc.Driver
jdbc.url=jdbc:mysql: //localhost:3306/db_mybatis?serverTimezone=UTC
jdbc.username=root
jdbc.password=root
```

然后在MyBatis配置文件mybatis-config.xml中配置<properties/>属性，并修改配置文件中数据库连接的信息，修改完成后mybatis-config.xml内容如下：

```
<?xml version="1.0" encoding="UTF-8"?>
<!DOCTYPE configuration PUBLIC "-//mybatis.org//DTD Config 3.0//EN"
    "http://mybatis.org/dtd/mybatis-3-config.dtd">
<configuration>
    <properties resource="db.properties" />
    <environments default="mysql">
        <environment id="mysql">
            <transactionManager type="JDBC" />
            <dataSource type="POOLED">
                <!--数据库驱动 -->
                <property name="driver" value="${jdbc.driver}" />
                <!--连接数据库的url -->
                <property name="url" value="${jdbc.url}" />
                <!--连接数据库的用户名 -->
                <property name="username" value="${jdbc.username}" />
                <!--连接数据库的密码-->
                <property name="password" value="${jdbc.password}" />
            </dataSource>
        </environment>
    </environments>
    <mappers>
        <mapper resource="com/ssm/mapper/UserMapper.xml" />
    </mappers>
</configuration>
```

步骤04 创建一个com.ssm.util包，在该包下创建工具类MybatisUtil，在其中定义获取SqlSession的方法getSession()，如文件8.1所示。

文件8.1 MybatisUtil.java

```
01  package com.ssm.util;
02  import java.io.IOException;
03  import java.io.Reader;
04  import org.apache.ibatis.io.Resources;
05  import org.apache.ibatis.session.SqlSession;
06  import org.apache.ibatis.session.SqlSessionFactory;
```

```
07  import org.apache.ibatis.session.SqlSessionFactoryBuilder;
08  public class MybatisUtils{
09      private static SqlSessionFactory sqlSessionFactory=null;
10      static{
11          try {
12          String resource = "mybatis-config.xml";
13              InputStream inputStream = Resources.getResourceAsStream(resource);
14              sqlSessionFactory=new SqlSessionFactoryBuilder().build(inputStream);
15          } catch (IOException e) {
16              e.printStackTrace();
17          }
18      }
19      // 获取 SqlSession 的方法
20      public static SqlSession getSession(){
21          return sqlSessionFactory.openSession();
22      }
23  }
```

项目创建后的项目文件结构如图 8.1 所示。

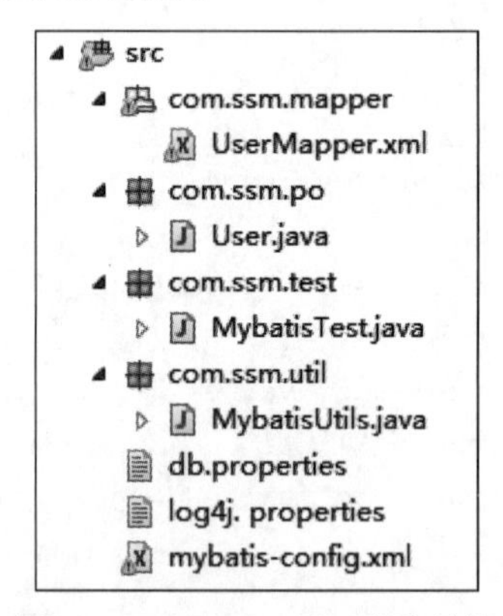

图 8.1 chapter08 项目文件结构

步骤 05 修改映射文件 UserMapper.xml，在映射文件中，使用<if>元素编写根据用户姓名（username）和职业（jobs）组合条件查询用户信息列表的动态 SQL，如文件 8.2 所示。

文件 8.2 UserMapper.xml

```
01  <?xml version="1.0" encoding="UTF-8"?>
02  <!DOCTYPE mapper PUBLIC "-//mybatis.org//DTD Mapper 3.0//EN"
03      "http://mybatis.org/dtd/mybatis-3-mapper.dtd">
04  <mapper namespace="com.ssm.mapper.UserMapper">
05      <!--<if>元素的使用 -->
06      <select id="findUserByNameAndJobs" parameterType="com.ssm.po.User"
07       resultType="com.ssm.po.User">
08        select * from t_user where 1=1
09        <if test="username !=null and username !=''">
10              and username like concat('%',#{username},'%')
```

```
11          </if>
12          <if test="jobs !=null and jobs !=''">
13             and jobs=#{jobs}
14          </if>
15       </select>
16    </mapper>
```

在文件 8.2 中，使用<if>元素的 test 属性分别对 username 和 jobs 进行了非空判断（test 属性多用于条件判断语句中，用于判断真假，大部分的场景中都是进行非空判断的，有时也需要判断字符串、数字和枚举等），如果传入的查询条件非空，就进行动态 SQL 组装。

步骤06 在测试类 MybatisTest 中，编写测试方法 findUserByNameAndJobsTest()，如文件 8.3 所示。

文件 8.3 MybatisTest.java

```
01  package com.ssm.test;
02  import java.io.InputStream;
03  import java.util.List;
04  import org.apache.ibatis.io.Resources;
05  import org.apache.ibatis.session.SqlSession;
06  import org.apache.ibatis.session.SqlSessionFactory;
07  import org.apache.ibatis.session.SqlSessionFactoryBuilder;
08  import org.junit.Test;
09  import com.ssm.po.User;
10  import com.ssm.util.MybatisUtils;
11  public class MybatisTest {
12  /*
13   * 根据用户姓名和职业组合条件查询用户信息列表
14   */
15      @Test
16      public void findUserByNameAndJobsTest() throws Exception {
17          //通过工具类生成 SqlSession 对象
18          SqlSession sqlSession = MybatisUtils.getSession();
19          //创建 User 对象，封装需要组合查询的条件
20          User user=new User();
21          user.setUsername("zhangsan");
22          user.setJobs("teacher");
23          //执行 SqlSession 的查询方法，返回结果集
24          List<User> users =
25  sqlSession.selectList("com.ssm.mapper.UserMapper.findUserByNameAndJobs", user);
26          //输出查询结果
27          for (User u : users) {
28             System.out.println(u.toString());
29          }
30          sqlSession.close();
31      }
32  }
```

在文件 8.3 的 findUserByNameAndJobsTest()方法中，首先通过 MybatisUtils 工具类获取了 SqlSession 对象，然后使用 User 对象封装了用户名为 zhangsan 且职业为 teacher 的查询条件，并通过 SqlSession 对象的 selectList()方法执行多条件组合的查询操作。最后，程序执行完毕时关闭了 SqlSession 对象。执行 findUserByNameAndJobsTest()方法后，控制台的输出结果如图 8.2 所示。

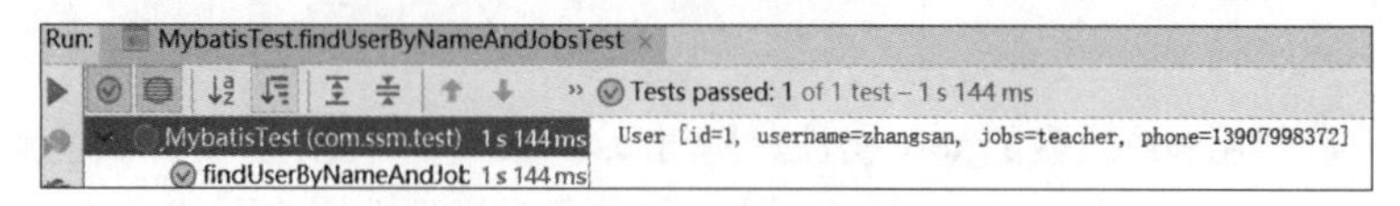

图 8.2 运行结果

从图 8.2 中可以看出，已经查询出了 username 为 zhangsan 且 jobs 为 teacher 的用户信息。先将封装到 User 对象中的 zhangsan 和 teacher 两行代码注释掉，然后再次执行 findUserByNameAndJobsTest()方法，控制台的输出结果如图 8.3 所示。

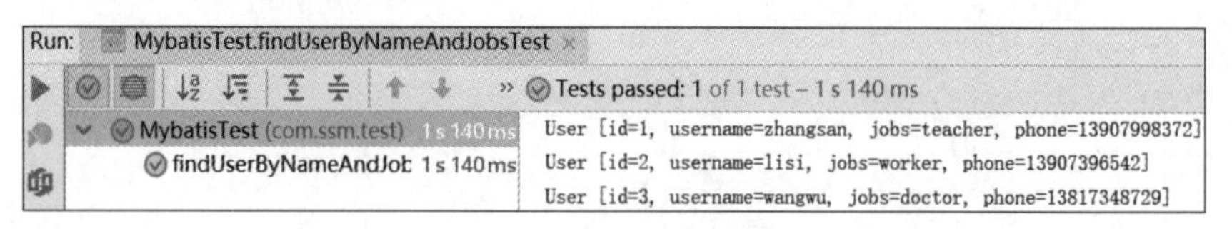

图 8.3 运行结果

从图 8.3 中可以看到，当未传递任何参数时，会将数据表中的所有数据查询出来，这就是<if>元素的使用。

8.2 <choose>、<when>和<otherwise>元素

在使用<if>元素时，只要 test 属性中的表达式为 true，就会执行元素中的条件语句，但是在实际应用中，有时只需要从多个选项中选择一个执行。例如，若用户姓名不为空，则只根据用户姓名进行筛选；若用户姓名为空而用户职业不为空，则只根据用户职业进行筛选；若用户姓名和用户职业都为空，则要求查询出所有电话号码不为空的用户信息。

此种情况下，使用<if>元素进行处理是非常不合适的，可以使用<choose>、<when>、<otherwise>元素进行处理，类似于在 Java 语言中使用 switch...case...default 语句。

【示例 8-2】使用<choose>、<when>、<otherwise>元素组合实现上面的情况。

步骤 01 在映射文件 UserMapper.xml 中，使用<choose>、<when>、<otherwise>元素执行上述情况的动态 SQL 代码如下：

```
01  <!--<choose>(<when>、<otherwise>)元素的使用 -->
02  <select id="findUserByNameOrJobs" parameterType="com.ssm.po.User"
03      resultType="com.ssm.po.User">
04       select * from t_user where 1=1
05       <choose>
06           <when test="username !=null and username !=''">
07               and username like concat('%',#{username},'%')
```

```
08          </when>
09          <when test="jobs !=null and jobs !=''">
10              and jobs=#{jobs}
11          </when>
12          <otherwise>
13              and phone is not null
14          </otherwise>
15      </choose>
16  </select>
```

在上述代码中，使用了<choose>元素进行SQL拼接，若第一个<when>元素中的条件为真，则只动态组装第一个<when>元素内的SQL片段；否则继续向下判断第二个<when>元素中的条件是否为真，以此类推；若前面所有when元素中的条件都不为真，则只组装<otherwise>元素内的SQL片段。

步骤02 在测试类MybatisTest中，编写测试方法findUserByNameOrJobsTest()，代码如下：

```
01  /*
02   * 根据用户姓名或者职业组合条件查询用户信息列表
03   */
04  @Test
05  public void findUserByNameOrJobsTest() throws Exception {
06      SqlSession sqlSession = MybatisUtils.getSession();
07      User user=new User();
08      user.setUsername("zhangsan");
09      user.setJobs("teacher");
10      //执行SqlSession的查询方法，返回结果集
11      List<User> users =
12  sqlSession.selectList("com.ssm.mapper.UserMapper.findUserByNameOrJobs", user);
13      for (User u : users) {
14          System.out.println(u.toString());
15      }
16      sqlSession.close();
17  }
```

执行上述方法后，结果与图8.2所示的结果相同，不过虽然同时传入了姓名和职业两个查询条件，但MyBatis生成的SQL是动态组装用户姓名进行条件查询的。如果将上述代码中的"user.setUsername("zhangsan");"删除或者注释掉，然后再次执行，这时MyBatis生成的SQL组装用户职业进行条件查询，同样会查询出用户信息。如果将设置用户姓名和职业参数值的两行代码都注释掉，那么程序的执行结果如图8.3所示，MyBatis的SQL组装<otherwise>元素中的SQL片段进行条件查询。

8.3 <where>、<trim>元素

在前两节的案例中，映射文件中编写的SQL后面都加入了"where1=1"的条件，这是为了保证

当条件不成立时拼接起来的 SQL 语句在执行时不会报错，即使得 SQL 不出现语法错误。那么在 MyBatis 中，有没有什么办法不用加入“1=1”这样的条件也能使拼接后的 SQL 成立呢？针对这种情况，MyBatis 提供了<where>元素。

【示例 8-3】以【示例 8-1】为例，将映射文件中的“where 1=1”条件删除，使用<where>元素替换后的代码如下：

```
01  <!--<if>、<where>元素的使用 -->
02  <select id="findUserByNameAndJobs" parameterType="com.ssm.po.User"
03      resultType="com.ssm.po.User">
04          select * from t_user
05      <where>
06          <if test="username !=null and username !=''">
07              and username like concat('%',#{username},'%')
08          </if>
09          <if test="jobs !=null and jobs !=''">
10              and jobs=#{jobs}
11          </if>
12      </where>
13  </select>
```

上述配置代码中，使用<where>元素对“where 1=1”条件进行了替换，<where>元素会自动判断组合条件下拼装的 SQL 语句，只有<where>元素内的条件成立时，才会在拼接 SQL 中加入 where 关键字，否则将不会添加；即使 where 之后的内容有多余的“AND”或“OR”，<where>元素也会自动将它们去除。除了使用<where>元素外，还可以通过<trim>元素来定制需要的功能，上述代码可以修改为如下形式：

```
01  <!--<if>、<trim>元素的使用 -->
02  <select id="findUserByNameAndJobs" parameterType="com.ssm.po.User"
03      resultType="com.ssm.po.User">
04          select * from t_user
05      <trim prefix="where" prefixOverrides="and">
06          <if test="username !=null and username !=''">
07              and username like concat('%',#{username},'%')
08          </if>
09          <if test="jobs !=null and jobs !=''">
10              and jobs=#{jobs}
11          </if>
12      </trim>
13  </select>
```

上述配置代码中，同样使用<trim>元素对“where 1=1”条件进行了替换，<trim>元素的作用是去除一些特殊的字符串，它的 prefix 属性代表的是语句的前缀（这里使用 where 来连接后面的 SOL 片段），而 prefixOverrides 属性代表的是需要去除的那些特殊字符串（这里定义了要去除 SQL 中的 and），上面的写法和使用<where>元素基本是等效的。

8.4 <set>元素

在 Hibernate 中，如果想要更新某一个对象，就需要发送所有的字段给持久化对象，然而实际应用中有时只需要更新某一个或几个字段。为了让程序只更新需要更新的字段，MyBatis 提供了<set>元素来完成这一项工作。<set>元素主要用于更新操作，其作用是在动态包含的 SQL 语句前输出一个 SET 关键字，并去除 SQL 语句中最后一个多余的逗号。

【示例 8-4】以更新操作为例，使用<set>元素对映射文件中更新用户信息的 SQL 语句进行修改，代码如下：

```
01  <!-- <set>元素 -->
02  <update id="updateUser" parameterType="com.ssm.po.User">
03      update t_user
04      <set>
05          <if test="username !=null and username !=''">
06              username=#{username},
07          </if>
08          <if test="jobs !=null and jobs !=''">
09              jobs=#{jobs},
10          </if>
11          <if test="phone !=null and phone !=''">
12              phone=#{phone},
13          </if>
14      </set>
15      where id=#{id}
16  </update>
```

在上述配置的 SQL 语句中，使用<set>元素和<if>元素相结合的方式来组装 update 语句。其中<set>元素会动态前置 SET 关键字，同时消除 SQL 语句中最后一个多余的逗号；<if>元素用于判断相应的字段是否传入值，如果传入的更新字段非空，那么就将此字段进行动态 SQL 组装，并更新此字段，否则此字段不执行更新。

注意：在映射文件中使用<set>元素和<if>元素组合进行 update 语句动态 SQL 组装时，如果<set>元素内包含的内容都为空，那么就会出现 SQL 语法错误，所以在使用<set>元素进行字段信息更新时，要确保传入的更新字段不都为空。

8.5 <foreach>元素

MyBatis 中已经提供了一种用于数组和集合循环遍历的方式，那就是使用<foreach>元素。假设在一个用户表中有 1000 条数据，现在需要将 id 值小于 100 的用户信息全部查询出来，就可以通过<foreach>元素来解决。

<foreach>元素通常在构建 IN 条件语句时使用，使用方式如下：

```
01  <!--<foreach>元素的使用 -->
02  <select id="findUserByIds" parameterType="List" resultType="com.ssm.po.User">
03      select * from t_user where id in
04      <foreach item="id" index="index" collection="list" open="(" separator="," close=")">
05         #{id}
06      </foreach>
07  </select>
```

在上述代码中，使用<foreach>元素对传入的集合进行遍历和动态 SQL 组装。<foreach>元素中使用的几种属性的描述具体如下：

- item：配置的是循环中的当前元素。
- index：配置的是当前元素在集合中的位置下标。
- collection：配置的 list 是传递过来的参数类型（首字母小写），可以是一个 array、list（或 collection）、Map 集合的键、POJO 包装类中的数组或集合类型的属性名等。
- open 和 close：配置的是以什么符号将这些集合元素包装起来。
- separator：配置的是各个元素的间隔符。

注意：可以将任何可迭代对象（如列表、集合等）、字典或者数组对象传递给<foreach>作为集合参数。当使用可迭代对象或者数组时，index 是当前迭代的次数，item 的值是本次迭代获取的元素。当使用字典（或者 MapEntry 对象的集合）时，index 是键，item 是值。

【示例 8-5】对<foreach>元素的使用进行测试。在测试类 MybatisTest 中，编写测试方法 findUserByIdsTest()，其代码如下：

```
01 /* 根据用户编号批量查询用户信息 */
02  @Test
03  public void findUserByIdsTest(){
04      SqlSession sqlSession = MybatisUtils.getSession();
05      //创建 List 集合，封闭查询 id
06      List<Integer> ids=new ArrayList<Integer>();
07      ids.add(1);
08      ids.add(2);
09      //执行 SqlSession 的查询方法，返回结果集
10      List<User> users =
11               sqlSession.selectList("com.ssm.mapper.UserMapper.findUserByIds", ids);
12      for (User user : users) {
13          System.out.println(user.toString());
14      }
15      sqlSession.close();
16 }
```

在上述代码中，执行查询操作时传入了一个客户编号集合 ids。执行 findUserByIdsTest()方法后，控制台的输出结果如图 8.4 所示。从图中可以看出，已成功地批量查询出对应的用户信息。

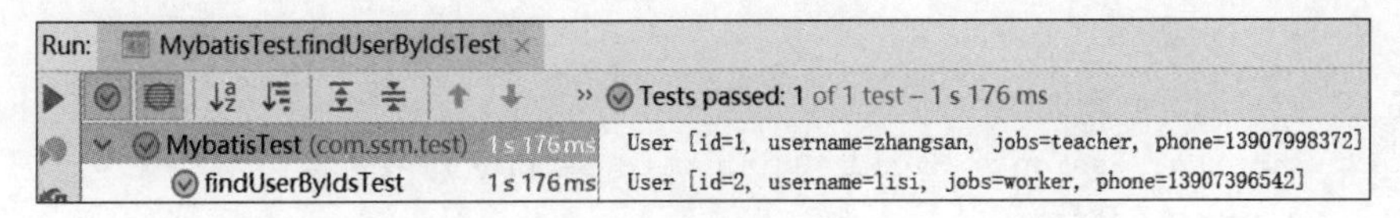

图 8.4　运行结果

在使用<foreach>元素时，最关键也最容易出错的就是 collection 属性，该属性是必须指定的，而且在不同情况下该属性的值不一样，主要有以下 3 种情况：

- 如果传入的是单参数且参数类型是一个数组或者列表，那么 collection 属性值分别为 array、list（或 collection）。
- 如果传入的参数有多个，就需要把它们封装成一个 Map，当然单参数也可以封装成 Map 集合，这时 collection 属性值就为 Map 的键。
- 如果传入的参数是 POJO 包装类，那么 collection 属性值就为该包装类中需要进行遍历的数组或集合的属性名。

在设置 collection 属性值的时候，必须按照实际情况配置，否则程序就会出现异常。

8.6　<bind>元素

在对模糊查询编写 SQL 语句的时候，若使用“${}”进行字符串拼接，则无法防止 SQL 注入问题；若使用 concat 函数进行拼接，则只针对 MySQL 数据库有效；若使用的是 Oracle 数据库，则要使用连接符号“||”。这样，映射文件中的 SQL 就要根据不同的情况提供不同形式的实现，显然是比较麻烦的，且不利于项目的移植。为此，MyBatis 提供了<bind>元素来解决这一问题。我们完全不必使用数据库语言，只需使用 MyBatis 的语言即可与所需参数连接。

MyBatis 的<bind>元素可以通过 OGNL 表达式来创建一个上下文变量，其使用方式如下：

```
01  <!--<bind>元素的使用：根据用户姓名模糊查询用户信息 -->
02  <select id="findUserByName2" parameterType="com.ssm.po.User"
03      resultType="com.ssm.po.User">
04      <!--_parameter.getUsername()也可以直接写成传入的字段属性名，即 username  -->
05      <bind name="p_username" value="'%'+_parameter.getUsername()+'%'"/>
06          select * from t_user
07          where username like #{p_username}
08  </select>
```

上述配置代码中，使用<bind>元素定义了一个 name 为 p_username 的变量，<bind>元素中 value 的属性值就是拼接的查询字符串，其中_parameter.getUsername()表示传递进来的参数（也可以直接写成对应的参数变量名，如 username）。在 SQL 语句中，直接引用<bind>元素的 name 属性值即可进行动态 SQL 组装。

【示例 8-6】在测试类 MybatisTest 中，编写测试方法 findUserByName2()进行测试，代码如下：

```
01  /* 根据用户姓名模糊查询用户信息 */
```

```
02  @Test
03  public void findUserByName2(){
04      SqlSession sqlSession = MybatisUtils.getSession();
05      User user=new User();
06      user.setUsername("s");
07      List<User> users =
08      sqlSession.selectList("com.ssm.mapper.UserMapper.findUserByName2", user);
09      for (User u : users) {
10          System.out.println(u.toString());
11      }
12      sqlSession.close();
13  }
```

程序执行后，控制台的输出结果如图 8.5 所示。从图中可以看出，使用 MyBatis 的<bind>元素已经完成了动态 SQL 组装，并成功地模糊查询出了用户信息。

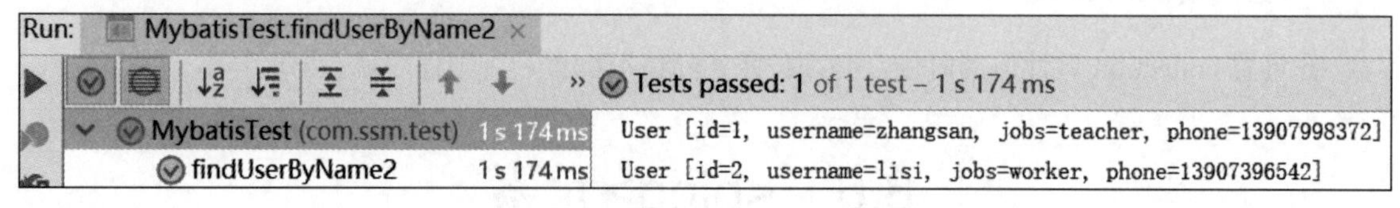

图 8.5 运行结果

第 9 章

MyBatis 的关联映射

在实际应用中，对数据库的操作会涉及多张表，这在面向对象中就涉及对象与对象之间的关联关系。针对多表之间的操作，MyBatis 提供了关联映射，通过关联映射来处理对象与对象之间的关联关系。

本章主要涉及的知识点如下：

- 数据表之间以及对象之间的 3 种关联关系。
- 一对一、一对多和多对多关联映射的使用。
- 关联关系中的嵌套查询和嵌套结果。

9.1 关联关系概述

在关系数据库中，多表之间存在 3 种关联关系，分别为一对一、一对多和多对多。

- 一对一：在任意一方引入对方主键作为外键。
- 一对多：在“多”的一方添加“一”的一方的主键作为外键。
- 多对多：产生中间关系表，引入两张表的主键作为外键，两个主键成为联合主键或使用新的字段作为主键。

对象之间也存在 3 种关联关系：

- 一对一的关系：在本类中定义对方类型的对象，比如在 A 类中定义 B 类类型的属性 b、在 B 类中定义 A 类类型的属性 a。
- 一对多的关系：一个 A 类类型对应多个 B 类类型的情况，需要在 A 类中以集合的方式引入 B 类类型的对象，在 B 类中定义 A 类类型的属性 a。

- 多对多的关系：在A类中定义B类类型的集合，在B类中定义A类类型的集合。

9.2 MyBatis中的关联关系

本节将对如何使用MyBatis处理对象中的3种关联关系进行详细讲解。

9.2.1 一对一

在现实生活中，一对一关联关系十分常见。例如，一个学生只有一本学生证，同时一本学生证也只对应一个学生。

那么MyBatis是怎么处理这种一对一关联关系的呢？在本书7.3.5节讲解的<resultMap>元素中包含一个<association>子元素，MyBatis就是通过该元素来处理一对一关联关系的。

在<association>元素中，通常可以配置以下属性：

- property：指定映射到的实体类对象属性，与表字段一一对应。
- column：指定表中对应的字段。
- javaType：指定映射到实体对象属性的类型。
- select：指定引入嵌套查询的子SQL语句，用于关联映射中的嵌套查询。
- fetchType：指定在关联查询时是否启用延迟加载，有lazy和eager两个属性值，默认值为lazy（默认关联映射延迟加载）。

<association>元素有如下两种配置方式：

```
<!--方式一：嵌套查询-->
<association property="card" column="card_id" javaType="com.ssm.po.StudentIdCard"
select="com.ssm.mapper.StudentIdCardMapper.findCodeById"/>
<!--方式二：嵌套结果-->
<association property="card" javaType="com.ssm.po.StudentIdCard">
    <id property="id" column=""card_id"/>
    <result property="code" column="code"/>
</association>
```

注意：MyBatis在映射文件中主要通过两种方式加载关联关系对象：嵌套查询和嵌套结果。嵌套查询是指通过执行另一条SQL映射语句来返回预期的复杂类型，嵌套结果是使用嵌套结果映射来处理重复的联合结果的子集。

【示例9-1】接下来以学生和学生证之间的一对一关联关系为例来进一步进行讲解。

查询学生及其关联的学生证信息的方法是先通过查询学生表中的主键来获取学生信息，然后通过表中的外键来获取学生证表中的学生证号信息。具体实现步骤如下：

步骤01 创建数据表。在db_mybatis数据库中分别创建名为tb_studentidcard和tb_student的数据表，同时预先插入几条数据。执行的SQL语句如下：

```
#使用数据库 db_mybatis
USE db_mybatis;
# 创建一个名称为 tb_studentidcard 的表
CREATE TABLE tb_studentidcard(
       id INT PRIMARY KEY AUTO_INCREMENT,
       CODE VARCHAR(8)
);
# 插入两条数据
INSERT INTO tb_studentidcard(CODE) VALUES('18030128');
INSERT INTo tb_studentidcard(CODE) VALUES ('18030135');
# 创建一个名称为 tb_student 的表（暂时添加少量字段）
CREATE TABLE tb_student(
       id INT PRIMARY KEY AUTO_INCREMENT,
       name VARCHAR(32),
       sex CHAR(1),
       card_id INT UNIQUE,
       FOREIGN KEY (card_id) REFERENCES tb_studentidcard(id)
);
# 插入两条数据
INSERT INTO tb_student(name, sex, card_id) VALUES('limin','f',1);
INSERT INTO tb_student(name, sex, card_id) VALUES('jack','m',2);
```

步骤 02 在 IntelliJ IDEA 中创建一个名为 chapter09 的 Web 项目，然后引入相关 JAR 包、MybatisUtils 工具类以及 mybatis-config.xml 核心配置文件。

步骤 03 在项目的 com.ssm.po 包下创建持久化类：学生证类 StudentIdCard 和学生类 Student。编辑后的代码如文件 9.1 和文件 9.2 所示。

文件 9.1 StudentIdCard.java

```
01  package com.ssm.po;
02  //学生证类
03  public class StudentIdCard {
04      private Integer id;
05      private String code;
06      public Integer getId() {
07          return id;
08      }
09      public void setId(Integer id) {
10          this.id = id;
11      }
12      public String getCode() {
13          return code;
14      }
15      public void setCode(String code) {
16          this.code = code;
17      }
18      public String toString() {
```

```
19          return "StudentIdCard [id=" + id + ", code=" + code + "]";
20      }
21  }
```

文件 9.2 Student.java

```
01  package com.ssm.po;
02  //学生类
03  public class Student {
04      private Integer id;
05      private String name;
06      private String sex;
07      private StudentIdCard studentIdCard;
08      public Integer getId() {
09          return id;
10      }
11      public void setId(Integer id) {
12          this.id = id;
13      }
14      public String getName() {
15          return name;
16      }
17      public void setName(String name) {
18          this.name = name;
19      }
20      public String getSex() {
21          return sex;
22      }
23      public void setSex(String sex) {
24          this.sex = sex;
25      }
26      public StudentIdCard getStudentIdCard() {
27          return studentIdCard;
28      }
29      public void setStudentIdCard(StudentIdCard studentIdCard) {
30          this.studentIdCard = studentIdCard;
31      }
32      public String toString() {
33          return "Student [id=" + id + ", name=" + name + ", sex=" + sex +
34            ", studentIdCard=" + studentIdCard + "]";
35      }
36  }
```

步骤 04 在com.ssm.mapper包中创建学生证映射文件StudentIdCardMapper.xml和学生映射文件StudentMapper.xml，并在两个映射文件中编写一对一关联映射查询的配置信息，如文件9.3和文件9.4所示。

文件 9.3 StudentIdCardMapper.xml

```
01 <?xml version="1.0" encoding="UTF-8"?>
02 <!DOCTYPE mapper PUBLIC "-//mybatis.org//DTD Mapper 3.0//EN"
03     "http://mybatis.org/dtd/mybatis-3-mapper.dtd">
04 <mapper namespace="com.ssm.mapper.StudentIdCardMapper">
05     <!--根据 id 获取学生证信息 -->
06     <select id="findStudentIdCardById" parameterType="Integer" resultType= "StudentIdCard">
07         select * from tb_studentidcard where id=#{id}
08     </select>
09 </mapper>
```

文件 9.4 StudentMapper.xml

```
01 <?xml version="1.0" encoding="UTF-8"?>
02 <!DOCTYPE mapper PUBLIC "-//mybatis.org//DTD Mapper 3.0//EN"
03     "http://mybatis.org/dtd/mybatis-3-mapper.dtd">
04 <mapper namespace="com.ssm.mapper.StudentMapper">
05     <!--嵌套查询，通过执行一条 SQL 映射语句来返回预期的特殊类型 -->
06     <select id="findStudentById" parameterType="Integer"
07       resultMap="StudentIdCardWithStudentResult">
08           select * from tb_student where id=#{id}
09     </select>
10     <resultMap type="Student" id="StudentIdCardWithStudentResult">
11         <id property="id" column="id"/>
12         <result property="name" column="name"/>
13         <result property="sex" column="sex"/>
14         <!-- 一对一，association 使用 select 属性引入另一条 SQL 语句 -->
15         <association property="studentIdCard" column="card_id" javaType="StudentIdCard"
16          select="com.ssm.mapper.StudentIdCardMapper.findStudentIdCardById"/>
17     </resultMap>
18 </mapper>
```

在上述两个映射文件中，使用了 MyBatis 中的嵌套查询方式进行学生及其关联的学生证信息查询，因为返回的学生对象中除了基本属性外，还有一个关联的 studentIdCard 属性，所以需要手动编写结果映射。从映射文件 StudentMapper.xml 中可以看出，嵌套查询的方法是先执行一个简单的 SQL 语句，然后在进行结果映射时将关联对象在<association>元素中使用 select 属性执行另一条 SQL 语句（StudentIdCardMapper.xml 中的 SQL）。

步骤 05 在核心配置文件 mybatis-config.xml 中引入 Mapper 映射文件并定义别名，如文件 9.5 所示。

文件 9.5 mybatis-config.xml

```
01 <?xml version="1.0" encoding="UTF-8"?>
02 <!DOCTYPE configuration PUBLIC "-//mybatis.org//DTD Config 3.0//EN"
03     "http://mybatis.org/dtd/mybatis-3-config.dtd">
04 <configuration>
05     <!-- 引入数据库连接配置文件 -->
```

```
06      <properties resource="db.properties" />
07      <!-- 使用扫描包的形式定义别名 -->
08      <typeAliases>
09          <package name="com.ssm.po"/>
10      </typeAliases>
11      <environments default="mysql">
12          <environment id="mysql">
13              <transactionManager type="JDBC" />
14              <dataSource type="POOLED">
15                  <!--数据库驱动 -->
16                  <property name="driver" value="${jdbc.driver}" />
17                  <!--连接数据库的 url -->
18                  <property name="url" value="${jdbc.url}" />
19                  <!--连接数据库的用户名 -->
20                  <property name="username" value="${jdbc.username}" />
21                  <!--连接数据库的密码-->
22                  <property name="password" value="${jdbc.password}" />
23              </dataSource>
24          </environment>
25      </environments>
26      <!-- 配置 Mapper 的位置 -->
27      <mappers>
28          <mapper resource="com/ssm/mapper/StudentIdCardMapper.xml" />
29          <mapper resource="com/ssm/mapper/StudentMapper.xml" />
30          <mapper resource="com/ssm/mapper/UserMapper.xml" />
31      </mappers>
32  </configuration>
```

在上述核心配置文件中，首先引入了数据库连接的配置文件，然后使用扫描包的形式自定义别名，接下来进行环境的配置，最后配置了 Mapper 映射文件的位置信息。

步骤 06 在 com.ssm.test 包中创建测试类 MybatisAssociatedTest，并在类中编写测试方法 findStudentByIdTest()，如文件 9.6 所示。

文件 9.6 MybatisAssociatedTest.java

```
01  package com.ssm.test;
02  import org.apache.ibatis.session.SqlSession;
03  import org.junit.Test;
04  import com.ssm.po.Student;
05  import com.ssm.util.MybatisUtils;
06  public class MybatisAssociatedTest {
07      /* 嵌套查询 */
08      @Test
09      public void findStudentByIdTest(){
10          SqlSession sqlSession = MybatisUtils.getSession();
11          //使用 MyBatis 嵌套查询的方法查询 id 为 1 的学生信息
12          Student student=
```

```
13            sqlSession.selectOne("com.ssm.mapper.StudentMapper.findStudentById",1);
14            System.out.println(student.toString());
15            sqlSession.close();
16        }
17    }
```

在文件 9.6 的 findStudentByIdTest()方法中，首先通过 MybatisUtils 工具类获取了 SqlSession 对象，然后通过 SqlSession 对象的 selectOne()方法获取了学生信息，最后关闭了 SqlSession。执行方法后，控制台的输出结果如图 9.1 所示。从图中可以看出使用 MyBatis 嵌套查询的方式查询出了学生及其关联的学生证信息，这就是 MyBatis 中的一对一关联查询。

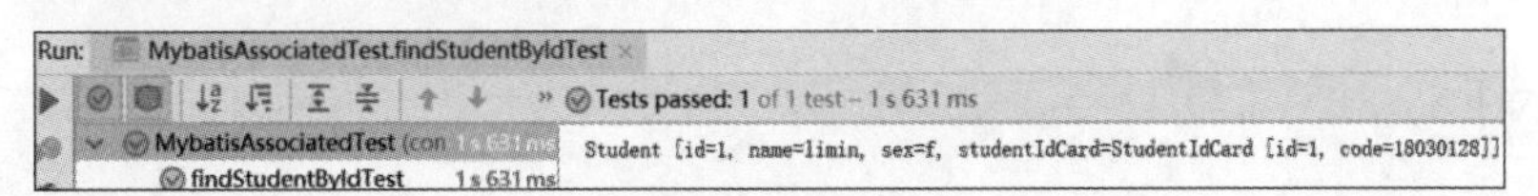

图 9.1　运行结果

虽然使用嵌套查询的方式比较简单，但是嵌套查询的方式要执行多条 SQL 语句，这对于大型数据集合和列表来说不是很好，因为这样可能会导致成百上千条关联的 SQL 语句被执行，从而极大地消耗了数据库性能，并且会降低查询效率。为此，MyBatis 提供了嵌套结果的方式进行关联查询。

在 StudentMapper.xml 中，使用 MyBatis 嵌套结果的方式进行学生及其关联的学生证信息查询，所添加的代码如下：

```
<!--嵌套结果，通过嵌套结果映射来处理重复的联合结果的子集 -->
<select id="findStudentById2" parameterType="Integer"
resultMap="StudentIdCardWithStudentResult2">
select s.*,sidcard.code
    from tb_student s,tb_studentidcard sidcard
    where s.card_id=sidcard.id and s.id=#{id}
</select>
<resultMap type="Student" id="StudentIdCardWithStudentResult2">
    <id property="id" column="id" />
    <result property="name" column="name" />
    <result property="sex" column="sex" />
    <association property="studentIdCard" javaType="StudentIdCard">
        <id property="id" column="card_id" />
        <result property="code" column="code" />
    </association>
</resultMap>
```

从上述代码中可以看出，MyBatis 嵌套结果的方式只编写了一条复杂的多表关联的 SQL 语句，并且在<association>元素中继续使用相关子元素进行数据库表字段和实体类属性的一一映射。执行结果与图 9.1 所示的结果相同，但使用 MyBatis 嵌套结果的方式只执行了一条 SQL 语句。

注意：在使用 MyBatis 嵌套查询方式进行关联查询映射时，使用 MyBatis 的延迟加载在一定程度上可以降低运行消耗并提高查询效率。MyBatis 默认没有开启延迟加载，需要在核心配置文件 mybatis-config.xml 中的<settings>元素内进行配置，具体配置方式如下：

```
<settings>
<!--打开延迟加载的开关-->
<setting name="lazyLoadingEnabled" value="true"/>
<!--将积极加载改为消极加载，即按需加载-->
<setting name="aggressiveLazyLoading" value="false" />
</settings>
```

在映射文件中，MyBatis 关联映射的<association>元素和<collection>元素中都已经默认配置了延迟加载属性，即默认属性 fetchType="lazy"（属性 fetchType="eager"表示立即加载），所以在配置文件中开启延迟加载后，无须在映射文件中再做配置。

9.2.2 一对多

在实际应用中，应用得更多的关联关系是一对多（或多对一）。例如，一个班级有多个学生，即多个学生属于一个班级。使用 MyBatis 是怎么处理这种一对多关联关系的呢？在本书第 7.3.5 节讲解的<resultMap>元素中包含一个<collection>子元素，MyBatis 就是通过这个<collection>子元素来处理一对多关联关系的。<collection>子元素的属性大部分与<collection>元素相同，但它还包含一个特殊属性——ofType。ofType 属性与 javaType 属性对应，用于指定实体对象中集合类属性所包含的元素类型。

<collection>元素可以参考如下两种方式进行配置：

```
<!--方式一：嵌套查询-->
<collection property="studentList" column="id" ofType="com.ssm.po.Student"
    select= "com.ssm.mapper.StudentMapper.selectStudent"/>
<!--方式二：嵌套结果-->
<collection property="studentList" ofType="com.ssm.po.Student">
    <id property="id" column="student_id"/>
    <result property="username" column="username"/>
</collection>
```

【示例 9-2】在了解了 MyBatis 处理一对多关联关系的元素和方式后，接下来以班级和学生之间的这种一对多关联关系为例，详细讲解如何在 MyBatis 中处理一对多关联关系，具体步骤如下：

步骤 01 在 db_mybatis 数据库中创建两个数据表：tb_banji 和 tb_student，同时在表中预先插入几条数据，执行的 SQL 语句如下：

```
# 创建一个名称为 tb_banji 的表（暂时添加少量字段）
CREATE TABLE tb_banji(
    id INT PRIMARY KEY AUTO_INCREMENT,
    name VARCHAR(32)
);
#插入两条数据
INSERT INTO tb_banji VALUES(1,'16 软件技术 1 班');
INSERT INTO tb_banji VALUES(2,'16 软件技术 2 班');
# 创建一个名称为 tb_student 的表（暂时添加少量字段）
```

```
CREATE TABLE tb_student(
    id INT PRIMARY KEY AUTO_INCREMENT,
    name VARCHAR(32),
    sex CHAR(1),
    banji_id INT ,
    FOREIGN KEY (banji_id) REFERENCES tb_banji(id)
);
#插入 3 条数据
INSERT INTO tb_student VALUES(1,'孙淼','m',1);
INSERT INTO tb_student VALUES(2,'刘梦奕','f',1);
INSERT INTO tb_student VALUES(3,'无为','m',2);
```

步骤 02 在 com.ssm.po 包中创建持久化类：班级类（Banji）和学生类（Student），并在两个类中定义相关属性和方法，如文件 9.7 和文件 9.8 所示。

文件 9.7　Banji.java

```
01  package com.ssm.po;
02  import java.util.List;
03  //班级类
04  public class Banji {
05      private Integer id;
06      private String name;
07      private List<Student> studentList;
08      public Integer getId() {
09          return id;
10      }
11      public void setId(Integer id) {
12          this.id = id;
13      }
14      public String getName() {
15          return name;
16      }
17      public void setName(String name) {
18          this.name = name;
19      }
20      public List<Student> getStudentList() {
21          return studentList;
22      }
23      public void setStudentList(List<Student> studentList) {
24          this.studentList = studentList;
25      }
26      public String toString() {
27          return "Banji [id="+ id + ", name=" + name + ", studentList=" + studentList + "]";
28      }
29  }
```

文件 9.8 Student.java

```
01  package com.ssm.po;
02  //学生类
03  public class Student {
04      private Integer id;
05      private String name;
06      private String sex;
07      public Integer getId() {
08          return id;
09      }
10      public void setId(Integer id) {
11          this.id = id;
12      }
13      public String getName() {
14          return name;
15      }
16      public void setName(String name) {
17          this.name = name;
18      }
19      public String getSex() {
20          return sex;
21      }
22      public void setSex(String sex) {
23          this.sex = sex;
24      }
25      public String toString() {
26          return "Student [id=" + id + ", name=" + name + ", sex=" + sex + "]";
27      }
28  }
```

步骤 03 在 com.ssm.mapper 包中创建班级实体映射文件 BanjiMapper.xml，并在文件中编写一对多关联映射查询的配置，如文件 9.9 所示。

文件 9.9 BanjiMapper.xml

```
01  <?xml version="1.0" encoding="UTF-8"?>
02  <!DOCTYPE mapper PUBLIC "-//mybatis.org//DTD Mapper 3.0//EN"
03      "http://mybatis.org/dtd/mybatis-3-mapper.dtd">
04  <mapper namespace="com.ssm.mapper.BanjiMapper">
05      <!--一对多：查看某一班级及其关联的学生信息
06      注意：若关联查询出的列名相同，则默认使用前者，所以需要使用别名进行区分  -->
07      <select id="findBanjiWithStudent" parameterType="Integer"
08             resultMap="BanjiWithStudentResult">
09         select b.*,s.id as student_id,s.name as sname,s.sex
10             from tb_banij b,tb_student s
11             where b.id=s.banji_id and b.id=#{id}
12      </select>
```

```
13      <resultMap type="Banji" id="BanjiWithStudentResult">
14          <id property="id" column="id" />
15          <result property="name" column="name" />
16          <!--一对多关联映射：collection
17  ofType 表示属性集合中元素的类型 List<Student>属性，即 Student 类  -->
18          <collection property="studentList" ofType="Student">
19              <id property="id" column="student_id" />
20              <result property="name" column="sname" />
21              <result property="sex" column="sex" />
22          </collection>
23      </resultMap>
24  </mapper>
```

在文件 9.9 中，使用 MyBatis 嵌套结果的方式定义一个根据班级 id 查询班级及其关联的学生信息的 select 语句。因为返回的班级对象中包含 Student 集合对象属性，所以需要手动编写结果映射信息。

步骤 04 将映射文件 BanjiMapper.xml 的路径配置到核心配置文件 mybatis-config.xml 中，代码如下：

```
<mapper resource="com/ssm/mapper/BanjiMapper.xml" />
```

步骤 05 在测试类 MyBatisAssociatedTest 中编写测试方法 findBanjiTest()，代码如下：

```
@Test
public void findBanjiTest(){
    SqlSession sqlSession = MybatisUtils.getSession();
    //查询班级 id 为 1 的班级信息（及其关联的学生集合信息）
    Banji banji=
            sqlSession.selectOne("com.ssm.mapper.BanjiMapper.findBanjiWithStudent",1);
    System.out.println(banji.toString());
    sqlSession.close();
}
```

执行方法后，控制台输出结果如图 9.2 所示。从图中可以看出，使用 MyBatis 嵌套结果的方式查询出了班级及其关联的学生集合信息，这就是 MyBatis 一对多的关联查询。

```
Tests passed: 1 of 1 test – 1 s 325 ms
Banji [id=1, name=16软件技术1班, studentList=[Student [id=1, name=孙淼, sex=m], Student [id=2, name=刘梦奕, sex=f]]]
```

图 9.2　运行结果

注意：上述案例从班级的角度出发，班级与学生之间是一对多的关联关系，但如果从单个学生的角度出发，一个学生只能属于一个班级，即为一对一的关联关系。

9.2.3　多对多

在实际项目开发中，多对多的关联关系是非常常见的。以学生和课程为例，一个学生可以选修多门课程，而一门课程又可以被多个学生选修，学生和课程就属于多对多的关联关系。

在数据库中，多对多的关联关系通常使用一个中间表（选课表）来维护，选课表（electiveCourse）中的学生 id（student_id）作为外键参照学生表的 id，课程 id（course_id）作为外键参照课程表的 id。三个表的关联关系如图 9.3 所示。

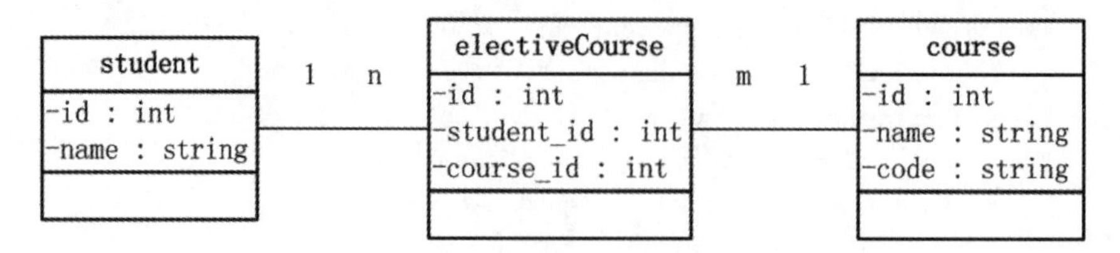

图 9.3 学生表、中间表与课程表之间的关联关系

【示例 9-3】了解了数据库中学生表与课程表之间的多对多关联关系后，下面我们通过具体的案例来讲解如何使用 MyBatis 处理这种多对多的关系，具体实现步骤如下：

步骤 01 创建数据表。在 db-mybatis 数据库中新建名称为 tb_course 和 tb_electiveCourse 的两个数据表（tb_student 表已在前面的案例中创建，这里直接引用），同时在表中预先插入几条数据。执行的 SQL 语句如下：

```
# 创建一个名称为 tb_course 的表
CREATE TABLE tb_course(
    id INT PRIMARY KEY AUTO_INCREMENT,
    name VARCHAR (32),
    code VARCHAR (32)
)
# 插入两条数据
INSERT INTO tb_course VALUES (1,'Java 程序设计语言', '08113226');
INSERT INTO tb_course VALUES (2,'JavaWeb 程序开发入门','08113228');
# 创建一个名称为 tb_electiveCourse 的中间表
CREATE TABLE tb_electiveCourse (
    id INT PRIMARY KEY AUTO_INCREMENT,
    student_id INT,
    course_id INT,
    FOREIGN KEY(student_id) REFERENCES tb_student(id),
    FOREIGN KEY(course_id) REFERENCES tb_course(id)
)
# 插入 3 条数据
INSERT INTO tb_electiveCourse VALUES (1,1,1);
INSERT INTo tb_electiveCourse VALUES (2,1,2);
INSERT INTO tb_electiveCourse VALUES (3,2,2);
```

步骤 02 在 com.ssm.po 包中创建持久化类课程类 Course，并在类中定义相关属性和方法，如文件 9.10 所示。

文件 9.10 Course.java

```
01  package com.ssm.po;
02  import java.util.List;
```

```
03  //课程类
04  public class Course {
05      private Integer id;
06      private String name;
07      private String code;
08      private List<Student> studentlist;//与学生集合的关联属性
09      public Integer getId() {
10          return id;
11      }
12      public void setId(Integer id) {
13          this.id = id;
14      }
15      public String getName() {
16          return name;
17      }
18      public void setName(String name) {
19          this.name = name;
20      }
21      public String getCode() {
22          return code;
23      }
24      public void setCode(String code) {
25          this.code = code;
26      }
27      public List<Student> getStudentlist() {
28          return studentlist;
29      }
30      public void setStudentlist(List<Student> studentlist) {
31          this.studentlist = studentlist;
32      }
33      public String toString() {
34          return "Course [id=" + id + ", name=" + name + ", code=" + code + "]";
35      }
36  }
```

除了需要在课程持久化类（Course.java）中添加学生集合的属性外，还需要在学生持久化类（Student.java）中增加课程集合的属性及其对应的 getter()方法和 setter()方法；同时为了方便查看输出结果，需要重写 tostring()方法。Student 类中添加的代码如下：

```
//关联课程集合信息
private List<Course> courseList;
//省略 getter()、setter()方法以及重写的 tostring()方法
```

步骤 03 在 com.ssm.mapper 包中创建课程实体映射文件 CourseMapper.xml 和学生实体映射文件 StudentMapper.xml，两个映射文件编辑后分别如文件 9.11 和文件 9.12 所示。

文件 9.11　CourseMapper.xml

```
01  <?xml version="1.0" encoding="UTF-8"?>
02  <!DOCTYPE mapper PUBLIC "-//mybatis.org//DTD Mapper 3.0//EN"
03      "http://mybatis.org/dtd/mybatis-3-mapper.dtd">
04  <mapper namespace="com.ssm.mapper.CourseMapper">
05      <!--多对多嵌套查询：通过执行一条 SQL 映射语句来返回预期的特殊类型  -->
06      <select id="findCourseWithStudent" parameterType="Integer"
07          resultMap="CourseWithStudentResult">
08          select * from tb_course where id=#{id}
09      </select>
10      <resultMap type="Course" id="CourseWithStudentResult">
11          <id property="id" column="id" />
12          <result property="name" column="name" />
13          <result property="code" column="code" />
14          <collection property="studentList" column="id" ofType="Student"
15   select="com.ssm.mapper.StudentMapper.findStudentById">
16          </collection>
17      </resultMap>
18  </mapper>
```

在文件 9.11 中，使用嵌套查询的方式定义了一个 id 为 findCourseWithStudent 的 select 语句来查询课程及其关联的学生信息。在<resultMap>元素中使用了<collection>元素来映射多对多的关联关系，其中 property 属性表示课程持久化类中的学生属性，ofType 属性表示集合中的数据为 Student 类型，而 column 的属性值会作为参数执行 StudentMapper.xml 中定义的 id 为 findStudentById 的执行语句，来查询课程中的学生信息。

文件 9.12　StudentMapper.xml

```
01  <?xml version="1.0" encoding="UTF-8"?>
02  <!DOCTYPE mapper PUBLIC "-//mybatis.org//DTD Mapper 3.0//EN"
03      "http://mybatis.org/dtd/mybatis-3-mapper.dtd">
04  <mapper namespace="com.ssm.mapper.StudentMapper">
05      <select id="findStudentById" parameterType="Integer" resultType="Student">
06          select * from tb_student where id in(
07          select student_id from tb_electivecourse where course_id=#{id}
08          )
09      </select>
10  </mapper>
```

在文件 9.12 中定义了一个 id 为 findStudentById 的执行语句，该执行语句中的 SQL 会根据课程 id 查询与该课程关联的学生信息。由于课程和学生是多对多的关联关系，因此需要通过中间表来查询学生信息。

步骤 04 将新创建的映射文件 CourseMapper.xml 和 StudentMapper.xml 的文件路径配置到核心配置文件 mybatis-config.xml 中，代码如下：

```
<mapper resource="com/ssm/mapper/CourseMapper.xml" />
```

```
<mapper resource="com/ssm/mapper/StudentMapper.xml" />
```

步骤 05 在测试类 MyBatisAssociatedTest 中编写多对多关联查询的测试方法 findCourseByIdTest()，代码如下：

```
<!--多对多嵌套查询-->
@Test
public void findCourseByIdTest(){
    SqlSession sqlSession = MybatisUtils.getSession();
    //查询课程 id 为 1 的课程中的学生信息
    Course course=
        sqlSession.selectOne("com.ssm.mapper.CourseMapper.findCourseWithStudent", 1);
    System.out.println(course);
    sqlSession.close();
}
```

执行方法后，控制台的输出结果如图 9.4 所示。从图中可以看出，使用 MyBatis 嵌套查询的方式查询出了课程及其关联的学生信息，这就是 MyBatis 多对多的关联查询。

```
Tests passed: 1 of 1 test – 1 s 414 ms
Course [id=1, name=Java程序设计语言, code=08113226, studentList=[Student [id=1, name=孙淼, sex=m]]]
```

图 9.4 运行结果

如果读者对多表关联查询的 SQL 语句比较熟悉，那么就可以在 CourseMapper.xml 中使用嵌套结果查询的方式，代码如下：

```
<!--多对多嵌套结果查询：查询某课程及其关联的学生信息 -->
<select id="findCourseWithStudent2" parameterType="Integer"
    resultMap="CourseWithStudentResult2">
    select c.*,s.id as sid,s.name as sname
    from tb_course c,tb_student s,tb_electivecourse ec
    where ec.course_id=c.id
    and ec.student_id=s.id and c.id=#{id}
</select>
<resultMap type="Course" id="CourseWithStudentResult2">
    <id property="id" column="id" />
    <result property="name" column="name" />
    <result property="code" column="code" />
    <collection property="studentList" ofType="Student">
        <id property="id" column="sid" />
        <result property="name" column="sname" />
    </collection>
</resultMap>
```

第 10 章

Spring 与 MyBatis 的整合

前面章节分别讲解了 Spring 和 MyBatis 的相关知识，在实际的项目开发中，需要将 Spring 与 MyBatis 整合在一起使用，本章就对 Spring 和 MyBatis 的整合进行讲解。

本章主要涉及的知识点如下：

- Spring 和 MyBatis 整合环境搭建。
- 传统 DAO 方式的开发整合。
- Mapper 接口方式的开发整合。

10.1 整合环境搭建

Spring 和 MyBatis 的整合环境搭建主要涉及准备所需的 JAR 包和编写配置文件，下面将详细介绍。

10.1.1 准备所需的 JAR 包

要实现 Spring 与 MyBatis 的整合，需要这两个框架相关的 JAR 包，除此之外，还需要其他的 JAR 包来配合使用。

1. Spring 框架所需的 JAR 包

Spring 框架所需要准备的 JAR 包共 10 个，其中包括 4 个核心模块 JAR 包、AOP 开发使用的 JAR 包、JDBC 和事务的 JAR 包（其中核心容器依赖的 commons-logging 的 JAR 包在 MyBatis 框架的 lib 目录中已经包含），具体如下：

- aopalliance-1.0.jar
- aspectjweaver-1.9.9.1.jar
- spring-aop-5.2.9.RELEASE.jar
- spring-aspects-5.2.9.RELEASE.jar
- spring-beans-5.2.9.RELEASE.jar
- spring-context-5.2.9.RELEASE.jar
- spring-core-5.2.9.RELEASE.jar
- spring-expression-5.2.9.RELEASE.jar
- spring-jdbc-5.2.9.RELEASE.jar
- spring-tx-5.2.9.RELEASE.jar

2. MyBatis 框架所需的 JAR 包

MyBatis 框架需要准备的 JAR 包共 12 个，其中包括核心包 mybatis-3.5.10.jar 及其解压文件夹中 lib 目录中的所有 JAR 包，具体如下：

- ant-1.10.12.jar
- ant-launcher-1.10.12.jar
- asm-9.3.jar
- cglib-3.3.0.jar
- commons-logging-1.2.jar
- javassist-3.29.0-GA.jar
- log4j-api-2.18.0.jar
- log4j-core-2.18.0.jar
- mybatis-3.5.10.jar
- ognl-3.3.2.jar
- slf4j-api-2.0.0-alpha1.jar
- slf4j-log4j12-2.0.0-alpha1.jar

3. Spring 与 MyBatis 整合所需的中间 JAR 包

为了满足 MyBatis 用户对 Spring 框架的需求，MyBatis 社区开发了一个用于整合 MyBatis 和 Spring 两个框架的中间件——mybatis-spring。

本书使用的中间件是 mybatis-spring-2.0.7.jar。此版本的 JAR 包可以通过下面这个链接获取：https://mvnrepository.com/artifact/org.mybatis/mybatis-spring/2.0.7。

4. 数据库驱动 JAR 包

本书所使用的数据库驱动包为 mysql-connector-java-8.0.23.jar。

5. 数据源所需的 JAR 包

整合时所使用的是 DBCP 数据源，所以需要准备 DBCP 和连接池的 JAR 包，具体如下：

- commons-dbcp2-2.9.0.jar

- commons-pool2-2.11.1.jar

10.1.2 编写配置文件

在 IntelliJ IDEA 中创建一个名称为 chapter10 的 Web 项目，将 10.1.1 节准备的全部 JAR 包添加到项目的 lib 目录中，并发布到类路径下。

【示例 10-1】参照前面章节的内容和案例，在项目的 src 目录下分别创建 db.properties 文件、Spring 的配置文件 applicationContext.xml 以及 MyBatis 的配置文件 mybatis-config.xml，如文件 10.1~文件 10.3 所示。

文件 10.1 db.properties

```
01  jdbc.driver=com.mysql.cj.jdbc.Driver
02  jdbc.url=jdbc:mysql://localhost:3306/db_mybatis?serverTimezone=UTC
03  jdbc.username=root
04  jdbc.password=root
05  jdbc.maxTotal=30
06  jdbc.maxIdle=10
07  jdbc.initialSize=5
```

在文件 10.1 中，除了配置连接数据库的基本 4 项外，还配置数据库连接池的最大连接数（maxTotal）、最大空闲连接数（maxIdle）以及初始化连接数（initialSize）。

文件 10.2 applicationContext.xml

```
01  <?xml version="1.0" encoding="UTF-8"?>
02  <beans xmlns="http://www.springframework.org/schema/beans"
03      xmlns:xsi="http://www.w3.org/2001/XMLSchema-instance"
04      xmlns:aop="http://www.springframework.org/schema/aop"
05      xmlns:tx="http://www.springframework.org/schema/tx"
06      xmlns:context="http://www.springframework.org/schema/context"
07      xsi:schemaLocation="http://www.springframework.org/schema/beans
08        http://www.springframework.org/schema/beans/spring-beans.xsd
09        http://www.springframework.org/schema/tx
10        http://www.springframework.org/schema/tx/spring-tx.xsd
11        http://www.springframework.org/schema/context
12        http://www.springframework.org/schema/context/spring-context.xsd
13        http://www.springframework.org/schema/aop
14        http://www.springframework.org/schema/aop/spring-aop.xsd">
15      <!--读取 db.properties-->
16      <context:property-placeholder location="classpath:db.properties"/>
17      <!--配置数据源 -->
18      <bean id="dataSource" class="org.apache.commons.dbcp2.BasicDataSource">
19          <!--数据库驱动 -->
20          <property name="driverClassName" value="${jdbc.driver}" />
21          <!--连接数据库的url -->
```

```
22            <property name="url" value="${jdbc.url}" />
23            <!--连接数据库的用户名 -->
24            <property name="username" value="${jdbc.username}" />
25            <!--连接数据库的密码-->
26            <property name="password" value="${jdbc.password}" />
27            <!--最大连接数-->
28            <property name="maxTotal" value="${jdbc.maxTotal}" />
29            <!--最大空闲连接-->
30            <property name="maxIdle" value="${jdbc.maxIdle}" />
31            <!--初始化连接数-->
32            <property name="initialSize" value="${jdbc.initialSize}" />
33        </bean>
34        <!--事务管理器，依赖于数据源 -->
35        <bean id="transactionManager"
36    class="org.springframework.jdbc.datasource.DataSourceTransactionManager">
37            <property name="dataSource" ref="dataSource"/>
38        </bean>
39        <!--注册事务管理器驱动，开启事务注解 -->
40        <tx:annotation-driven transaction-manager="transactionManager"/>
41        <!--配置 MyBatis 工厂 -->
42        <bean id="sqlSessionFactory" class="org.mybatis.spring.SqlSessionFactoryBean">
43            <!--注入数据源 -->
44            <property name="dataSource" ref="dataSource" />
45            <!--指定核心配置文件位置 -->
46            <property name="configLocation" value="classpath:mybatis-config.xml" />
47        </bean>
48        <bean id="userDao" class="com.ssm.dao.impl.UserDaoImpl">
49            <property name="sqlSessionFactory" ref="sqlSessionFactory"></property>
50        </bean>
51    </beans>
```

在文件 10.2 中，首先定义了读取 properties 文件的配置，然后配置了数据源，接下来配置了事务管理器并开启了事务注解，最后配置了 MyBatis 工厂来与 Spring 整合。其中，MyBatis 工厂的作用是构建 SqlSessionFactory，它通过 mybatis-spring 包中提供的 org.mybatis.Spring.SqlSessionFactoryBean 类来配置。通常在配置时需要提供两个参数：一个是数据源，另一个是 MyBatis 的配置文件路径。这样 Spring 的 IoC 容器就会在初始化 id 为 sqlSessionFactory 的 Bean 时，解析 MyBatis 的配置文件，并与数据源一同保存到 Spring 的 Bean 中。

文件 10.3 mybatis-config.xml

```
01    <?xml version="1.0" encoding="UTF-8"?>
02    <!DOCTYPE configuration PUBLIC "-//mybatis.org//DTD Config 3.0//EN"
03        "http://mybatis.org/dtd/mybatis-3-config.dtd">
04    <configuration>
05        <!--配置别名 -->
06        <typeAliases>
07            <package name="com.ssm.po"/>
```

```
08        </typeAliases>
09        <!--配置 Mapper 的位置 -->
10        <mappers>
11            ...
12        </mappers>
13    </configuration>
```

由于在 Spring 中已经配置了数据源信息，因此在 MyBatis 的配置文件中不再需要配置数据源信息。这里只需要使用<typeAliases>和<mappers>元素来配置文件别名以及指定 mapper 文件位置即可。

此外，还需在项目的 src 目录下创建 log4j.properties 文件，该文件的编写可参考第 6 章的入门案例，也可将前面章节创建的 log4j.properties 文件复制到此项目中使用。

10.2 整　合

10.1 节已经完成了 MyBatis 与 Spring 整合环境的搭建工作，可以说完成这些配置后就已经完成了这两个框架大部分的整合工作。接下来，本节将对传统 DAO 方式和 Mapper 接口方式的开发整合进行介绍。

10.2.1 传统 DAO 方式的开发整合

采用传统 DAO 开发方式进行 MyBatis 与 Spring 框架的整合时，我们需要编写 DAO 接口以及接口的实现类，并且需要向 DAO 实现类中注入 SqlSessionFactory，然后在方法体内通过 SqlSessionFactory 创建 SqlSession。为此，我们可以使用 mybatis-spring 包中提供的 SqlSessionTemplate 类或 SqlSessionDaoSupport 类来实现此功能。这两个类的描述如下：

- SqlSessionTemplate：mybatis-spring 的核心类，它负责管理 MyBatis 的 SqlSession，调用 MyBatis 的 SQL 方法。当调用 SQL 方法时，SqlSessionTemplate 会保证使用的 SqlSession 和当前 Spring 的事务是相关的。它还管理 SqlSession 的生命周期，包含必要的关闭、提交和回滚操作。
- SqlSessionDaoSupport：一个抽象支持类，它继承自 DaoSupport 类，主要是作为 DAO 的基类来使用。可以通过 SqlSessionDaoSupport 类的 getSqlSession()方法来获取所需的 SqlSession。

【示例 10-2】在了解了传统 DAO 方式开发的整合可以使用的两个类后，下面以 SqlSessionDaoSupport 类的使用为例，讲解传统的 DAO 方式开发的整合的实现，具体步骤如下：

1. 实现持久层

步骤 01 在 src 目录下创建一个 com.ssm.po 包，并在包中创建持久化类 User，在 User 类中定义相关属性和方法，如文件 10.4 所示。

文件 10.4 User.java

```
01  package com.ssm.po;
```

```
02  //用户类
03  public class User {
04      private Integer id;
05      private String username;
06      private String jobs;
07      private String phone;
08      public Integer getId() {
09          return id;
10      }
11      public void setId(Integer id) {
12          this.id = id;
13      }
14      public String getUsername() {
15          return username;
16      }
17      public void setUsername(String username) {
18          this.username = username;
19      }
20      public String getJobs() {
21          return jobs;
22      }
23      public void setJobs(String jobs) {
24          this.jobs = jobs;
25      }
26      public String getPhone() {
27          return phone;
28      }
29      public void setPhone(String phone) {
30          this.phone = phone;
31      }
32      public String toString() {
33          return "User [id=" + id + ", username=" + username + ", jobs=" + jobs +
34   ", phone=" + phone + "]";
35      }
36  }
```

步骤02 在com.ssm.po包中创建映射文件UserMapper.xml，在该文件中编写根据id查询用户信息的映射语句，如文件10.5所示。

文件10.5 UserMapper.xml

```
01  <?xml version="1.0" encoding="UTF-8"?>
02  <!DOCTYPE mapper PUBLIC "-//mybatis.org//DTD Mapper 3.0//EN"
03      "http://mybatis.org/dtd/mybatis-3-mapper.dtd">
04  <mapper namespace="com.ssm.po.UserMapper">
05      <!--根据用户编号获取用户信息 -->
06      <select id="findUserById" parameterType="Integer" resultType="User">
```

```
07          select * from t_user where id=#{id}
08      </select>
09  </mapper>
```

步骤03 在 MyBatis 的配置文件 mybatis-config.xml 中配置映射文件 UserMapper.xml 的位置，具体如下：

```
<mapper resource="com/ssm/po/UserMapper.xml" />
```

2. 实现 DAO 层

步骤01 在 src 目录下创建一个 com.ssm.dao 包，并在包中创建接口 UseDao，在接口中编写一个通过 id 查询用户的方法 findUserById()，如文件 10.6 所示。

文件 10.6 UserDao.iava

```
01  package com.ssm.dao;
02  import com.ssm.po.User;
03  //用户接口类
04  public interface UserDao {
05      //根据用户 id 查询用户的方法
06      public User findUserById(Integer id);
07  }
```

步骤02 在 src 目录下创建一个 com.ssm.dao.impl 包，并在包中创建 UserDao 接口的实现类 UserDaoImpl，如文件 10.7 所示。

文件 10.7 UserDaoImpl.java

```
01  package com.ssm.dao.impl;
02  import org.mybatis.spring.support.SqlSessionDaoSupport;
03  import com.ssm.dao.UserDao;
04  import com.ssm.po.User;
05  //用户接口实现类
06  public class UserDaoImpl extends SqlSessionDaoSupport implements UserDao {
07      //根据用户 id 查询用户的实现方法
08      public User findUserById(Integer id) {
09          return this.getSqlSession().selectOne("com.ssm.po.UserMapper.findUserById", id);
10      }
11  }
```

在文件 10.7 中，UserDaoImpl 类继承了 SqlSessionDaoSupport 并实现了 UserDao 接口。其中，SqlSessionDaoSupport 类在使用时需要一个 SqlSessionFactory 或一个 SqlSessionTemplate 对象，所以需要通过 Spring 给 SqlSessionDaoSupport 类的子类对象注入一个 SqlSessionFactory 或 SqlSessionTemplate。这样，在子类中就能通过调用 SqlSessionDaoSupport 类的 getSqlSession()方法来获取 SqlSession 对象，并使用 SqlSession 对象中的方法了。

步骤03 在 Spring 的配置文件 applicationContext.xml 中编写实例化 UserDaoImpl 的配置，代码如下：

```
<!-- 实例化 Dao -->
<bean id="userDao" class="com.ssm.dao.impl.UserDaoImpl">
    <property name="sqlSessionFactory" ref="sqlSessionFactory"></property>
</bean>
```

上述代码创建了一个 id 为 userDao 的 Bean，并将 SqlSessionFactory 对象注入该 Bean 的实例化对象中。

3. 整合测试

在 src 目录下创建一个 com.ssm.test 包，在包中创建测试类 UserDaoTest，并在类中编写测试方法 findUserByIdDaoTest()，如文件 10.8 所示。

文件 10.8　UserDaoTest.java

```
01  package com.ssm.test;
02  import org.junit.Test;
03  import org.springframework.context.ApplicationContext;
04  import org.springframework.context.support.ClassPathXmlApplicationContext;
05  import com.ssm.dao.UserDao;
06  import com.ssm.po.User;
07  public class UserDaoTest {
08      @Test
09       public void findUserByIdDaotest(){
10           //1.初始化 Spring 容器，加载配置文件
11           ApplicationContext applicationContext=
12  new ClassPathXmlApplicationContext("applicationContext.xml");
13           //2.通过容器获取 userDao 实例
14           UserDao userDao=(UserDao)applicationContext.getBean("userDao");
15           //调用 UserDao 接口的查询用户方法（用户 id 值为 1）
16           User user=userDao.findUserById(1);
17           System.out.println(user);
18      }
19  }
```

在上述方法中，我们采用的是根据容器中 Bean 的 id 来获取指定 Bean 的方式。执行上述方法后，控制台的输出结果如图 10.1 所示。从图中可以看出，通过 UserDao 实例的 findUserById()方法，已成功经查询出了 id 为 1 的用户信息，这就说明 MyBatis 与 Spring 整合成功。

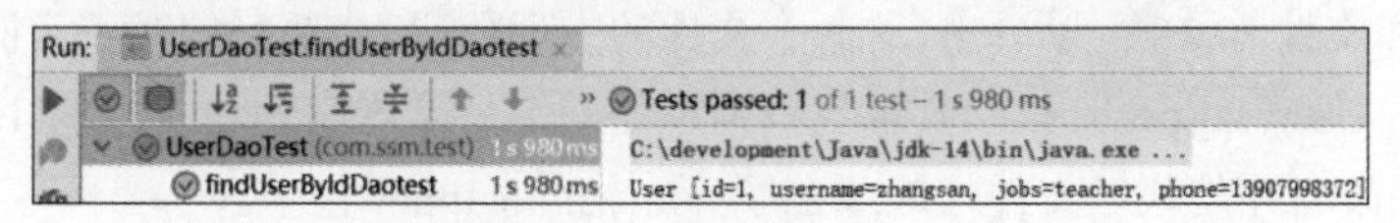

图 10.1　运行结果

10.2.2　Mapper 接口方式的开发整合

在 MyBatis+Spring 的项目中，虽然使用传统的 DAO 开发方式可以实现所需功能，但是采用这种方式在实现类中会出现大量的重复代码，在方法中也需要指定映射文件中执行语句的 id，并且不能保

证编写时 id 的正确性（运行时才能知道）。为此，我们可以使用 MyBatis 提供的另一种编程方式，即使用 Mapper 接口编程。接下来将讲解如何使用 Mapper 接口方式实现 MyBatis 与 Spring 的整合。

1. 基于 MapperFactoryBean 的整合

MapperFactoryBean 是 mybatis-spring 团队提供的一个用于根据 Mapper 接口生成 Mapper 对象的类，该类在 Spring 配置文件中使用时可以配置以下参数：

- mapperInterface：用于指定接口。
- SqlSessionFactory：用于指定 SqlSessionFactory。
- SqlSessionTemplate：用于指定 SqlSessionTemplate。若与 SqlSessionFactory 同时设定，则只会启用 SqlsessionTemplate。

【示例 10-3】在了解了 MapperFactoryBean 类后，接下来通过一个具体的案例来演示如何通过 MapperFactoryBean 实现 MyBatis 与 Spring 的整合，具体步骤如下：

步骤 01 在 src 目录下创建一个 com.ssm.mapper 包，然后在该包中创建 UserMapper 接口以及对应的映射文件，如文件 10.9 和文件 10.10 所示。

文件 10.9 UserMapper.java

```
01  package com.ssm.mapper;
02  import com.ssm.po.User;
03  public interface UserMapper {
04      public User findUserById(Integer id);
05  }
```

文件 10.10 UserMapper.xml

```
01  <?xml version="1.0" encoding="UTF-8"?>
02  <!DOCTYPE mapper PUBLIC "-//mybatis.org//DTD Mapper 3.0//EN"
03      "http://mybatis.org/dtd/mybatis-3-mapper.dtd">
04  <mapper namespace="com.ssm.mapper.UserMapper">
05      <!--根据用户编号获取用户信息 -->
06      <select id="findUserById" parameterType="Integer" resultType="User">
07          select * from t_user where id=#{id}
08      </select>
09  </mapper>
```

从文件 10.9 和文件 10.10 可以看出，这两个文件的实现代码与 10.2.1 节的【示例 10-2】中 UserDao 接口以及映射文件 UserMapper.xml 的实现代码基本相同，只是本案例与【示例 10-2】的区别在于，本案例将接口文件改名并与映射文件一起放在了 com.ssm.mapper 包中。

步骤 02 在 MyBatis 的配置文件中引入新的映射文件，代码如下：

```
<!-- Mapper 接口开发方式 -->
<mapper resource="com/ssm/mapper/UserMapper.xml" />
```

步骤 03 在 Spring 的配置文件中创建一个 id 为 userMapper 的 Bean，代码如下：

```
<!-- Mapper 代理开发（基于 MapperFactoryBean） -->
<bean id="userMapper" class="org.mybatis.spring.mapper.MapperFactoryBean">
    <property name="mapperInterface" value="com.ssm.mapper.UserMapper" />
    <property name="sqlSessionFactory" ref="sqlSessionFactory" />
</bean>
```

上述配置代码为 MapperFactoryBean 指定了接口以及 SqlSessionFactory。

步骤 04 在测试类 UserDaoTest 中编写测试方法 findUserByIdMapperTest()，代码如下：

```
@Test
public void findUserByIdMapperTest(){
    ApplicationContext applicationContext=
                new ClassPathXmlApplicationContext("applicationContext.xml");
    UserMapper userMapper=(UserMapper)applicationContext.getBean("userMapper");
    User user=userMapper.findUserById(1);
    System.out.println(user);
}
```

上述方法中，通过 Spring 容器获取了 UserMapper 实例，并调用了实例中的 findUserById()方法来查询 id 为 1 的用户信息。执行方法后，控制台的输出结果与图 10.1 所示的结果相同。

注意：Mapper 接口编程方式只需要程序员编写 Mapper 接口（相当于 DAO 接口），然后由 MyBatis 框架根据接口的定义创建接口的动态代理对象，这个代理对象的方法体等同于【示例 10-2】中 DAO 接口的实现类方法。

虽然使用 Mapper 接口编程的方式很简单，但是在具体使用时还是需要遵循以下规范：

（1）Mapper 接口的名称和对应的 Mapper.xml 映射文件的名称必须一致。

（2）Mapper.xml 文件中的 namespace 与 Mapper 接口的类路径相同（即接口文件和映射文件需要放在同一个包中）。

（3）Mapper 接口中的方法名和 Mapper.xml 中定义的每条执行语句的 id 相同。

（4）Mapper 接口中方法的输入参数类型要和 Mapper.xml 中定义的每个 SQL 的 parameterType 的类型相同。

（5）Mapper 接口方法的输出参数类型要和 Mapper.xml 中定义的每个 SQL 的 resultType 的类型相同。

只要遵循了这些开发规范，MyBatis 就可以自动生成 Mapper 接口实现类的代理对象，从而简化开发。

2. 基于 MapperScannerConfigurer 的整合

在实际的项目中，DAO 层会包含很多接口，如果每一个接口都像【示例 10-2】中那样在 Spring 配置文件中配置，那么不但会增加工作量，还会使得 Spring 配置文件非常臃肿。为此，mybsatis-spring 团队提供了一种自动扫描的形式来配置 MyBatis 中的映射器——采用 MapperScannerConfigurer 类。

MapperScannerConfigurer 类在 Spring 配置文件中使用时，可以配置以下属性：

- basePackage：指定映射接口文件所在的包路径，当需要扫描多个包时可以使用分号或逗号作为分隔符。指定包路径后，会扫描该包及其子包中的所有文件。
- annotationClass：指定要扫描的注解名称，只有被注解标识的类才会被配置为映射器。
- sqlSessionFactoryBeanName：指定在 Spring 中定义的 SqlSessionFactory 的 Bean 名称。
- sqlSessionTemplateBeanName：指定在 Spring 中定义的 SqlSessionTemplate 的 Bean 名称。若定义此属性，则 sqlSessionFactoryBeanName 将不起作用。
- markerInterface：指定创建映射器的接口。

MapperScannerConfigurer 的使用非常简单，只需要在 Spring 的配置文件中编写如下代码：

```
<!-- Mapper 代理开发（基于 MapperScannerConfigurer） -->
<bean class="org.mybatis.spring.mapper.MapperScannerConfigurer">
    <property name="basePackage" value="com.ssm.mapper" />
</bean>
```

在通常情况下，MapperScannerConfigurer 在使用时只需通过 basePackage 属性指定需要扫描的包即可。Spring 会自动地通过包中的接口生成映射器。这使得开发人员可以在编写很少代码的情况下完成对映射器的配置，从而提高开发效率。

要验证上面的配置很容易，只需将上述配置代码写入 Spring 的配置文件，并将基于 MapperFactoryBean 的整合案例（【示例 10-3】）中的步骤 02 和步骤 03 的代码注释掉，再次执行 findUserByIdMapperTest()方法进行测试即可。方法执行后的结果与之前一致。

第 11 章 Spring MVC 入门

在学习了 Spring 框架和 MyBatis 框架的使用及整合后，接下来将讲解 Spring MVC 框架。

本章主要涉及的知识点如下：

- Spring MVC 的特点。
- Spring MVC 入门程序。
- Spring MVC 核心类。
- Spring MVC 注解。

11.1 Spring MVC 概述

Spring MVC 是 Spring 提供的一个轻量级 Web 框架，它实现了 Web MVC 设计模式。Spring MVC 在使用和性能等方面比 Struts 2 框架更加优异。

Spring MVC 具有如下特点：

- 它是 Spring 框架的一部分，可以方便地利用 Spring 所提供的其他功能。
- 灵活性强，易于与其他框架集成。
- 提供了一个前端控制器 DispatcherServlet，使开发人员无须额外开发控制器对象。
- 可自动绑定用户输入，并能正确地转换数据类型。
- 内置了常见的校验器，可以校验用户输入。如果校验不能通过，就会重定向到输入表单。
- 支持国际化，可以根据用户区域显示多国语言。
- 支持多种视图技术，如 JSP、Velocity 和 FreeMarker 等视图技术。
- 使用基于 XML 的配置文件，在编辑后，不需要重新编译应用程序。

11.2 应用案例——第一个 Spring MVC 应用

了解了什么是 Spring MVC 及其优点后，接下来通过一个简单的入门案例来演示 Spring MVC 的使用。

1. 创建项目，引入 JAR 包

在 IntelliJ IDEA 中创建一个名称为 chapter11 的 Web 项目，在项目的 lib 目录中添加运行 Spring MVC 程序所需要的 JAR 包，并发布到类路径下，如图 11.1 所示。

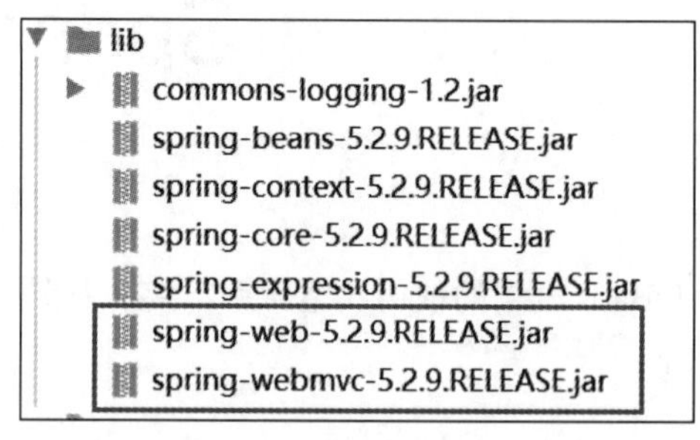

图 11.1 所需要的 JAR 包

从图 11.1 中可以看到，在项目中添加了 4 个核心 JAR 包、1 个 commons-logging 的 JAR 包以及 2 个 Web 相关的 JAR 包，这 2 个 Web 相关的 JAR 包就是 Spring MVC 框架所需的 JAR 包。

2. 配置前端控制器

在 web.xml 中配置 Spring MVC 的前端控制器 DispatcherServlet，如文件 11.1 所示。

文件 11.1 web.xml

```
01 <?xml version="1.0" encoding="UTF-8"?>
02 <web-app xmlns="http://xmlns.jcp.org/xml/ns/javaee"
03         xmlns:xsi="http://www.w3.org/2001/XMLSchema-instance"
04         xsi:schemaLocation="http://xmlns.jcp.org/xml/ns/javaee
05         http://xmlns.jcp.org/xml/ns/javaee/web-app_4_0.xsd"
06         id="WebApp_ID" version="4.0">
07     <servlet>
08         <!-- 配置前端过滤器 -->
09         <servlet-name>springmvc</servlet-name>
10         <servlet-class>org.springframework.web.servlet.DispatcherServlet </servlet-class>
11         <!-- 初始化时加载配置文件 -->
12         <init-param>
13             <param-name>contextConfigLocation</param-name>
14             <param-value>classpath:springmvc-config.xml</param-value>
15         </init-param>
16             <!-- 表示容器在启动时立即加载 Servlet -->
17             <load-on-startup>1</load-on-startup>
18     </servlet>
```

```
19      <servlet-mapping>
20          <servlet-name>springmvc</servlet-name>
21          <url-pattern>/</url-pattern>
22      </servlet-mapping>
23  </web-app>
```

在文件 11.1 中，主要对<servlet>和<servlet-mapping>元素进行了配置。在<servlet>中配置了 Spring MVC 的前端控制器 DispatcherServlet，并通过其子元素<init-param>配置了 Spring MVC 配置文件的位置，<load-on-startup>元素中的 1 表示容器在启动时立即加载这个 Servlet；在<servlet-mapping>中，通过<url-pattern>元素的“/”拦截所有 URL，并交由 DispatcherServlet 处理。

3. 创建 Controller 类

在 src 目录下创建一个 com.ssm.controller 包，并在包中创建控制器类 ControllerTest，该类需要实现 Controller 接口，如文件 11.2 所示。

文件 11.2　ControllerTest.java

```
01  package com.ssm.controller;
02  import javax.servlet.http.HttpServletRequest;
03  import javax.servlet.http.HttpServletResponse;
04  import org.springframework.web.servlet.ModelAndView;
05  import org.springframework.web.servlet.mvc.Controller;
06  public class ControllerTest implements Controller {
07      @Override
08      public ModelAndView handleRequest(HttpServletRequest arg0, HttpServletResponse arg1)
09      throws Exception {
10          //创建 ModelAndView 对象
11          ModelAndView m=new ModelAndView();
12          //向模型对象中添加一个名称为 msg 的字符串对象
13          m.addObject("msg","第一个 Spring MVC 程序");
14          //设置返回的视图路径
15          m.setViewName("/WEB-INF/jsp/welcome.jsp");
16          return m;
17      }
18  }
```

在文件 11.2 中，handleRequest()是 Controller 接口的实现方法，ControllerTest 类会调用该方法处理请求，并返回一个包含视图名或包含视图名与模型的 ModelAndview 对象。本案例向模型对象中添加了一个名称为 msg 的字符串对象，并设置返回的视图路径为 WEB-INF/jsp/welcome. jsp，这样请求就会被转发到 welcome. jsp 页面。

4. 创建 Spring MVC 的配置文件，配置控制器映射信息

在 src 目录下创建配置文件 springmvc-config.xml，并在文件中配置控制器信息，如文件 11.3 所示。

文件 11.3　springmvc-config.xml

```
01  <?xml version="1.0" encoding="UTF-8"?>
```

```
02  <beans xmlns="http://www.springframework.org/schema/beans"
03         xmlns:xsi="http://www.w3.org/2001/XMLSchema-instance"
04         xsi:schemaLocation="http://www.springframework.org/schema/beans
05         http://www.springframework.org/schema/beans/spring-beans.xsd">
06      <!--配置处理器 Handle，映射"controllerTest"请求 -->
07      <bean name="/controllerTest" class="com.ssm.controller.ControllerTest" />
08      <!--处理器映射，将处理器 Handle 的 name 作为 url 进行查找 -->
09      <bean class="org.springframework.web.servlet.handler.BeanNameUrlHandlerMapping" />
10      <!--处理器适配器，配置对处理器中 handleRequest()方法的调用 -->
11      <bean class="org.springframework.web.servlet.mvc.SimpleControllerHandlerAdapter" />
12      <!--视图解析器 -->
13      <bean class="org.springframework.web.servlet.view.InternalResourceViewResolver" />
14  </beans>
```

在文件 11.3 中，首先定义了一个名称为“/ControllerTest”的 Bean，该 Bean 会将控制器类 ControllerTest 映射到“/ControllerTest”请求中；然后配置了处理器映射器 BeanNameUrlHandlerMapping 和处理器适配器 SimpleControllerHandlerAdapter，其中处理器映射器用于将处理器 Bean 中的 name（即 url）进行处理器查找，而处理器适配器用于完成对 ControllerTest 处理器中 handleRequest()方法的调用；最后配置了视图解析器 InternalResourceViewResolver 来解析结果视图，并将结果呈现给用户。

注意：在旧版本的 Spring 中，配置文件内必须配置处理器映射器、处理器适配器和视图解析器。但在 Spring 4.0 以后，如果不配置处理器映射器、处理器适配器和视图解析器，就会使用 Spring 内部默认的配置来完成相应的工作。这里显示的配置处理器映射器、处理器适配器和视图解析器是为了让读者能够更加清晰地了解 Spring MVC 的工作流程。

5. 创建视图（View）页面

在 WEB-INF 目录下创建一个 jsp 文件夹，并在文件夹中创建一个页面文件 welcome.jsp，在该页面中使用 EL 表达式获取 msg 中的信息，如文件 11.4 所示。

文件 11.4　welcome. jsp

```
01  <%@ page contentType="text/html;charset=UTF-8" language="java" %>
02  <html>
03  <head>
04  <title>入门程序</title>
05  <script src="https://unpkg.com/vue@next"></script>
06  </head>
07  <body>
08  <div id="app">
09  {{msg}}
10  </div>
11  <script type="text/javascript">
12  const app=Vue.createApp({
13      data(){
14          return{
```

```
15                msg:''
16            }
17        }
18    });
19    const vm = app.mount('#app')
20    </script>
21    </body>
22    </html>
```

6. 启动项目，测试应用

将 chapter11 项目发布到 Tomcat 服务器中并启动服务器。在浏览器中访问 http://localhost:8080/chapter11/controllerTest，显示的页面效果如图 11.2 所示。从图中可以看到，浏览器页面中已经显示出模型对象中的字符串信息，这就说明第一个 Spring MVC 程序执行成功了。

图 11.2　运行结果的页面效果

通过入门案例的学习，我们总结一下 Spring MVC 程序的执行流程。

（1）用户通过浏览器向服务器发送请求，请求会被 Spring MVC 的前端控制器 DispatcherServlet 拦截。

（2）DispatcherServlet 拦截到请求后，会调用 HandlerMapping 处理器映射器。

（3）处理器映射器根据请求 URL 找到具体的处理器，生成处理器对象及处理器拦截器（如果有就生成）并一起返回给 DispatcherServlet。

（4）DispatcherServlet 会通过返回信息选择合适的 HandlerAdapter（处理器适配器）。

（5）HandlerAdapter 会调用并执行 Handler（处理器），这里的处理器就是程序中编写的 Controller 类，它也被称为后端控制器。

（6）Controller 执行完成后，会返回一个 ModelAndView 对象，该对象中包含视图名或包含模型与视图名。

（7）HandlerAdapter 将 ModelAndView 对象返回给 DispatcherServlet。

（8）DispatcherServlet 会根据 ModelAndView 对象选择一个合适的 ViewResolver（视图解析器）。

（9）ViewResolver 解析后，会向 DispatcherServlet 中返回具体的 View（视图）。

（10）DispatcherServlet 对 View 进行渲染（即将模型数据填充至视图中）。

（11）视图渲染结果会返回给客户端浏览器显示。

在上述执行过程中，DispatcherServlet、HandlerMapping、HandlerAdapter 和 ViewResolver 对象的工作是在框架内部执行的，开发人员并不需要关心这些对象内部的实现过程，只需要配置前端控制器（DispatcherServlet），完成 Controller 中的业务处理，并在视图中（View）中展示相应信息即可。

11.3 Spring MVC 的注解

在 Spring 2.5 之前，只能使用实现 Controller 接口的方式来开发一个控制器，11.2 节的案例使用的就是这种方式。在 Spring 2.5 之后，新增加了基于注解的控制器以及其他一些常用注解，这些注解的使用极大地减少了程序员的开发工作。本节将详细讲解 Spring MVC 中的常用核心类及其常用注解。

11.3.1 DispatcherServlet

DispatcherServlet 的全名是 org.Springframework.web.servlet.DispatcherServlet，它在程序中充当前端控制器的角色。

【示例 11-1】在使用 DispatcherServlet 时，只需将它配置在项目的 web.xml 文件中，其配置代码如下：

```
01  <servlet>
02      <!-- 配置前端过滤器 -->
03      <servlet-name>springmvc</servlet-name>
04      <servlet-class>org.srpingframework.web.servlet.DispatcherServlet</servlet-class>
05      <!-- 初始化时加载配置文件 -->
06      <init-param>
07          <param-name>contextConfigLocation</param-name>
08          <param-value>classpath:springmvc-config.xml</param-value>
09      </init-param>
10      <!-- 表示容器在启动时立即加载 Servlet -->
11      <load-on-startup>1</load-on-startup>
12  </servlet>
13  <servlet-mapping>
14      <servlet-name>springmvc</servlet-name>
15      <url-pattern>/</url-pattern>
16  </servlet-mapping>
```

在上述代码中，<load-on-startup>元素和<init-param>元素都是可选的。如果<load-on-startup>元素的值为 1，那么应用程序在启动时就会立即加载 Servlet；如果<load-on-startup>元素不存在，那么应用程序就会在第一个 Servlet 请求时加载该 Servlet。如果<init-param>元素存在，并且通过它的子元素配置了 Spring MVC 配置文件的路径，那么应用程序在启动时就会加载配置路径下的配置文件；如果没有通过<init-param>元素配置 Spring MVC 配置文件的路径，那么应用程序就会默认到 WEB-INF 目录下寻找如下方式命名的配置文件。

```
servletName-servlet.xml
```

其中，servletName 指的是部署在 web.xml 中的 DispatcherServlet 的名称，在上面 web.xml 的配置代码中即为 springmvc；而-servlet.xml 是配置文件名的固定写法，所以应用程序会在 WEB-INF 下寻找 springmvc-servlet.xml。

11.3.2 Controller 注解类型

org.springframework.stereotype.Controller 注解类型用于指示 Spring 类的实例是一个控制器，其注解形式为@Controller。该注解在使用时不需要再实现 Controller 接口，只需要将@Controller 注解加入控制器类上，然后通过 Spring 的扫描机制找到标注了该注解的控制器即可。

【示例 11-2】@Controller 注解在控制器类中的使用示例如下：

```
package com.ssm.controller;
import org.springframework.stereotype.Controller;
...
//Controller 注解
@Controller
public class ControllerTest{
    ...
}
```

为了保证 Spring 能够找到控制器类，还需要在 Spring MVC 的配置文件中添加相应的扫描配置信息，具体如下：

（1）在配置文件的声明中引入 spring-context。

（2）使用<context: component-scan>元素指定需要扫描的类包。

完整的配置文件如文件 11.5 所示。

文件 11.5　springmvc-config.xml

```
01  <?xml version="1.0" encoding="UTF-8"?>
02  <beans xmlns="http://www.springframework.org/schema/beans"
03      xmlns:xsi="http://www.w3.org/2001/XMLSchema-instance"
04      xmlns:context="http://www.springframework.org/schema/context"
05      xsi:schemaLocation="http://www.springframework.org/schema/beans
06        http://www.springframework.org/schema/beans/spring-beans.xsd
07        http://www.springframework.org/schema/context
08        http://www.springframework.org/schema/context/spring-context.xsd">
09      <!--指定需要扫描的包 -->
10      <context:component-scan base-package="com.ssm.controller" />
11  </beans>
```

在文件 11.5 中，<context: component-scan>元素的属性 base-package 指定了需要扫描的类包为 com.ssm.controller。在运行时，该类包及其子包下所有标注了注解的类都会被 Spring 处理。与实现了 Controller 接口的方式相比，使用注解的方式显然更加简单。同时，Controller 接口的实现类只能处理一个单一的请求动作，而基于注解的控制器可以同时处理多个请求动作，在使用上更加灵活。因此，在实际开发中通常都会使用基于注解的形式。

注意：使用注解方式时，程序的运行需要依赖 Spring 的 AOP 包，因此需要向 lib 目录中添加 spring.aop-5.2.9 RELEASE.jar，否则程序运行时会报错。

11.3.3 RequestMapping 注解类型

1. @RequestMapping 注解的使用

Spring 通过@Controller 注解找到相应的控制器类后，还需要知道控制器内部对每一个请求是如何处理的，这就需要使用 org.springframework.web.bind.annotation.RequestMapping 注解类型。RequestMapping 用于映射一个请求或一个方法，其注解形式为@RequestMapping，可以使用该注解标注在一个方法或一个类上。

（1）标注在方法上

当标注在一个方法上时，该方法将成为一个请求处理方法，它会在程序接收到对应的 URL 请求时被调用。

【示例 11-3】使用@RequestMapping 注解标注在方法上的示例如下：

```
01  package com.ssm.controller;
02  import org.springframework.stereotype.Controller;
03  import org.springframework.web.bind.annotation.RequestMapping;
04  …
05  //Controller 注解
06  @Controller
07  public class AnnotationControllerTest {
08      //@RequestMapping 注解标注在方法上
09      @RequestMapping(value="/annotationController")
10      public ModelAndView handleRequest(HttpServletRequest arg0,
11                                 HttpServletResponse arg1) throws Exception {
12          …
13          return m;
14      }
15  }
```

使用@RequestMapping 注解后，上述代码中的 handleRequest()方法就可以通过地址 http://localhost:8080/chapter11/annotationController 进行访问。

（2）标注在类上

当标注在一个类上时，该类中的所有方法都将映射为相对于类级别的请求，表示该控制器所处理的所有请求都被映射到 value 属性值所指定的路径下。

【示例 11-4】使用@RequestMapping 注解标注在类上的示例如下：

```
01  package com.ssm.controller;
02  import org.springframework.stereotype.Controller;
03  import org.springframework.web.bind.annotation.RequestMapping;
```

```
04 …
05 //Controller注解
06 @Controller
07 //@RequestMapping注解标注在类上
08 @RequestMapping(value="/controll")
09 public class AnnotationControllerTest {
10     @RequestMapping(value="/annotationController")
11     public ModelAndView handleRequest(HttpServletRequest arg0,
12                             HttpServletResponse arg1) throws Exception {
13         …
14         return m;
15     }
16 }
```

由于在类上添加了@RequestMapping注解，并且其value属性值为“/controll”，因此上述代码方法的请求路径将变为 http://localhost:8080/chapter11/controll/annotationController。如果该类中还包含其他方法，那么在其他方法的请求路径中也需要加入“/controll”。

2. @RequestMapping注解的属性

@RequestMapping注解除了可以指定value属性外，还可以指定一些其他属性，如表11.1所示。

表11.1 @RequestMapping注解的属性

属性名	类型	描述
name	String	可选属性，用于为映射地址指定别名
value	String[]	可选属性，同时也是默认属性，用于映射一个请求和一种方法，可以标注在一个方法或一个类上
method	RequestMethod[]	可选属性，用于指定该方法用于处理哪种类型的请求方式，其请求方式包括 GET、POST、HEAD、OPTIONS、PUT、PATCH、DELETE和TRACE，例如method= RequestMethod. GET表示只支持GET请求，如果需要支持多个请求方式，那就需要通过“{}”写成数组的形式，并且多个请求方式之间用英文逗号（，）分隔
params	String[]	可选属性，用于指定 Request中必须包含某些参数的值才可以通过它标注的方法来处理
headers	String[]	可选属性，用于指定 Request中必须包含某些指定的header的值才可以通过它标注的方法来处理
consumes	String[]	可选属性，用于指定处理请求的提交内容类型（Content-Type），比如application/json、text/html等
produces	String[]	可选属性，用于指定返回的内容类型，返回的内容类型必须是request请求头（Accept）中所包含的类型

在表11.1中，所有属性都是可选的，但它们的默认属性都是value。当value是它们的唯一属性时，可以省略属性名，例如下面两种标注的含义相同。

```
@RequestMapping(value="/annotationController")
```

```
@RequestMapping("/annotationController")
```

3. 组合注解

前面已经对@RequestMapping 注解及其属性进行了详细讲解，而在 Spring 4.3 及以后的版本中引入了组合注解来帮助简化常用的 HTTP 方法的映射，并更好地表达被注解方法的语义。组合注解如下：

- @GetMapping：匹配 GET 方式的请求。
- @PostMapping：匹配 POST 方式的请求。
- @PutMapping：匹配 PUT 方式的请求。
- @DeleteMapping：匹配 DELETE 方式的请求。
- @PatchMapping：匹配 PATCH 方式的请求。

以@GetMapping 为例，该组合注解是@RequestMapping(method=RequestMethod.GET)的缩写，它会将 Http Get 映射到特定的处理方法上。在实际开发中，传统的@RequestMapping 注解使用方式如下：

```
@RequestMapping(value="/user/{id}",method=RequestMethod.GET)
public String selectUserById(String id){
    …
}
```

而使用新注解@GetMapping 后，可以省略 method 属性，从而简化代码，使用方式如下：

```
@GetMapping(value="/user/{id}")
public String selectUserById(String id){
    …
}
```

4. 请求处理方法的参数类型和返回类型

在控制器类中，每一个请求处理方法都可以有多个不同类型的参数，以及一个多种类型的返回结果。例如，在 11.2 节的入门案例中，handleRequest()方法的参数就是对应请求的 HttpServletRequest 和 HttpServletResponse 两种类型参数。除此之外，还可以使用其他类型的参数，例如，在请求处理方法中需要访问 HttpSession 对象，就可以添加 HttpSession 作为参数，Spring 会将对象正确地传递给方法，使用示例如下：

```
@RequestMapping(value="/annotationController ")
public ModelAndView(HttpSession session)
{
    …
    return m;
}
```

在请求处理方法中，可以出现的参数类型如下：

- javax.servlet.ServletRequest/javax.servlet.http.HttpServletRequest

- javax.servlet.ServletResponse/javax.servlet.http.HttpServletResponse
- javax.servlet.http.HttpSession
- org.springframework.web.context.request.WebRequest 或 org.springframework.web.context.request.NativeWebRequest
- java.util.Locale
- java.util.TimeZone (Java 6+)/java.time.Zoneld(Java 8+)
- java.io.InputStream/java.io.Reader
- Java.io.OutputStream/java.io.Writer
- org.springframework.http.HttpMethod
- java.security.Principal
- 注解：@PathVariable、@MatrixVariable、@RequestParam、@RequestHeader、@RequestBody、@RequestPart、@SessionAttribute、@RequestAttributeHttpEntity<?>
- java.util.Map/org.springframework.ui.Model/lorg.springframework.ui.ModelMap
- org.springframework.web.servlet.mvc.support.RedirectAttributes
- org.springframework.validation.Errors/org.springframework.validation.BindingResult
- org.springframework.web.bind.support.SessionStatus
- org.springframework.web.util.UriComponentsBuilder

注意：org.springframework.ui.Model 类型不是一个 Servlet API 类型，而是一个包含 Map 对象的 Spring MVC 类型。如果方法中添加了 Model 参数，那么每次调用该请求处理方法时，Spring MVC 都会创建 Model 对象，并将它作为参数传递给方法。

在 11.2 节的入门案例中，请求处理方法返回的是一个 ModelAndView 类型的数据。除了这种类型外，请求处理方法还可以返回其他类型的数据。Spring MVC 所支持的常见方法返回类型如下：

- ModelAndView
- Model
- Map
- View
- String
- void
- HttpEntity<?>或 ResponseEntity<?>
- Callable<?>
- DeferredResult<?>

在上述列举的返回类型中，常见的返回类型有 ModelAndView、String 和 void。其中，ModelAndView 类型中可以添加 Model 数据，并指定视图；String 类型的返回值可以跳转视图，但不能携带数据；void 类型主要在异步请求时使用，它只返回数据，而不会跳转视图。

由于 ModelAndView 类型未能实现数据与视图之间的解耦，因此在开发时，方法的返回类型通常都会使用 String。既然 String 类型的返回值不能携带数据，那么在方法中是如何将数据带入视图页面的呢？这就用到了上面所讲解的 Model 参数类型，通过该参数类型即可添加需要在视图中显示的属性。

返回 String 类型的方法的示例代码如下：

```
@RequestMapping(value="/ annotationController")
public String handleRequest(HttpServletRequest arg0, HttpServletResponse arg1,
Model model) throws Exception {
    model.addAttribute("msg","第一个 Spring MVC 程序");
    return "/WEB-INF/jsp/welcome.jsp";
}
```

在上述方法代码中增加了一个 Model 类型的参数，通过该参数实例的 addAttribute()方法即可添加所需数据。String 类型除了可以返回上述代码中的视图页面外，还可以进行重定向与请求转发，具体方式如下：

（1）redirect 重定向

例如，在修改用户信息后，将请求重定向到用户查询方法的实现代码如下：

```
@RequestMapping(value="/update")
public string update(httpServletRequest request,HttpServletRespoNse response,
Model model){
    ...
    //重定向请求路径
    return "redirect: queryUser";
}
```

（2）forward 请求转发

例如，用户执行修改操作时，转发到用户修改页面的实现代码如下：

```
@RequestMapping(value="/toEdit")
public string toEdit(httpServletRequest request,HttpServletRespoNse response,
Model model){
    ...
    //请求转发
    return "forward:editUser";
}
```

关于重定向和转发的具体使用，在本书后面章节中会有具体的应用案例。

11.3.4 ViewResolver 视图解析器

Spring MVC 中的视图解析器负责解析视图，可以在配置文件中定义一个 ViewResolver 来配置视图解析器，配置示例如下：

```
<!-- 定义视图解析器 -->
<bean id="viewResolver "
class=" org.springframework.web.servlet.view.InternalResourceViewResolver ">
    <!-- 设置前缀 -->
    <property name="prefix" value="/WEB-INF/jsp/" />
```

```
    <!-- 设置后缀 -->
    <property name="suffix" value=".jsp" />
</bean>
```

在上述代码中定义了一个 id 为 viewResolver 的视图解析器，并设置了视图的前缀和后缀属性。这样设置后，方法中所定义的 view 路径将可以简化。例如，11.2 节的入门案例中的逻辑视图名只需设置为“welcome”，而不再需要设置为“/WEB-INF/jsp/welcome.jsp”，在访问时视图解析器会自动地增加前缀和后缀。

11.4 应用案例——基于注解的 Spring MVC 应用

为了帮助读者掌握 Spring MVC 的核心类和注解的使用，接下来我们将以注解的方式对入门案例进行改写，具体实现步骤如下：

1. 搭建项目环境

在 IntelliJ IDEA 中创建一个名为 chapter11_2 的 Web 项目，将 chapter11 项目中的所有 JAR 包以及编写的所有文件复制到 chapter11_2 项目，并向 lib 目录添加 Spring AOP 所需的 JAR 包（spring-aop-5.2.9.RELEASE.Jar）。

2. 修改配置文件

在 springmvc-config.xml 中添加注解扫描配置，并定义视图解析器，如文件 11.6 所示。

文件 11.6 springmvc-config.xml

```
01  <?xml version="1.0" encoding="UTF-8"?>
02  <beans xmlns="http://www.springframework.org/schema/beans"
03      xmlns:xsi="http://www.w3.org/2001/XMLSchema-instance"
04      xmlns:context="http://www.springframework.org/schema/context"
05      xsi:schemaLocation="http://www.springframework.org/schema/beans
06        http://www.springframework.org/schema/beans/spring-beans.xsd
07        http://www.springframework.org/schema/context
08        http://www.springframework.org/schema/context/spring-context.xsd">
09      <!--指定需要扫描的包 -->
10      <context:component-scan base-package="com.ssm.controller" />
11      <!-- 定义视图解析器 -->
12      <bean id="viewResoler"
13          class="org.springframework.web.servlet.view.InternalResourceViewResolver">
14          <!-- 设置前缀 -->
15          <property name="prefix" value="/WEB-INF/jsp/" />
16          <!-- 设置后缀 -->
17          <property name="suffix" value=".jsp" />
18      </bean>
19  </beans>
```

在文件 11.6 中，首先通过组件扫描器指定了需要扫描的包，然后定义了视图解析器，并在视图解析器中设置了视图文件的路径前缀和文件后缀名。

3. 修改 Controller Test 类

修改 ControllerTest 类，在类和方法上添加相应注解，如文件 11.7 所示。

文件 11.7 ControllerTest.java

```
01  package com.ssm.controller;
02  import javax.servlet.http.HttpServletRequest;
03  import javax.servlet.http.HttpServletResponse;
04  import org.springframework.stereotype.Controller;
05  import org.springframework.ui.Model;
06  import org.springframework.web.bind.annotation.RequestMapping;
07  //@Controller 注解
08  @Controller
09  //@RequestMapping 注解
10  @RequestMapping(value = "/controll")
11  public class ControllerTest {
12      @RequestMapping(value = "/annotationController")
13      public String handleRequest(HttpServletRequest arg0, HttpServletResponse arg1,
14                      Model model) throws Exception {
15          model.addAttribute("msg", "第一个 Spring MVC 程序");
16          return "welcome";
17      }
18  }
```

在文件 11.7 中使用@Controller 注解来标注控制器类，并使用@RequestMapping 注解标注在类名和方法名上来映射请求方法。在项目启动时，Spring 就会扫描到此类以及此类中标注了@RequestMapping 注解的方法。由于标注在类上的@RequestMapping 注解的 value 值为"/controll"，因此，类中所有请求方法的路径都需要加上"/controll"。由于类中的 handleRequest()方法的返回类型为 String，而 String 类型的返回值又无法携带数据，因此需要通过参数 Model 对象的 addAttribute()方法来添加数据信息。因为在配置文件的视图解析器中定义了视图文件的前缀和后缀名，所以 handleRequest()方法只需返回视图名"welcome"即可，在访问此方法时，系统会自动访问"WEB-INF/jsp/"路径下名称为 welcome 的 JSP 文件。

4. 启动项目，测试应用

将项目发布到 Tomcat 服务器中并启动，在浏览器中访问地址 http://localhost:8080/chapter11_2/controll/annotationController，页面显示效果如图 11.3 所示。从图中可以看出，通过注解的方式同样实现了第一个 Spring MVC 程序的运行。

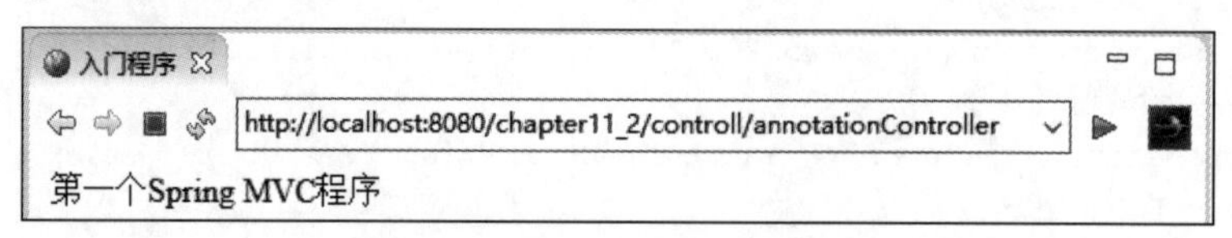

图 11.3 运行结果的页面效果

第12章

Spring MVC 数据绑定

在实际开发中，客户端在请求后台处理方法时可以传递参数，且参数包含多种类型。那么 Spring MVC 框架是如何绑定并获取请求参数的呢？本章将详细讲解这个数据绑定问题。

本章主要涉及的知识点如下：

- Spring MVC 中的数据绑定的相关概念。
- Spring MVC 中的数据绑定类型。
- Spring MVC 数据绑定的使用。

12.1 数据绑定概述

在执行程序时，Spring MVC 根据客户端请求参数的不同，将请求消息中的信息以一定的方式转换并绑定到控制器类的方法参数中。这种将请求消息数据与后台方法参数建立连接的过程就是 Spring MVC 中的数据绑定。

在数据绑定过程中，Spring MVC 框架会通过数据绑定组件（DataBinder）将请求参数中的内容进行类型转换，然后将转换后的值赋给控制器类中方法的形参，这样后台方法就可以正确绑定并获取客户端请求携带的参数。具体的信息处理过程如下：

（1）Spring MVC 将 ServletRequest 对象传递给 DataBinder。

（2）将处理方法的入参对象传递给 DataBinder。

（3）DataBinder 调用 ConversionService 组件进行数据类型转换、数据格式化等工作，并将 ServletRequest 对象中的消息填充到参数对象中。

（4）调用 Validator 组件对已经绑定了请求消息数据的参数对象进行数据合法性校验。

（5）校验完成后会生成数据绑定结果 BindingResult 对象，Spring MVC 会将 BindingResult 对象中的内容赋给处理方法的相应参数。

根据客户端请求参数类型和个数的不同，我们把 Spring MVC 中的数据绑定分为简单数据绑定和复杂数据绑定，接下来将讲解这两种类型的数据绑定。

12.2 简单数据绑定

简单数据绑定包括绑定默认数据类型、绑定简单数据类型、绑定 POJO 类型、绑定包装 POJO 等。

12.2.1 绑定默认数据类型

当前端请求的参数比较简单时，可以在后台方法的形参中直接使用 Spring MVC 提供的默认参数类型进行数据绑定。

常用的默认参数类型如下：

- HttpServletRequest：通过 request 对象获取请求信息。
- HttpServletResponse：通过 response 对象处理响应信息。
- HttpSession：通过 session 对象得到 session 中存储的对象。
- Model/ModelMap：Model 是一个接口，ModelMap 是一个接口实现，作用是将 Model 数据填充到 request 域。

【示例 12-1】下面以 HttpServletRequest 类型的使用为例进行演示说明，具体步骤如下：

步骤 01 在 IntelliJ IDEA 中创建一个名为 chapter12 的 Web 项目，然后将 Spring MVC 相关 JAR 包添加到项目的 lib 目录下，并发布到类路径中。添加 JAR 包后的目录如图 12.1 所示。

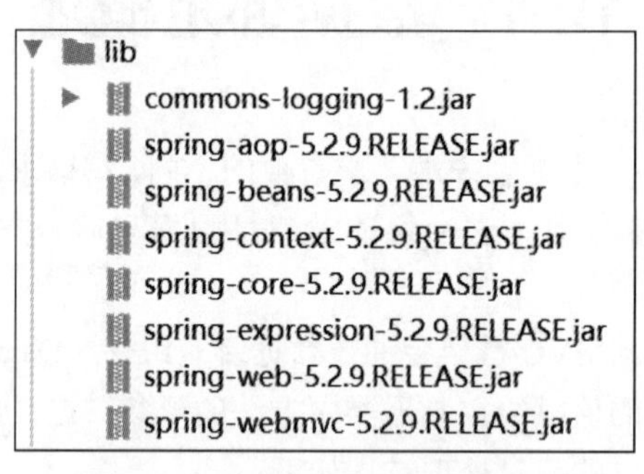

图 12.1 Spring MVC 相关 JAR 包

步骤 02 在 web.xml 中配置 Spring MVC 的前端控制器等信息，如文件 12.1 所示。

文件 12.1 web.xml

```
01  <?xml version="1.0" encoding="UTF-8"?>
02  <web-app xmlns:xsi="http://www.w3.org/2001/XMLSchema-instance"
03              xmlns="http://xmlns.jcp.org/xml/ns/javaee"
04              xsi:schemaLocation="http://xmlns.jcp.org/xml/ns/javaee
05          http://xmlns.jcp.org/xml/ns/javaee/web-app_4_0.xsd"
```

```
06          id="WebApp_ID" version="4.0">
07      <servlet>
08          <!-- 配置前端过滤器 -->
09          <servlet-name>springmvc</servlet-name>
10          <servlet-class>org.srpingframework.web.servlet.DispatcherServlet </servlet-class>
11          <!-- 初始化时加载配置文件 -->
12          <init-param>
13              <param-name>contextConfigLocation</param-name>
14              <param-value>classpath:springmvc-config.xml</param-value>
15          </init-param>
16          <!-- 表示容器在启动时立即加载 Servlet -->
17          <load-on-startup>1</load-on-startup>
18     </servlet>
19     <servlet-mapping>
20          <servlet-name>springmvc</servlet-name>
21          <url-pattern>/</url-pattern>
22     </servlet-mapping>
23  </web-app>
```

步骤03 在 src 目录下创建 Spring MVC 的核心配置文件 springmvc-config.xml，在该文件中配置组件扫描器和视图解析器，如文件 12.2 所示。

文件 12.2　springmvc-config.xml

```
01  <?xml version="1.0" encoding="UTF-8"?>
02  <beans xmlns="http://www.springframework.org/schema/beans"
03      xmlns:xsi="http://www.w3.org/2001/XMLSchema-instance"
04      xmlns:context="http://www.springframework.org/schema/context"
05      xsi:schemaLocation="http://www.springframework.org/schema/beans
06        http://www.springframework.org/schema/beans/spring-beans.xsd
07        http://www.springframework.org/schema/context
08        http://www.springframework.org/schema/context/spring-context.xsd">
09      <!--指定需要扫描的包 -->
10      <context:component-scan base-package="com.ssm.controller" />
11      <!-- 定义视图解析器 -->
12      <bean id="viewResolver"
13          class="org.springframework.web.servlet.view.InternalResourceViewResolver">
14          <!-- 设置前缀 -->
15          <property name="prefix" value="/WEB-INF/jsp/" />
16          <!-- 设置后缀 -->
17          <property name="suffix" value=".jsp" />
18      </bean>
19  </beans>
```

步骤04 在 src 目录下创建一个 com.ssm.controller 包，在该包下创建一个用于用户操作的控制器类 UserController，如文件 12.3 所示。

文件 12.3　UserController.java

```
01  package com.ssm.controller;
02  import javax.servlet.http.HttpServletRequest;
03  import org.springframework.stereotype.Controller;
04  import org.springframework.web.bind.annotation.RequestMapping;
05  //@Controller 注解
06  @Controller
07  public class UserController {
08      //@RequestMapping 注解在方法上
09      @RequestMapping(value="/selectUser")
10      public String selectUser(HttpServletRequest request) {
11          //获取请求地址中的参数 id 的值
12          String id=request.getParameter("id");
13          System.out.println("id="+id);
14          return "success";
15      }
16  }
```

在文件 12.3 中，使用注解方式定义了一个控制器类，同时定义了方法的访问路径。在方法参数中使用了 HttpServletRequest 类型，并通过该对象的 getParameter()方法获取了指定的参数。为了方便查看结果，将获取的参数进行输出打印，最后返回一个名为“success”的视图，Spring MVC 会通过视图解析器在 WEB-INF/jsp 路径下寻找 success.jsp 文件。

注意： 后台在编写控制器类时，通常会根据需要操作的业务对控制器类进行规范命名。例如，如果要编写一个对用户操作的控制器类，那么可以将控制器类命名为 UserController，然后在该控制器类中编写任何有关用户操作的方法。

步骤 05 在 WEB-INF 目录下创建一个名为 jsp 的文件夹，然后在该文件夹中创建页面文件 success.jsp，该页面只作为正确执行操作后的响应页面，没有其他业务逻辑，如文件 12.4 所示。

文件 12.4　success.jsp

```
01  <%@ page contentType="text/html;charset=UTF-8" language="java" %>
02  <html>
03  <head>
04  <title>数据绑定</title>
05  <script src="https://unpkg.com/vue@next"></script>
06  </head>
07  <body>
08  <div id="app">
09  {{msg}}
10  </div>
11  <script type="text/javascript">
12  const app=Vue.createApp({
13      data(){
14          return{
```

```
15            msg:'ok! ,执行成功。'
16         }
17     }
18 });
19 const vm = app.mount('#app')
20 </script>
21 </body>
22 </html>
```

步骤 06 将 chapter12 项目发布到 Tomcat 服务器中并启动，在浏览器中访问地址 http://localhost:8080/chapter12/selectUser?id=1，显示效果如图 12.2 所示。

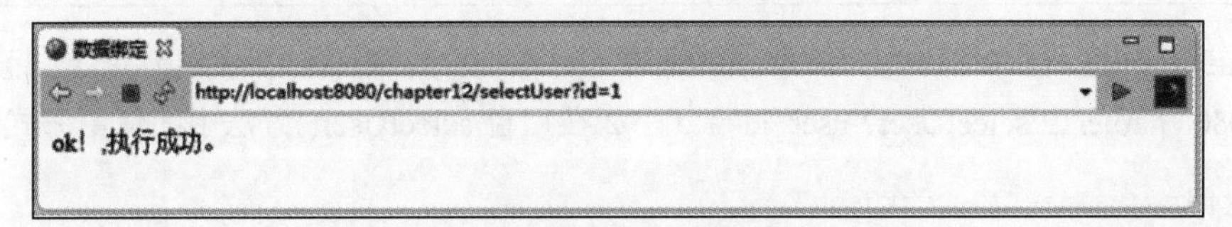

图 12.2 运行结果（success.jsp 页面）

此时的控制台输出信息如图 12.3 所示。从图中可以看出，后台方法已经从请求地址中正确地获取到了 id 的参数信息，这说明使用默认的 HttpServletRequest 参数类型已经成功地完成了数据绑定。

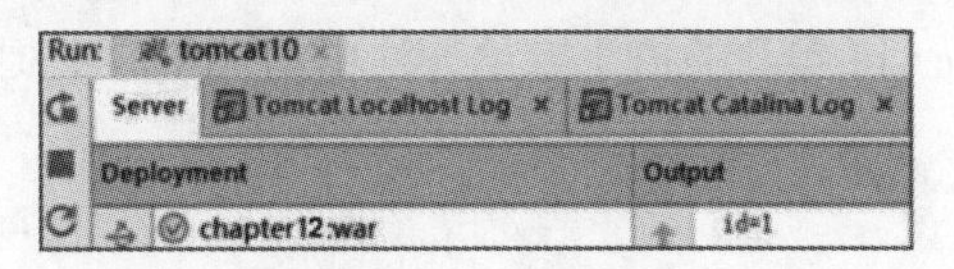

图 12.3 运行结果

12.2.2 绑定简单数据类型

简单数据类型的绑定就是指 Java 中几种基本数据类型的绑定，如 int、String、Double 等类型。

【示例 12-2】这里仍然以【示例 12-1】中参数 id 为 1 的请求为例来讲解简单数据类型的绑定。

首先修改控制器类，将控制器类 UserController 中的 selectUser()方法的参数修改为使用简单数据类型的形式，修改后的代码如下：

```
@RequestMapping(value="/selectUser")
public String selectUser(Integer id) {
   System.out.println("id="+id);
   return "success";
}
```

与默认数据类型案例中的 selectUser()方法相比，此方法中只是将 HttpServletRequest 参数类型替换为了 Integer 类型。启动项目重新运行，会得到与图 12.2 和图 12.3 所示相同的运行结果，这说明使用简单的数据类型同样完成了数据绑定。

注意：有时候前端请求中参数名和后台控制器类方法中的形参名不一样，就会导致后台无法正确绑定并接收前端请求的参数。为此，Spring MVC 提供了@RequestParam 注解来进行间接数据绑定。

@RequestParam 注解主要用于定义请求中的参数，在使用时可以指定它的 4 个属性，如表 12.1 所示。

表 12.1 @RequestParam 注解的属性及说明

属 性	说 明
value	name 属性的别名，这里指参数的名字，即入参的请求参数名字，如 value="user_id"表示传入请求的参数中名字为 user_id 的参数的值。如果只使用 vaule 属性，就可以省略 value 属性名
name	指定请求头绑定的名称
required	用于指定参数是否必需，默认是 true，表示请求中一定要有相应的参数
defaultValue	默认值，表示请求中没有同名参数的情况下的默认值

@RequestParam 注解的使用非常简单，假设浏览器中的请求地址为 http://localhost:8080/chapter12/selectUser? user_id=1，那么在后台 selectUser()方法中的使用方式如下：

```
@RequestMapping(value="/selectUser")
public String selectUser(@RequestParam (value="user_id")Integer id) {
    System.out.println("id="+id);
    return "success";
}
```

上述代码会将请求中 user_id 参数的值 1 赋给方法中的 id 形参，这样通过输出语句就可以输出 id 形参中的值。

12.2.3 绑定 POJO 类型

在使用简单数据类型绑定时，可以很容易地根据具体需求来定义方法中的形参类型和个数，然而在实际应用中，客户端请求可能会传递多个不同类型的参数数据，如果还使用简单数据类型进行绑定，就需要手动编写多个不同类型的参数，这种操作显然比较烦琐。此时就可以使用 POJO 类型进行数据绑定。

POJO 类型的数据绑定就是将所有关联的请求参数封装在一个 POJO 中，然后在方法中直接使用该 POJO 作为形参来完成数据绑定。

【示例 12-3】下面通过一个用户注册案例来演示 POJO 类型数据的绑定，具体实现步骤如下：

步骤 01 在 src 目录下创建一个 com.ssm.po 包，在该包下创建一个 User 类来封装用户注册的信息参数，如文件 12.5 所示。

文件 12.5 User.java

```
01  package com.ssm.po;
02  //用户类 User
03  public class User {
04      private Integer id;
05      private String username;
06      private String password;
```

```
07        public Integer getId() {
08           return id;
09        }
10        public void setId(Integer id) {
11           this.id = id;
12        }
13        public String getUsername() {
14           return username;
15        }
16        public void setUsername(String username) {
17           this.username = username;
18        }
19        public String getPassword() {
20           return password;
21        }
22        public void setPassword(String password) {
23           this.password = password;
24        }
25    }
```

步骤02 在控制器 UserController 类中编写向注册页面跳转和接收用户注册信息的方法，代码如下：

```
//向注册页面跳转
@RequestMapping("/toRegister")
public String toRegister() {
    return "register";
}
//接收用户注册信息
@RequestMapping("/registerUser")
public String registerUser(User user) {
    String username=user.getUserName();
    String password=user.getPassword();
    System.out.println("username="+username);
    System.out.println("password="+password);
    return "success";
}
```

步骤03 在 WEB-INF 目录下创建一个用户注册页面 register.jsp，在该页面中编写用户注册表单，表单需要以 POST 方式提交，并且在提交时发送一条以“registerUser”结尾的请求消息，如文件 12.6 所示。

文件 12.6 register.jsp

```
01    <%@ page contentType="text/html;charset=UTF-8" language="java" %>
02    <html>
03    <head>
04    <title>注册</title>
```

```
05  <script src="https://unpkg.com/axios/dist/axios.min.js"></script>
06  <script src="https://unpkg.com/vue@next"></script>
07  <meta http-equiv="Content-Type" content="text/html; charset=UTF-8">
08  </head>
09  <body>
10  <div id="register">
11  用户名: <input v-model="username" > <br>
12  密  码: <input v-model="password" /> <br />
13  <button @click="doregister">注册</button>
14  </div>
15  <script type="text/javascript" src="@{/plugins/jquery/jquery.min.js}">
16  const app=Vue.createApp({
17      data(){
18          return{
19              username:'',
20              password:'',
21          }
22      }
23  });
24  const vm = app.mount('#register')
25  </script>
26  </body>
27  </html>
```

注意：在使用 POJO 类型数据绑定时，前端请求的参数名（本例中指 form 表单内各元素的 name 属性值）必须与要绑定的 POJO 类中的属性名一样，这样才会自动将请求数据绑定到 POJO 对象中，否则后台接收的参数值为 null。

步骤04 发布并启动项目，在浏览器中访问地址 http://localhost:8080/chapter12/toRegister，就会跳转到用户注册页面 register.jsp，如图 12.4 所示。

图 12.4　注册页面（register.jsp）

在图 12.4 所示的页面中填写对应的用户名和密码，然后单击“注册”按钮即可完成模拟注册功能。这里假设用户注册的用户名和密码分别为“jack”和“jack_123”，当单击“注册”按钮后，浏览器会跳转到结果页面，此时控制台的输出结果如图 12.5 所示。

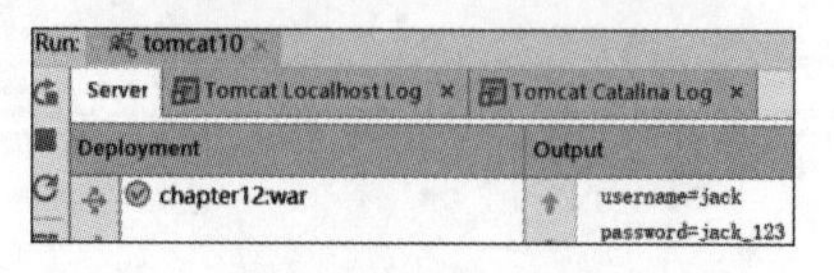

图 12.5　运行结果

从图12.5可以看出，使用POJO类型同样可以获取前端请求传递过来的数据信息，这就是POJO类型的数据绑定。

注意：在前端请求中，难免会有中文信息的传递，例如，在图12.4所示的用户名和密码输入框中输入用户名“张三”和密码“123”时，虽然浏览器可以正确跳转到结果页面，但是在控制台中输出的中文信息会出现乱码。为了防止前端传入的中文数据出现乱码问题，可以使用Spring提供的编码过滤器来统一编码。如果要使用编码过滤器，只需要在web.xml中添加如下代码：

```
<!--配置编码过滤器-->
<filter>
    <filter-name>CharacterEncodingFilter</filter-name>
    <filter-class>org.springframework.web.filter.CharacterEncodingFilter</filter-class>
    <init-param>
        <param-name>encoding</param-name>
        <param-value>UTF-8</param-value>
    </init-param>
</filter>
<filter-mapping>
    <filter-name>CharacterEncodingFilter</filter-name>
    <url-pattern>/*</url-pattern>
</filter-mapping>
```

上述代码中，<filter-mapping>元素的配置会拦截前端页面中的所有请求，并交由名称为CharacterEncodingFilter的编码过滤器类进行处理。在<filter>元素中，首先配置了编码过滤器类org.springframework.web.filter.CharacterEncodingFilter，然后通过初始化参数设置统一的编码为UTF-8，这样所有的请求信息都会以UTF-8的编码格式进行解析。

12.2.4　绑定包装POJO

使用简单POJO类型已经可以完成大多数的数据绑定，但有时客户端请求中传递的参数比较复杂，例如在老师查询学生信息时，页面传递的参数可能包括班级名称和学生号等信息，这就包含班级和学生两个对象的信息。如果将班级和学生的所有查询条件都封装在一个简单POJO中，显然会比较混乱，这时就可以考虑使用包装POJO类型的数据绑定。

所谓包装POJO，就是在一个POJO中包含另一个简单的POJO，例如在学生对象中包含班级对象。这样在使用时，就可以通过学生查询到班级信息。

【示例12-4】下面通过一个学生查询的案例来演示包装POJO数据绑定的使用，具体步骤如下：

步骤01 在chapter13项目的com.ssm.po包中创建班级类Banji和学生类Student，Student类用于封装学生和班级信息。两个类的代码分别如文件12.7和文件12.8所示。

文件12.7　Banji.java

```
01  package com.ssm.po;
02  //班级类Banji
```

```
03  public class Banji {
04      private Integer banji_id;           //班级 id
05      private String banji_name;          //班级名
06      //其他的属性省略
07      public Integer getBanji_id() {
08          return banji_id;
09      }
10      public void setBanji_id(Integer banji_id) {
11          this.banji_id = banji_id;
12      }
13      public String getBanji_name() {
14          return banji_name;
15      }
16      public void setBanji_name(String banji_name) {
17          this.banji_name = banji_name;
18      }
19  }
```

文件 12.8 Student.java

```
01  package com.ssm.po;
02  //学生类
03  public class Student {
04      private Integer stu_id;         //学生 id
05      private String stu_name;        //学生姓名
06      private Banji banji;            //学生所在班级
07      //其他属性省略
08      public Integer getStu_id() {
09          return stu_id;
10      }
11      public void setStu_id(Integer stu_id) {
12          this.stu_id = stu_id;
13      }
14      public String getStu_name() {
15          return stu_name;
16      }
17      public void setStu_name(String stu_name) {
18          this.stu_name = stu_name;
19      }
20      public Banji getBanji() {
21          return banji;
22      }
23      public void setBanji(Banji banji) {
24          this.banji = banji;
25      }
26  }
```

在上述包装POJO类Student中，不仅可以封装学生的基本属性参数，还可以封装Banji类型的属性参数。

步骤02 在com.ssm.controller包中创建一个学生控制器类StudentController,在该类中编写一个跳转到学生查询页面的方法和一个查询学生及班级信息的方法，如文件12.9所示。

文件 12.9 StudentController.java

```
01  package com.ssm.controller;
02  import org.springframework.stereotype.Controller;
03  import org.springframework.web.bind.annotation.RequestMapping;
04  import com.ssm.po.Banji;
05  import com.ssm.po.Student;
06  @Controller
07  public class StudentController {
08      //向学生查询页面跳转
09      @RequestMapping("/tofindStudentWithBanji")
10      public String tofindStudentWithBanji(){
11          return "student";
12      }
13      //查询学生和班级信息
14      @RequestMapping("/findStudentWithBanji")
15      public String findStudentWithBanji(Student student){
16          Integer stu_id=student.getStu_id();
17          Banji banji=student.getBanji();
18          String banji_name=banji.getBanji_name();
19          System.out.println("stu_id="+stu_id);
20          System.out.println("banji_name="+banji_name);
21          return "success";
22      }
23  }
```

在文件12.9中，通过访问页面跳转方法即可跳转到student.jsp页面，而通过查询学生和班级信息方法即可通过传递的参数条件调用Service中的相应方法来查询数据。这里只是为了讲解POJO的使用，所以只需要输出传递过来的参数。

步骤03 在WEB-INF/jsp目录下创建一个班级学生查询页面student.jsp，在页面中编写将学生编号和所属班级作为查询条件来查询学生信息的代码，如文件12.10所示。

文件 12.10 student.jsp

```
01  <%@ page contentType="text/html;charset=UTF-8" language="java" %>
02  <html>
03  <head>
04  <title>学生查询</title>
05  <script src="https://unpkg.com/axios/dist/axios.min.js"></script>
06  <script src="https://unpkg.com/vue@next"></script>
07  <meta http-equiv="Content-Type" content="text/html; charset=UTF-8">
```

```
08  </head>
09  <body>
10  <div id="student">
11  学生编号：<input v-model="stu_id" > <br>
12  所属班级：<input v-model="banji.banji_name" /> <br />
13  <button @click="select">查询</button>
14  </div>
15  <script type="text/javascript" src="@{/plugins/jquery/jquery.min.js}">
16  const app=Vue.createApp({
17      data(){
18          return{
19              stu_id:'',
20              banji.banji_name:'',
21          }
22      }
23  });
24  const vm = app.mount('#student')
25  </script>
26  </body>
27  </html>
```

注意：在使用包装POJO类型数据绑定时，前端请求的参数名编写必须符合以下两种情况。

① 如果查询条件参数是包装类的直接基本属性，那么参数名就直接使用对应的属性名，如上述代码中的stu_id。

② 如果查询条件参数是包装类中POJO的子属性，那么参数名就必须为“对象.属性”，其中“对象”要和包装POJO中的对象属性名称一致，属性要和包装POJO中的对象子属性一致，如上述代码中的banji.banji_name。

步骤04 发布并启动项目，在浏览器中访问地址http://localhost:8080/chapter12/tofindStudentWithBanji，显示效果如图12.6所示。

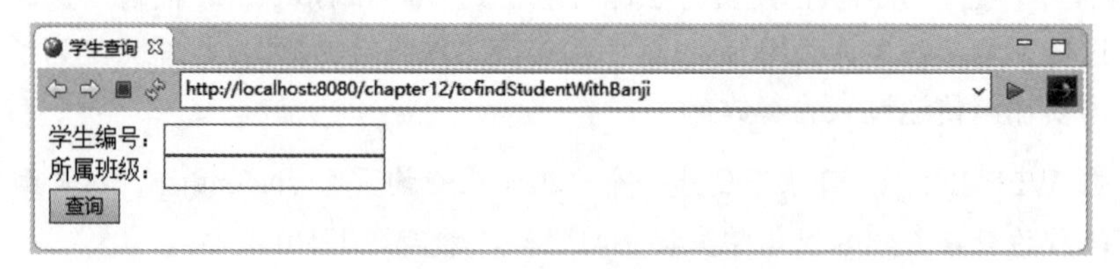

图12.6 student.jsp页面

在图12.6所示的页面中填写学生编号为“101011”，所属班级为“软件工程”，单击“查询”按钮后，浏览器会跳转到success.jsp页面，此时控制台中的打印信息如图12.7所示。从图12.7所示的结果中可以看出，使用包装POJO同样完成了数据绑定。

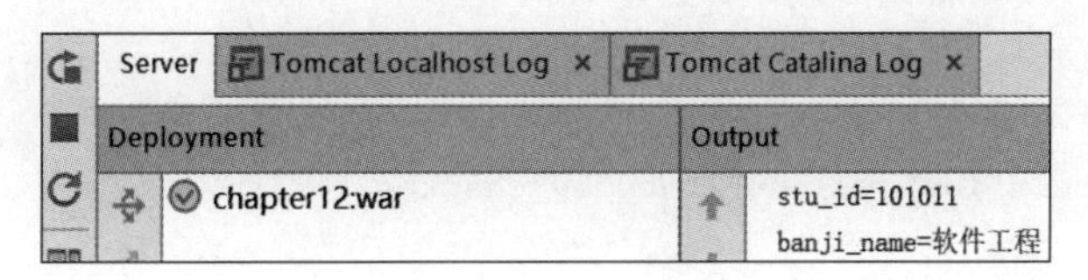

图 12.7　运行结果

除了上述几种简单的数据绑定外，有些特殊类型的参数无法在后台进行直接转换，例如日期数据需要开发者自定义转换器（Converter）或格式化（Formatter）进行数据绑定。相关内容读者可查阅相关资料自行学习。

12.3　复杂数据绑定

在实际项目开发中，除了简单数据类型外，还会经常遇到一些比较复杂的数据绑定问题，比如数组的绑定、集合的绑定，本节将具体讲解一下数组的绑定和集合的绑定的使用。

12.3.1　绑定数组

在实际开发时，可能会遇到前端请求需要传递到后台一个或多个相同名称参数的情况（如批量删除），此时不适合采用简单数据绑定而可以使用绑定数组的方式。

【示例 12-5】下面通过一个批量删除用户的例子来详细讲解绑定数组的操作。

步骤 01 在 chapter12 项目的 WEB-INF/jsp 目录下创建一个展示课程信息的列表页面 course.jsp，代码如文件 12.11 所示。

文件 12.11　course.jsp

```
01  <%@ page contentType="text/html;charset=UTF-8" language="java" %>
02  <html>
03  <head>
04  <title>课程列表</title>
05  <script src="https://unpkg.com/axios/dist/axios.min.js"></script>
06  <script src="https://unpkg.com/vue@next"></script>
07  </head>
08  <body>
09  <div id="course">
10  <table>
11  <tr>
12  <td>选择</td>
13  <td>课程名</td>
14  </tr>
15  <tr>
16  <td><input v-model="ids" value="1" type="checkbox"></td>
17  <td>JAVA 程序设计</td>
```

```
18  </tr>
19  <tr>
20  <td><input v-model="ids" value="2" type="checkbox"></td>
21  <td>MySQL 数据库</td>
22  </tr>
23  <tr>
24  <td><input v-model="ids" value="3" type="checkbox"></td>
25  <td>JavaEE 应用开发</td>
26  </tr>
27  </table>
28  <button @click="delete">删除</button>
29  </div>
30  <script type="text/javascript" src="@{/plugins/jquery/jquery.min.js}">
31  const app=Vue.createApp({
32      data(){
33          return{
34              ids:''
35          }
36      }
37  });
38  const vm = app.mount('#course')
39  </script>
40  </body>
41  </html>
```

在上述页面代码中定义了 3 个 name 属性相同而 value 属性值不同的复选框控件，并在每一个复选框对应的行中加上一个对应课程名称。在单击“删除”按钮执行删除操作时，表单会提交到一个以“/ deleteCourse”结尾的请求中。

步骤 02 在控制器类 CourseController 中编写接收批量删除课程的方法（同时为了方便向课程列表页面跳转，还需增加一个向 course.jsp 页面跳转的方法），其代码如文件 12.12 所示。

文件 12.12　CourseController.java

```
01  package com.ssm.controller;
02  import org.springframework.stereotype.Controller;
03  import org.springframework.web.bind.annotation.RequestMapping;
04  @Controller
05  public class CourseController {
06     //向课程页面跳转
07      @RequestMapping("/toCourse")
08      public String toCourse(){
09          return "course";
10      }
11      //删除课程
12      @RequestMapping("/deleteCourse")
13      public String deleteCourse(Integer[] ids){
```

```
14          if(ids!=null){
15              //使用输出语句模拟已经删除的课程
16              for(Integer id:ids){
17                  System.out.println("删除了 id 为"+id+"的课程");
18              }
19          }else{
20              System.out.println("ids=null");
21          }
22          return "success";
23      }
24  }
```

在上述代码中，先定义了一个向课程列表页面 course.jsp 跳转的方法 toCourse()，然后定义了一个接收前端批量删除用户的方法。在删除方法中使用了 Integer 类型的数组进行数据绑定，并通过 for 执行具体数据的删除操作。

步骤 03 发布并启动项目，在浏览器中访问地址 http://localhost:8080/chapter12/toCourse，显示效果如图 12.8 所示。

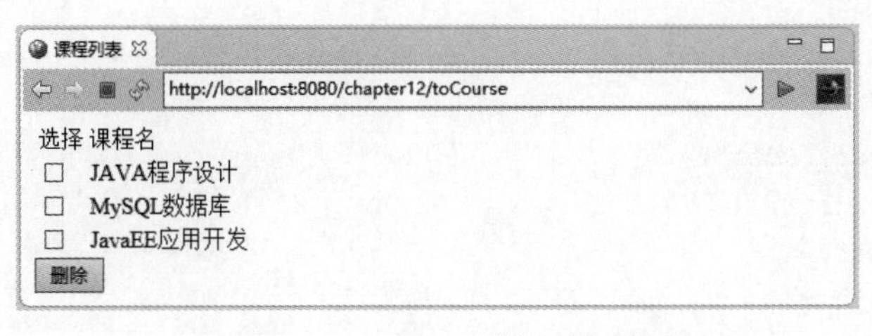

图 12.8 course.jsp 课程列表页面

勾选图 12.8 中的全部复选框，然后单击“删除”按钮，程序在正确执行后会跳转到 success.jsp 页面。此时控制台的打印信息如图 12.9 所示。从图中可以看出，已经成功执行了批量删除操作，这说明已成功实现了数组类型的数据绑定。

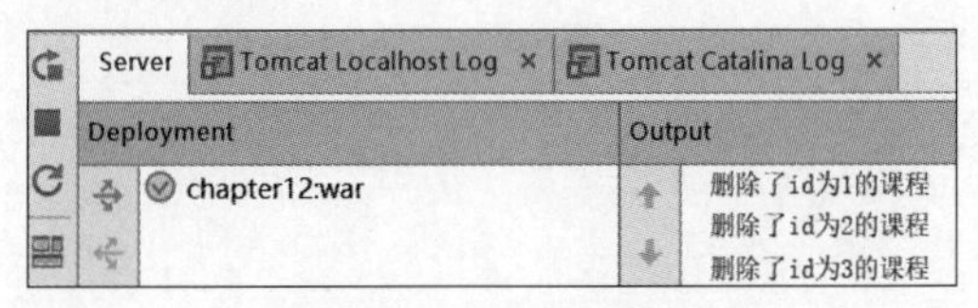

图 12.9 运行结果

12.3.2 绑定集合

在项目中，前端请求传递过来的数据可能会批量包含各种类型的数据，如 Integer、String 等。这种情况使用数组绑定是无法实现的。针对这种情况，可以使用集合数据绑定，即在包装类中定义一个包含对象类的集合，然后在接收方法中将参数类型定义为该包装类的集合。

【示例 12-6】下面以批量修改用户为例讲解一下集合数据绑定的使用。

步骤 01 在 src 目录下创建一个 com.ssm.vo 包，并在包中创建包装类 UserVo 来封装课程集合属性，代码如文件 12.13 所示。

文件 12.13 UserVo.java

```
01  package com.ssm.vo;
02  import java.util.List;
03  import com.ssm.po.User;
04  // 用户包装类
05  public class UserVo {
06      private List<User> users;   //用户列表
07      public List<User> getUsers() {
08          return users;
09      }
10      public void setUsers(List<User> users) {
11          this.users = users;
12      }
13  }
```

步骤 02 在控制器类 UserController 中编写接收批量修改用户的方法，以及向用户修改页面跳转的方法，代码如下：

```
//向用户批量修改页面跳转
@RequestMapping("/toUserEdit")
public String toUserEdit() {
    return "user_edit";
}
//接收批量修改用户的方法
@RequestMapping("/editUsers")
public String editUsers(UserVo userList){
    //将所有用户数据封装到集合中
    List<User> users=userList.getUsers();
    for(User user:users){
        if(user.getId()!=null){
            System.out.println("删除了 id 为"+user.getId()+"的用户名为"+user.getUsername());
        }
    }
    return "success";
}
```

在上述代码的两个方法中，通过 toUserEdit()方法将跳转到 user_edit.jsp 页面；通过 editUsers()方法将执行用户批量更新操作，该方法的 UserVo 类型参数用于绑定并获取页面传递过来的用户数据。

注意：在使用集合数据绑定时，后台方法中不支持直接使用集合形参进行数据绑定，所以需要使用包装 POJO 作为参数，然后在包装 POJO 中包装一个集合属性。

步骤 03 在项目的/WEB-INF/jsp 目录下创建页面文件 user_edit.jsp，如文件 12.14 所示。

文件 12.14　user_edit.jsp

```
01  <%@ page contentType="text/html;charset=UTF-8" language="java" %>
02  <html>
03  <head>
04  <title>修改用户</title>
05  <script src="https://unpkg.com/axios/dist/axios.min.js"></script>
06  <script src="https://unpkg.com/vue@next"></script>
07  </head>
08  <body>
09  <div id="user_edit">
10  <table>
11  <tr>
12  <td>选择</td>
13  <td>用户名</td>
14  </tr>
15  <tr>
16  <td><input  v-model="users[0].id" value="1" type="checkbox"></td>
17  <td>
18  <input  v-model="users[0].username" value="zhangsan" type="text">
19  </td>
20  </tr>
21  <tr>
22  <td><input  v-model="users[1].id" value="2" type="checkbox"></td>
23  <td>
24  <input  v-model="users[1].username" value="lisi" type="text">
25  </td>
26  </tr>
27  </table>
28  <button @click="update">修改</button>
29  </div>
30  <script type="text/javascript" src="@{/plugins/jquery/jquery.min.js}">
31  const app=Vue.createApp({
32      data(){
33          return{
34              users[0].id:'',
35              users[0].username:'',
36              users[1].id:'',
37              users[1].username:'',
38          }
39      }
40  });
41  const vm = app.mount('#course')
42  </script>
43  </body>
44  </html>
```

在上述页面代码中，模拟展示了 id 为 1、用户名为 zhangsan 和 id 为 2、用户名为 lisi 的两个用户。当单击“修改”按钮后，会将表单提交到一个以“editUsers”结尾的请求中。

步骤 04 发布并启动项目，在浏览器中访问地址 http://localhost:8080/chapter12/toUserEdit，显示效果如图 12.10 所示。

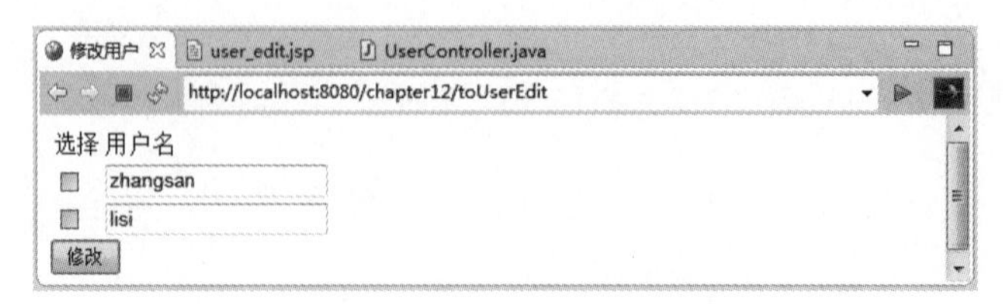

图 12.10 user_edit.jsp 页面

将图 12.10 所示页面中的用户名 zhangsan 改为 tom，lisi 改为 rose，并勾选两个数据前面的复选框，然后单击“修改”按钮，浏览器会跳转到 success.jsp 页面。此时控制台的打印信息如图 12.11 所示。从图中可以看出，已经成功输出了请求中批量修改的用户信息，这就是集合类型的数据绑定。

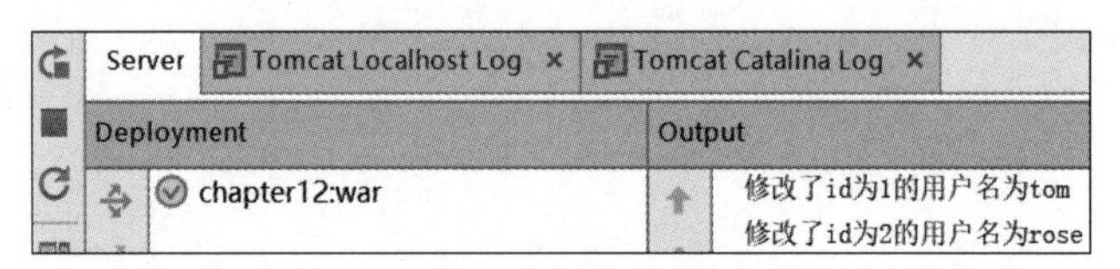

图 12.11 运行结果

第 13 章

JSON 数据交互和 RESTful 支持

Spring MVC 中的数据绑定需要对传递数据的格式和类型进行转换，包括第 12 章讲解的简单类型数据和复杂类型数据，实际上还包括 JSON 等其他类型的数据。本章将讲解 Spring MVC 中 JSON 类型的数据交互和 RESTful 支持。

本章主要涉及的知识点如下：

- Spring MVC 中 JSON 数据交互的使用。
- RESTful 风格的请求样式。
- Spring MVC 中 RESTful 风格请求的使用。

13.1 JSON 数据交互

JSON 与 XML 非常相似，都用于存储数据，但 JSON 相对于 XML 来说解析速度更快，占用空间更小。

13.1.1 JSON 概述

JSON（JavaScript Object Notation，JS 对象标记）是一种轻量级的数据交换格式，最近几年才流行起来。JSON 是基于 JavaScript 的一个子集，使用了 C、C++、C#、Java、JavaScript、Perl、Python 等其他语言的约定，采用完全独立于编程语言的文本格式来存储和表示数据。这些特性使得 JSON 成为理想的数据交互语言，它易于阅读和编写，同时也易于机器解析和生成。

与 XML 一样，JSON 也是基于纯文本的数据格式。初学者可以使用 JSON 传输一个简单的 String、Number、Boolean，也可以传输一个数组或者一个复杂的 Object 对象。JSON 有两种数据结构：对象结构和数组结构。

1. 对象结构

对象结构以“{”开始，以“}”结束，中间部分由 0 个或多个以英文逗号（,）分隔的 key:value 对构成，其中 key 和 value 之间也是以英文冒号（:）间隔的。

对象结构的语法结构代码如下：

```
{
    key1: value1,
    key2: value2,
    …
}
```

其中关键字（key）必须为 String 类型，值（value）可以是 String、Number、Object、Array 等数据类型。例如，一个 address 对象包含城市、街道等信息，使用 JSON 的表示形式如下：

```
{"city":"Guangzhou","street":"Zhongsan Road" }
```

2. 数组结构

数组结构以“[”开始，以“]”结束，中间部分由 0 个或多个以英文逗号（,）分隔的值的列表组成。

数组结构的语法结构代码如下：

```
{
    value1,
    value2,
    …
}
```

例如，一个数组包含 String、Number、Boolean、null 类型数据，使用 JSON 的表示形式如下：

```
{"jack",27,false,null}
```

JSON 这两种数据结构可以分别组合构成更为复杂的数据结构。例如，一个 person 对象包含 name、hobby 和 address 对象，其代码表现形式如下：

```
{
    "name":"zhangsan",
    "hobby":["羽毛球","读书"],
    "address":{
        "city": "Guangzhou",
        "street": "Zhongsan Road"
    }
}
```

注意：如果使用 JSON 存储单个数据（如 abc），一定要使用数组的形式，不要使用对象形式，因为对象形式必须是“名称:值”的形式。

13.1.2 JSON 数据转换

为了实现浏览器与控制器类之间的数据交互，Spring 提供了一个 HttpMessageConverter<T>接口来完成此项工作。该接口主要用于将请求信息中的数据转换为一个类型为 T 的对象，并将类型为 T 的对象绑定到请求方法的参数中，或者将对象转换为响应信息传递给浏览器显示。

Spring 为 HttpMessageConverter<T>接口提供了很多实现类，这些实现类可以对不同类型的数据进行信息转换，其中 MappingJacksona2HttpMessageConverter 是 Spring MVC 默认处理 JSON 格式请求响应的实现类。该实现类利用 Jackson 开源包读写 JSON 数据，将 Java 对象转换为 JSON 对象和 XML 文档，同时可以将 JSON 对象和 XML 文档转换为 Java 对象。

如果要使用 MappingJacksona2HttpMessageConverter 对数据进行转换，就需要使用 Jackson 的开源包，开发时所需的开源包及其描述如下：

- jackson-annoations-2.13.3.jar：JSON 转换注解包。
- jackson-core-2.13.3.jar：JSON 转换核心包。
- jackson- databind-2.13.3.jar：JSON 转换数据绑定包。

读者也可以下载更高版本的 JAR 包。

在使用注解式开发时，需要用到两个重要的 JSON 格式转换，即注解@RequestBody 和@ResponseBody，这两个注解的说明如表 13.1 所示。

表 13.1 JSON 数据交互注解及其说明

注 解	说 明
@RequestBody	用于将请求体中的数据绑定到方法的形参中，该注解用在方法的形参上
@ResponseBody	用于直接返回 return 对象，该注解用在方法上

【示例 13-1】在了解了 Spring MVC 中 JSON 数据交互需要使用的类和注解后，下面通过一个案例来演示如何进行 JSON 数据的交互。

步骤 01 创建项目并导入相关 JAR 包。使用 IntelliJ IDEA 创建一个名为 chapter13 的 Web 项目，然后将 Spring MVC 相关 JAR 包、JSON 转换包添加到项目的 lib 目录中，并发布到类路径下。添加包后的 lib 目录如图 13.1 所示。

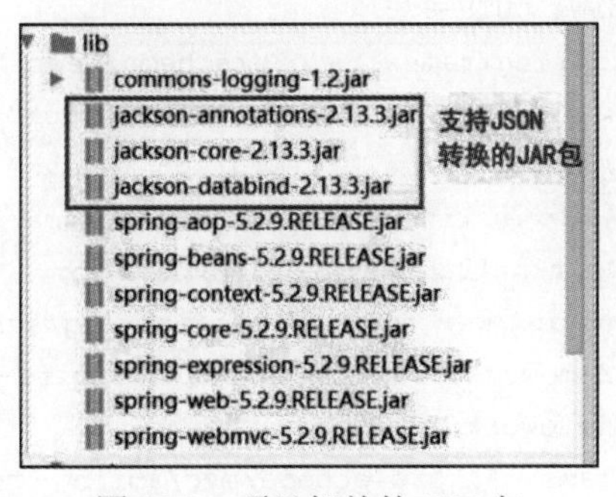

图 13.1 项目相关的 JAR 包

步骤 02 在 web.xml 中，对 Spring MVC 的前端控制器等信息进行配置，如文件 13.1 所示。

文件 13.1 web.xml

```
01  <?xml version="1.0" encoding="UTF-8"?>
02  <web-app xmlns:xsi="http://www.w3.org/2001/XMLSchema-instance"
03      xmlns="http://xmlns.jcp.org/xml/ns/javaee"
04      xsi:schemaLocation="http://xmlns.jcp.org/xml/ns/javaee
05                  http://xmlns.jcp.org/xml/ns/javaee/web-app_4_0.xsd"
06                      id="WebApp_ID" version="4.0">
07      <display-name>chapter13</display-name>
08      <welcome-file-list>
09          <welcome-file>index.jsp</welcome-file>
10      </welcome-file-list>
11      <!-- 配置前端控制器 -->
12      <servlet>
13          <!-- 配置前端过滤器 -->
14          <servlet-name>springmvc</servlet-name>
15          <servlet-class>org.springframework.web.servlet.DispatcherServlet </servlet-class>
16          <!-- 初始化时加载配置文件 -->
17          <init-param>
18              <param-name>contextConfigLocation</param-name>
19              <param-value>classpath:springmvc-config.xml</param-value>
20          </init-param>
21          <!-- 表示容器在启动时立即加载 Servlet -->
22          <load-on-startup>1</load-on-startup>
23      </servlet>
24      <servlet-mapping>
25          <servlet-name>springmvc</servlet-name>
26          <url-pattern>/</url-pattern>
27      </servlet-mapping>
28  </web-app>
```

步骤 03 在 src 目录下创建 Spring MVC 的核心配置文件 springmvc-config.xml，如文件 13.2 所示。

文件 13.2 springmvc-config.xml

```
01  <?xml version="1.0" encoding="UTF-8"?>
02  <beans xmlns="http://www.springframework.org/schema/beans"
03      xmlns:xsi="http://www.w3.org/2001/XMLSchema-instance"
04      xmlns:mvc="http://www.springframework.org/schema/mvc"
05      xmlns:context="http://www.springframework.org/schema/context"
06      xmlns:tx="http://www.springframework.org/schema/tx"
07      xsi:schemaLocation="http://www.springframework.org/schema/beans
08        http://www.springframework.org/schema/beans/spring-beans.xsd
09        http://www.springframework.org/schema/mvc
10        http://www.springframework.org/schema/mvc/spring-mvc.xsd
11        http://www.springframework.org/schema/context
12        http://www.springframework.org/schema/context/spring-context.xsd">
13      <!--指定需要扫描的包 -->
```

```
14        <context:component-scan base-package="com.ssm.controller" />
15        <!-- 配置注解驱动 -->
16        <mvc:annotation-driven />
17        <!-- 配置静态资源的访问映射，此配置中的文件将不被前端控制器拦截 -->
18        <mvc:resources location="/js/" mapping="/js/**"></mvc:resources>
19        <!-- 定义视图解析器 -->
20        <bean id="viewResoler"
21            class="org.springframework.web.servlet.view.InternalResourceViewResolver">
22            <!-- 设置前缀 -->
23            <property name="prefix" value="/WEB-INF/jsp/" />
24            <!-- 设置后缀 -->
25            <property name="suffix" value=".jsp" />
26        </bean>
27    </beans>
```

在文件 13.2 中，不仅配置了组件扫描器和视图解析器，还配置了 Spring MVC 的注解驱动<mvc:annotation-driven/>和静态资源访问映射<mvc:resources…/>。其中，<mvc:annotation-driven/>元素会自动注册 RequestMappingHandlerMapping 和 RequestMappingHandlerAdapter 两个 Bean，并提供对读写 XML 和读写 JSON 等功能的支持。<mvc:resources…/>元素用于配置静态资源的访问路径。由于在 web.xml 中配置的“/”会对页面中引入的静态文件进行拦截，而拦截后在页面中将找不到这些静态资源文件，因此会引起页面报错。增加了静态资源的访问映射配置后，程序会自动去配置路径下查找静态文件。

<mvc:resources…/>中有两个重要属性：location 和 mapping。关于这两个属性的说明如表 13.2 所示。

表 13.2 <mvc:resources>的属性及其说明

属 性	说 明
Location	用于定位需要访问的本地静态资源文件路径，具体到某个文件夹
Mapping	匹配静态资源全路径，其中“/”表示文件夹及其子文件夹下的某个具体文件

步骤 04 在 src 目录下创建一个 com.ssm.po 包，并在包中创建一个客户类 Customer，该类用于封装 User 类型的请求参数，如文件 13.3 所示。

文件 13.3 Customer.java

```
01    package com.ssm.po;
02    //客户类 Customer
03    public class Customer {
04        private Integer id;
05        private String loginname;     //客户登录名
06        private String nickname;      //昵称
07        private String password;      //密码
08        public Integer getId() {
09            return id;
10        }
11        public void setId(Integer id) {
```

```
12          this.id = id;
13      }
14      public String getLoginname() {
15          return loginname;
16      }
17      public void setLoginname(String loginname) {
18          this.loginname = loginname;
19      }
20      public String getNickname() {
21          return nickname;
22      }
23      public void setNickname(String nickname) {
24          this.nickname = nickname;
25      }
26      public String getPassword() {
27          return password;
28      }
29      public void setPassword(String password) {
30          this.password = password;
31      }
32      public String toString() {
33          return "Customer [id=" + id + ", loginname=" + loginname + ",nickname=" + nickname
+    ", password=" + password+ "]";
34      }
35  }
```

步骤 05 在 WebContent 目录下创建页面文件 json.jsp，用于测试 JSON 数据交互，如文件 13.4 所示。

文件 13.4 json.jsp

```
01  <%@ page language="java" contentType="text/html; charset=UTF-8"
02      pageEncoding="UTF-8"%>
03  <!DOCTYPE HTML>
04  <html>
05      <head>
06      <meta http-equiv="Content-Type" content="text/html; charset=UTF-8">
07      <title>测试 JSON 交互</title>
08      <script type="text/javascript"
09       src="${pageContext.request.contextPath}/js/jquery-3.5.0.min.js">
10      </script>
11      <script type="text/javascript">
12      function testJson(){
13          //获取输入的客户信息
14          var loginname=$("#loginname").val();
15          var password=$("#password").val();
16          $.ajax({
```

```
17              url:"${pageContext.request.contextPath}/testJson",
18              type:"post",
19              //data 表示发送的数据
20              data:JSON.stringify({loginname:loginname,password:password}),
21              // 定义发送请求的数据格式为 JSON 字符串
22              contentType:"application/json;charset=UTF-8",
23              //定义回调响应的数据格式为 JSON 字符串，该属性可以省略
24              dataType:"json",
25              //成功响应的结果
26              success:function(data){
27                  if(data!=null){
28                      alert("您输入的登录名为："+data.loginname+"密码为："+data.password);
29                  }
30              }
31          });
32      }
33      </script>
34      </head>
35      <body>
36          <form>
37          登录名：<input type="text" name="loginname" id="loginname" /> <br />
38          密码：<input type="password" name="password" id="password" /> <br />
39          <input type="button" value="测试 JSON 交互" onclick="testJson()" />
40          </form>
41      </body>
42  </html>
```

在文件 13.4 中编写了一个测试 JSON 交互的表单，当单击“测试 JSON 交互”按钮时，会执行页面中的 testJson()函数。在函数中使用了 jQuery 的 AJAX 方式，将 JSON 格式的登录名和密码传递到以“/testJson”结尾的请求中。

在 AJAX 中包含 3 个特别重要的属性，说明如下：

- data：请求时携带的数据，当使用 JSON 格式时需要注意编写规范。
- contentType：当请求数据为 JSON 格式时，值必须为 application/json。
- dataType：当响应数据为 JSON 格式时，可以定义 dataType 属性，并且值必须为 json。其中 dataType:"json" 可以省略不写，页面会自动识别响应的数据格式。

提示：json.jsp 还需要引入 jquery.js 文件，本例中引入了 WebContent 目录下 js 文件夹中的 jquery-3.6.0.min.js。

步骤 06 在 src 目录下创建一个 com.ssm.controller 包，在该包下创建一个用于客户操作的控制器类 CustomerController，如文件 13.5 所示。

文件 13.5　CustomerController.java

```
01  package com.ssm.controller;
02  import org.springframework.stereotype.Controller;
```

```
03  import org.springframework.web.bind.annotation.RequestBody;
04  import org.springframework.web.bind.annotation.ResponseBody;
05  import com.ssm.po.Customer;
06  @Controller
07  public class CustomerController {
08      // 接收页面请求的 JSON 格式数据，并返回 JSON 格式结果
09      @RequestMapping("/testJson")
10      @ResponseBody
11      public Customer testJson(@RequestBody Customer customer){
12          //打印接收到的 JSON 格式数据
13          System.out.println(customer);
14          return customer;
15      }
16  }
```

在文件 13.5 中，使用注解方式定义了一个控制器类，并编写了接收和响应 JSON 格式数据的 testJson()方法，在方法中接收并打印了接收到的 JSON 格式的用户数据，然后返回了 JSON 格式的用户对象。

方法中的@RequestBody 注解用于将前端请求体中的 JSON 格式数据绑定到形参 customer 上，@ResponseBody 注解用于直接返回 Customer 对象（当返回 POJO 对象时，会默认转换为 JSON 格式数据进行响应）。

步骤 07 发布并启动项目，在浏览器中访问地址 http://localhost:8080/chapter13/json.jsp，显示效果如图 13.2 所示。

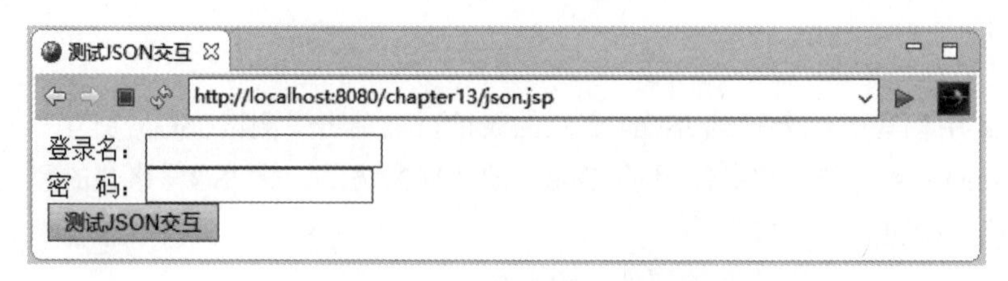

图 13.2 json.jsp 页面

在页面上的两个输入框中分别输入用户名“wujit”和密码“123456”，然后单击“测试 JSON 交互”按钮，当程序正确执行时，页面中会弹出显示用户名和密码的对话框，如图 13.3 所示。

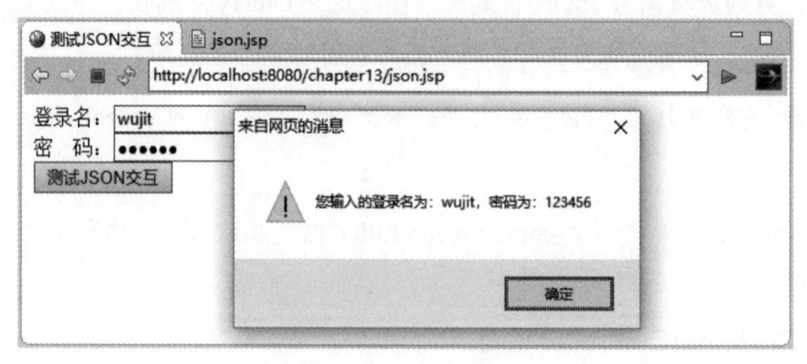

图 13.3 运行结果

从图 13.2 和图 13.3 中可以看出，编写的代码已经正确实现了 JSON 数据交互，可以将 JSON 格式的请求数据转换为方法中的 Java 对象，也可以将 Java 对象转换为 JSON 格式的响应数据。

13.2　RESTful 支持

Spring MVC 除了支持 JSON 数据交互外，还支持 RESTful 风格的编程。本节将介绍 RESTful 的概念和使用。

13.2.1　什么是 RESTful

RESTful 也称为 REST（Representational State Transfer），可以将它理解为一种软件架构风格或设计风格。

RESTful 风格就是把请求参数变成请求路径的一种风格。例如，传统的 URL 请求格式如下：

```
http://.../queryitems?id=1
```

而采用 RESTful 风格后，其 URL 请求为：

```
http://.../items/1
```

从上述两个请求中可以看出，RESTful 风格中的 URL 将请求参数 id=1 变成了请求路径的一部分，并且 URL 中的 queryltems 也变成了 items（RESTful 风格中的 URL 不存在动词形式的路径，如 queryltems 表示查询订单，是一个动词，而 tems 表示订单，为名词）。

RESTful 风格在 HTTP 请求中使用 PUT、DELETE、POST 和 GET 方式分别对应添加、删除、修改和查询的操作。不过目前国内开发还是只使用 POST 和 GET 方式进行增、删、改、查操作。

13.2.2　应用案例——查询客户信息

本案例将采用 RESTful 风格的请求实现对客户信息的查询，同时返回 JSON 格式的数据。具体实现步骤如下：

步骤 01 在控制器类 CustomerController 中编写客户查询方法 selectCustomer()，代码如下：

```
// 接收 RESTful 风格的请求，接收方式为 GET
@RequestMapping(value="/customer/{id}",method=RequestMethod.GET)
@ResponseBody
public Customer selectCustomer(@PathVariable("id") Integer id){
    //查看接收的数据
    System.out.println(id);
    Customer customer=new Customer();
    //模拟根据 id 查询出客户对象数据
    if(id==10){
        customer.setLoginname("wujit");
```

```
    }
    //返回 JSON 格式的数据
    return customer;
}
```

在上述代码中，@RequestMapping(value="customer/{id}", method= RequestMethod.GET)注解用于匹配请求路径（包括参数）和方式。其中 value="/user/{id}"表示可以匹配以“/user/{id}”结尾的请求，id 为请求中的动态参数；method= RequestMethod.GET 表示只接收 GET 方式的请求。方法中的@PathVariable("id")注解则用于接收并绑定请求参数，它可以将请求 URL 中的变量映射到方法的形参上。如果请求路径为“/user/{id}”，即请求参数中的 id 和方法形参名称 id 一样，则@PathVariable 后面的“("id")”就可以省略。

步骤 02 在 WebContent 目录下编写页面文件 restful.jsp，在页面中使用 AJAX 方式通过输入的客户编号来查询客户信息，如文件 13.6 所示。

文件 13.6 restful.jsp

```
01  <%@ page language="java" contentType="text/html; charset=UTF-8" pageEncoding="UTF-8"%>
02  <!DOCTYPE HTML>
03  <html>
04      <head>
05      <meta http-equiv="Content-Type" content="text/html; charset=UTF-8">
06      <title>RESTful 测试</title>
07      <script type="text/javascript"
08  src="${pageContext.request.contextPath }/js/jquery-3.5.0.min.js"></script>
09      <script type="text/javascript">
10      function search(){
11          //获取输入的查询编号
12          var id=$("#number").val();
13          $.ajax({
14              url:"${pageContext.request.contextPath }/customer/"+id, type:"GET",
15              //定义回调响应的数据格式为 JSON 字符串，该属性可以省略
16              dataType:"json",
17              //成功响应的结果
18              success:function(data){
19                  if(data.loginname!=null){
20                      alert("您查询的客户登录名为："+data.loginname);
21                  }else{
22                      alert("没有找到 id 为："+id+"的客户！");
23                  }
24              }
25          });
26      }
27      </script>
28      </head>
29      <body>
```

```
30          <form>
31          客户编号：<input type="text" name="number" id="number" /> <br />
32          <input type="button" value="查询" onclick="search()" />
33          </form>
34      </body>
35  </html>
```

在文件 13.6 中，在请求路径中使用了 RESTful 风格的 URL，并且定义了请求方式为 GET。

步骤 03 发布并启动项目，在浏览器中访问地址 http://localhost:8080/chapter13/restful.jsp，页面显示效果如图 13.4 所示。

图 13.4　restful. jsp 页面

在图 13.4 中的输入框中输入编号“10”，然后单击“查询”按钮，当程序正确执行时，页面中会弹出显示客户信息的对话框，如图 13.5 所示。

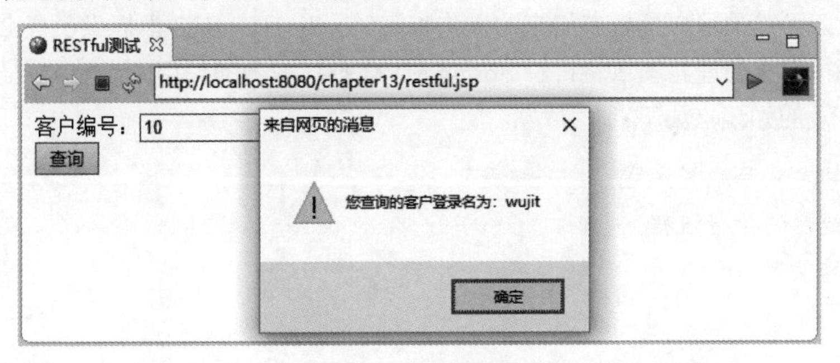

图 13.5　运行结果

从图 13.4 和图 13.5 中可以看出，我们已经成功地使用 RESTful 风格的请求查询出了客户信息。

第 14 章

拦截器

拦截器（Interceptor）的使用非常普遍，例如在 OA 系统中通过拦截器可以拦截未登录的用户，或者使用它来验证已登录用户是否有相应的操作权限等。Spring MVC 中提供了拦截器功能，通过配置即可对请求进行拦截处理。本章将讲解 Spring MVC 中拦截器的使用。

本章主要涉及的知识点如下：

- 拦截器的定义和配置方式。
- 熟悉拦截器的执行流程。
- 拦截器的使用。

14.1 拦截器概述

Spring MVC 中的拦截器类似于 Servlet 中的过滤器（Filter），它主要用于拦截用户请求并做出相应的处理。例如，通过拦截器可以进行权限验证、判断用户是否已登录等。

14.1.1 拦截器的定义

如果要使用 Spring MVC 中的拦截器，那么需要对拦截器类进行定义和配置。通常拦截器类可以通过以下两种方式来定义：

- 一种是通过实现 HandlerInterceptor 接口或者继承 HandlerInterceptor 接口的实现类（如 HandlerInterceptorAdapter）来定义。
- 另一种是通过实现 WebRequestInterceptor 接口或继承 WebRequestInterceptor 接口的实现类来定义。

以实现 HandlerInterceptor 接口的定义方式为例，自定义拦截器类的代码如下：

```
public class UserInterceptor implements HandlerInterceptor{
    @Override
    public boolean preHandle(HttpServletRequest arg0, HttpServletResponse arg1, Object arg2)
throws Exception {
        return false;
    }
    @Override
    public void postHandle(HttpServletRequest arg0, HttpServletResponse arg1, Object arg2,
ModelAndView arg3) throws Exception {
    }
    @Override
    public void afterCompletion(HttpServletRequest arg0, HttpServletResponse arg1, Object arg2,
Exception arg3) throws Exception {
    }
}
```

从上述代码可以看出，自定义的拦截器类实现了 HandlerInterceptor 接口，并实现了接口中的 3 个方法。关于这 3 个方法的具体描述如下：

- preHandle()方法：该方法会在调用控制器方法之前执行，它的返回值表示是否中断后续操作。当返回值为 true 时，表示继续向下执行；当返回值为 false 时，会中断后续的所有操作（包括调用下一个拦截器和控制器类中的方法执行等）。
- postHandle()方法：该方法会在调用控制器方法之后且解析视图之前执行。可以通过此方法对请求域中的模型和视图做出进一步的修改。
- afterCompletion()方法：该方法在整个请求完成即视图渲染结束之后执行。可以通过此方法实现一些资源清理、记录日志信息等工作。

14.1.2 拦截器的配置

如果要使自定义的拦截器类生效，则需要在 Spring MVC 的配置文件中进行配置，配置代码如下：

```
<!-- 配置拦截器 -->
<mvc:interceptors>
    <!-- 使用 bean 直接定义在<mvc:interceptors>下面的 Interceptor 将拦截所有请求 -->
    <bean class="com.ssm.interceptor.UserInterceptor" />
    <!-- 拦截器 1 -->
    <mvc:interceptor>
        <!-- 配置拦截器作用的路径 -->
        <mvc:mapping path="/**" />
        <!-- 配置不需要拦截器作用的路径 -->
        <mvc:exclude-mapping path="" />
        <!-- 定义在<mvc:interceptor>下面，表示对匹配路径的请求进行拦截 -->
```

```
        <bean class="com.ssm.interceptor.Interceptor1" />
    </mvc:interceptor>
    <!-- 拦截器 2 -->
    <mvc:interceptor>
        <mvc:mapping path="/hello" />
        <bean class="com.ssm.interceptor.Interceptor2" />
    </mvc:interceptor>
    ...
</mvc:interceptors>
```

在上述代码中，<mvc:interceptors>元素用于配置一组拦截器，其子元素<bean>中定义的是全局拦截器，它会拦截所有的请求；而<mvc:interceptor>元素中定义的是指定路径的拦截器，它会对指定路径下的请求生效。<mvc:interceptor>元素的子元素<mvc:mapping>用于配置拦截器作用的路径，该路径在其属性 path 中定义。如上述代码中 path 的属性值“/**”表示拦截所有路径，“/hello”表示拦截所有以“hello”结尾的路径。如果需要在请求路径中包含不需要拦截的内容，则可以通过<mvc:exclude-mapping>元素进行配置。

注意：<mvc: interceptor>中的子元素必须按照上述代码的配置顺序进行编写，即<mvc: mapping... />→<mvc: exclude-mapping.../>→<bean.../>的顺序，否则文件会报错。

14.2 拦截器的执行流程

拦截器的执行是有一定顺序的，该顺序与配置文件中定义的拦截器的顺序有关。本节将对单个拦截器的执行流程和多个拦截器的执行流程进行讲解。

14.2.1 单个拦截器的执行流程

如果在项目中只定义了一个拦截器，那么该拦截器在程序中的执行流程如图 14.1 所示。从图中可以看出，程序首先会执行拦截器类中的 preHandle()方法，如果该方法的返回值为 true，则程序继续向下执行处理器中的方法，否则将不再向下执行；在业务处理器（即控制器 Controller 类）处理完请求后，会执行 postHandle()方法，然后通过 DispatcherServlet 向客户端返回响应；在 DispatcherServlet 处理完请求后，才会执行 afterCompletion()方法。

【示例 14-1】下面通过一个案例来演示拦截器的执行流程。

步骤 01 在 IntelliJ IDEA 中创建一个名为 chapter14 的 Web 项目，将 Spring MVC 程序运行所需的 JAR 包复制到项目的 lib 目录中，并发布到类路径下。

步骤 02 在 web.xml 中配置 Spring MVC 的前端过滤器和初始化加载配置文件等信息。

步骤 03 在 src 目录下创建一个 com.ssm.controller 包，并在包中创建控制器类 HelloController，如文件 14.1 所示。

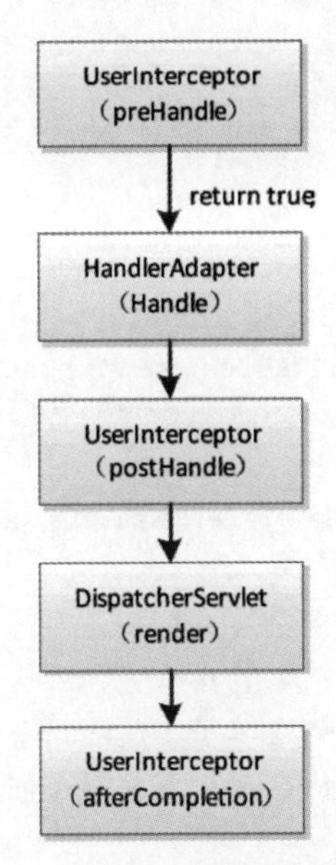

图 14.1 单个拦截器的执行流程

文件 14.1 HelloController.java

```
01 package com.ssm.controller;
02 import org.springframework.stereotype.Controller;
03 import org.springframework.web.bind.annotation.RequestMapping;
04 @Controller
05 public class HelloController {
06    @RequestMapping("/hello")
07    public String hello(){
08       System.out.println("Hello");
09       return "success";
10    }
11 }
```

步骤 04 在 src 目录下创建一个 com.ssm.interceptor 包，并在包中创建拦截器类 UserInterceptor。该类需要实现 HandlerInterceptor 接口，并且在实现方法中需要编写输出语句来输出信息，如文件 14.2 所示。

文件 14.2 UserInterceptor.java

```
01 package com.ssm.interceptor;
02 import javax.servlet.http.HttpServletRequest;
03 import javax.servlet.http.HttpServletResponse;
04 import org.springframework.web.servlet.HandlerInterceptor;
05 import org.springframework.web.servlet.ModelAndView;
06 public class UserInterceptor implements HandlerInterceptor{
07     @Override
08     public boolean preHandle(HttpServletRequest arg0, HttpServletResponse arg1, Object
   arg2) throws Exception {
09         System.out.println("UserInterceptor...preHandle");
```

```
10            //对拦截的请求进行放行处理
11            return true;
12        }
13        @Override
14        public void postHandle(HttpServletRequest arg0, HttpServletResponse arg1, Object arg2,
15          ModelAndView arg3) throws Exception {
16            System.out.println("UserInterceptor...postHandle");
17        }
18        @Override
19        public void afterCompletion(HttpServletRequest arg0,HttpServletResponse arg1,Object arg2,
20    Exception arg3) throws Exception {
21            System.out.println("UserInterceptor...afterCompletion");
22        }
23    }
```

步骤05 在 src 目录下创建 Spring MVC 的配置文件 springmvc-config.xml，如文件 14.3 所示。

文件 14.3 springmvc-config.xml

```
01 <?xml version="1.0" encoding="UTF-8"?>
02 <beans xmlns="http://www.springframework.org/schema/beans"
03     xmlns:xsi="http://www.w3.org/2001/XMLSchema-instance"
04     xmlns:mvc="http://www.springframework.org/schema/mvc"
05     xmlns:context="http://www.springframework.org/schema/context"
06     xsi:schemaLocation="http://www.springframework.org/schema/beans
07        http://www.springframework.org/schema/beans/spring-beans.xsd
08        http://www.springframework.org/schema/mvc
09        http://www.springframework.org/schema/mvc/spring-mvc.xsd
10        http://www.springframework.org/schema/context
11        http://www.springframework.org/schema/context/spring-context.xsd">
12     <!-- 指定需要扫描的包 -->
13     <context:component-scan base-package="com.ssm.controller" />
14     <!-- 定义视图解析器 -->
15     <bean id="viewResoler"
16         class="org.springframework.web.servlet.view.InternalResourceViewResolver">
17         <!-- 设置前缀 -->
18         <property name="prefix" value="/WEB-INF/jsp/" />
19         <!-- 设置后缀 -->
20         <property name="suffix" value=".jsp" />
21     </bean>
22     <!-- 配置拦截器 -->
23     <mvc:interceptors>
24         <!-- 使用 bean 直接定义在<mvc:interceptors>下面的 Interceptor 将拦截所有请求 -->
25         <bean class="com.ssm.interceptor.UserInterceptor" />
26     </mvc:interceptors>
27 </beans>
```

由于配置拦截器使用的是<mvc: interceptors>元素，因此需要配置 MVC 的 schema 信息。本案例演示的是单个拦截器的执行顺序，所以这里只配置了一个全局的拦截器。

步骤06 在 WEB-INF 目录下创建一个 jsp 文件夹，并在该文件夹中创建一个页面文件 success.jsp，然后在页面文件的<body>元素内编写任意显示信息，如“ok，执行成功！”。

步骤07 发布并启动项目，在浏览器中访问地址 http://localhost:8080/chapter14/hello，程序正确执行后，浏览器会跳转到 success.jsp 页面，此时控制台的输出结果如图 14.2 所示。从图中可以看出，程序先执行了拦截器类中的 preHandle()方法，然后执行了控制器类的 Hello()方法，最后分别执行了拦截器类中的 postHandle()方法和 afterCompletion()方法，这与前文描述的单个拦截器的执行顺序是一致的。

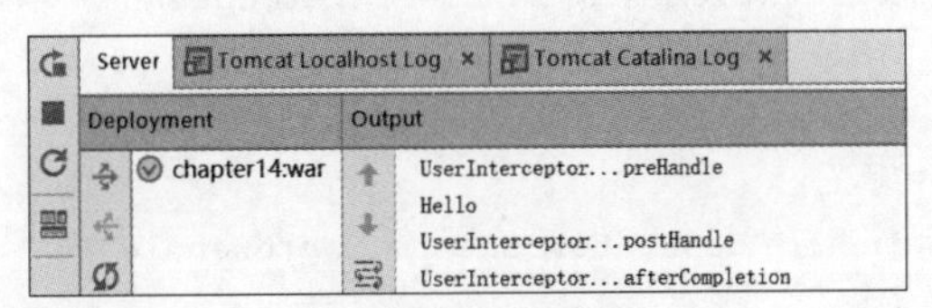

图 14.2 运行结果

14.2.2 多个拦截器的执行流程

在大型项目中，通常会定义很多拦截器来实现不同的功能。多个拦截器的执行顺序如图 14.3 所示。这里假设有两个拦截器 Interceptor1 和 Interceptor2，并且在配置文件中 Interceptor1 拦截器配置在前。

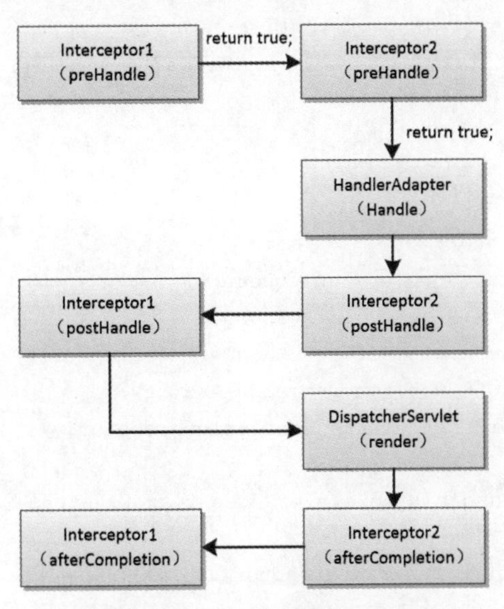

图 14.3 多个拦截器的执行流程

从图 14.3 可以看出，当有多个拦截器同时工作时，它们的 preHandle()方法会按照配置文件中拦截器的配置顺序执行，而它们的 postHandle()方法和 afterCompletion()方法则会按照配置顺序的反序执行。

【示例 14-2】为了验证上述描述，下面修改 14.2.1 节的【示例 14-1】来演示多个拦截器的执行。

步骤01 在 com.ssm.interceptor 包中创建两个拦截器类 Interceptor1 和 Interceptor2，这两个拦截器类均实现了 HandlerInterceptor 接口，并重写了其中的方法，如文件 14.4 和文件 14.5 所示。

文件 14.4 Interceptor1.java

```
01  package com.ssm.interceptor;
02  import javax.servlet.http.HttpServletRequest;
03  import javax.servlet.http.HttpServletResponse;
04  import org.springframework.web.servlet.HandlerInterceptor;
05  import org.springframework.web.servlet.ModelAndView;
06  public class Interceptor1 implements HandlerInterceptor{
07      @Override
08      public boolean preHandle(HttpServletRequest arg0, HttpServletResponse arg1, Object arg2)
09  throws Exception {
10          System.out.println("UserInterceptor1...preHandle");
11          return true;
12      }
13      @Override
14      public void postHandle(HttpServletRequest arg0, HttpServletResponse arg1, Object
    arg2, ModelAndView arg3) throws Exception {
15          System.out.println("UserInterceptor1...postHandle");
16      }
17      @Override
18      public void afterCompletion(HttpServletRequest arg0, HttpServletResponse arg1, Object
    arg2, Exception arg3) throws Exception {
19          System.out.println("UserInterceptor1...afterCompletion");
20      }
21  }
```

文件 14.5 Interceptor2.java

```
01  package com.ssm.interceptor;
02  import javax.servlet.http.HttpServletRequest;
03  import javax.servlet.http.HttpServletResponse;
04  import org.springframework.web.servlet.HandlerInterceptor;
05  import org.springframework.web.servlet.ModelAndView;
06  public class Interceptor2 implements HandlerInterceptor{
07      @Override
08      public boolean preHandle(HttpServletRequest arg0, HttpServletResponse arg1, Object
09  arg2)  throws Exception {
10          System.out.println("UserInterceptor2...preHandle");
11          return true;
12      }
13      @Override
14      public void postHandle(HttpServletRequest arg0, HttpServletResponse arg1, Object
15  arg2, ModelAndView arg3) throws Exception {
16          System.out.println("UserInterceptor2...postHandle");
17      }
```

```
18      @Override
19      public void afterCompletion(HttpServletRequest arg0, HttpServletResponse arg1,
20  Object arg2, Exception arg3) throws Exception {
21              System.out.println("UserInterceptor2...afterCompletion");
22      }
23  }
```

步骤02 在配置文件springmvc-config.xml中的<mvc:interceptors>元素内配置上面所定义的两个拦截器，配置代码如下：

```
<!-- 拦截器 1 -->
<mvc:interceptor>
    <!-- 配置拦截器作用的路径 -->
    <mvc:mapping path="/**" />
    <!-- 定义在<mvc:interceptor>下面，表示对匹配路径的请求才进行拦截 -->
    <bean class="com.ssm.interceptor.Interceptor1" />
</mvc:interceptor>
<!-- 拦截器 2 -->
<mvc:interceptor>
    <!-- 配置拦截器作用的路径 -->
    <mvc:mapping path="/hello" />
    <!-- 定义在<mvc:interceptor>下面，表示对匹配路径的请求才进行拦截 -->
    <bean class="com.ssm.interceptor.Interceptor2" />
</mvc:interceptor>
```

在上述拦截器的配置代码中，第一个拦截器会作用于所有路径下的请求，而第二个拦截器会作用于以“/hello”结尾的请求。

注意：为了不影响程序的输出结果，可以先注释掉【示例14-1】中的配置。

步骤03 发布并启动项目，在浏览器中访问地址http://localhost:8080/chapter14/hello，程序正确执行后，浏览器会跳转到success.jsp页面，此时控制台的输出结果如图14.4所示。

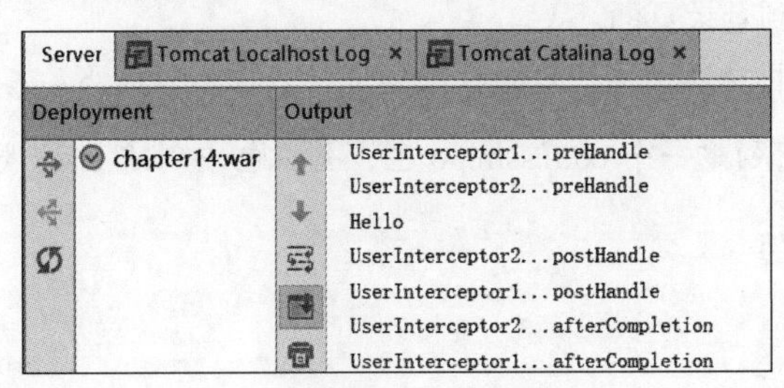

图14.4 运行结果

从图14.4中可以看出，程序先执行了前两个拦截器类中的preHandle()方法，这两个方法的执行顺序与配置文件中定义的顺序相同；然后执行了控制器类中的Hello()方法；最后执行了两个拦截器类中的postHandle()方法和afterCompletion()方法，且这两个方法的执行顺序与配置文件中定义的拦截器顺序相反。

14.3 应用案例——用户登录权限验证

本节将通过拦截器来完成一个用户登录权限验证的案例。该案例的整个执行流程如图 14.5 所示。

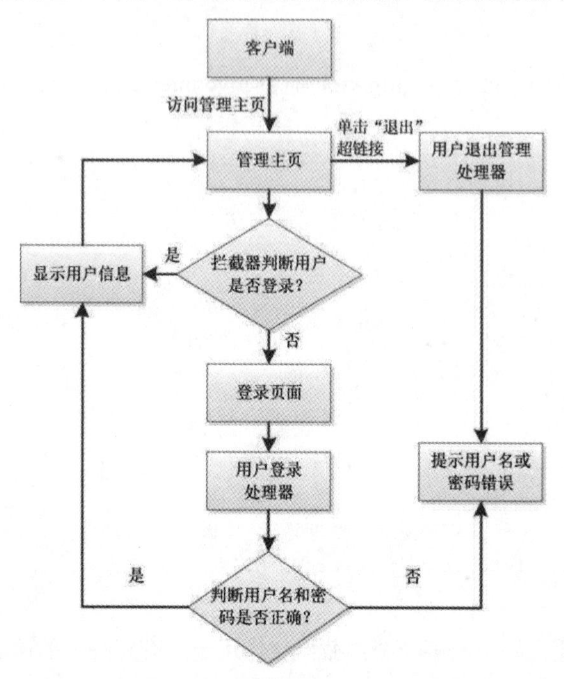

图 14.5 用户权限验证的执行流程图

从图 14.5 所示的流程图中可以看出，只有登录后的用户才能访问管理主页，如果没有登录而直接访问页面，拦截器就会拦截访问请求并转发到登录页面，同时在登录页面中给出提示信息。如果用户名或密码错误，也会在登录页面给出相应的提示信息。当已登录的用户在管理主页中单击“退出”超链接时，同样会回到登录页面。

接下来讲解如何在项目中实现用户登录权限验证，具体步骤如下：

步骤 01 在 src 目录下创建一个 com.ssm.po 包，并在包中创建 User 类，如文件 14.6 所示。

文件 14.6 User.java

```
01  package com.ssm.po;
02  public class User {
03      private Integer id;
04      private String username;
05      private String password;
06      public Integer getId() {
07          return id;
08      }
09      public void setId(Integer id) {
10          this.id = id;
11      }
```

```
12      public String getUsername() {
13          return username;
14      }
15      public void setUsername(String username) {
16          this.username = username;
17      }
18      public String getPassword() {
19          return password;
20      }
21      public void setPassword(String password) {
22          this.password = password;
23      }
24  }
```

步骤02 在 com.ssm.controller 包中创建控制器类 UserController，并在该类中定义向主页跳转、向登录页面跳转、执行用户登录等操作的方法，如文件 14.7 所示。

文件 14.7 UserController.java

```
01  package com.ssm.controller;
02  import javax.servlet.http.HttpSession;
03  import org.springframework.stereotype.Controller;
04  import org.springframework.ui.Model;
05  import org.springframework.web.bind.annotation.RequestMapping;
06  import org.springframework.web.bind.annotation.RequestMethod;
07  import com.ssm.po.User;
08  @Controller
09  public class UserController {
10      // 向用户登录页面跳转
11      @RequestMapping(value="/toLogin",method=RequestMethod.GET)
12      public String toLogin(){
13          return "login";
14      }
15      // 用户登录
16      @RequestMapping(value="/login",method=RequestMethod.POST)
17      public String login(User user,Model model,HttpSession session){
18          String username=user.getUsername();
19          String password=user.getPassword();
20          //模拟从数据库获取用户名和密码进行判断
21          if(username!=null && username.equals("wujit")){
22              if(password!=null && password.equals("123456")){
23                  //用户存在，将用户信息保存到 Session 中，并重定向到主页
24                  session.setAttribute("user_session", user);
25                  return "redirect:main";
26              }
27          }
28          //用户不存在，添加错误信息到model 中，并跳转到登录页面
```

```
29          model.addAttribute("msg","用户名或密码错误，请重新输入！");
30          return "login";
31      }
32      // 向管理主页跳转
33      @RequestMapping(value="/main")
34      public String toMain(){
35          return "main";
36      }
37      //退出
38      @RequestMapping(value="/logout")
39      public String logout(HttpSession session){
40          session.invalidate();
41          return "redirect:toLogin";
42      }
43  }
```

在文件 14.7 中的用户登录方法中，先通过 User 类型的参数获取用户名和密码，然后通过语句来模拟从数据库中获取到用户名和密码后进行判断。如果存在此用户，就将用户信息保存到 session 中，并重定向到主页，否则跳转到登录页面。

步骤 03 在 com.ssm.interceptor 包中创建拦截器类 LoginInterceptor，如文件 14.8 所示。

文件 14.8　LoginInterceptor.java

```
01  package com.ssm.interceptor;
02  import javax.servlet.http.HttpServletRequest;
03  import javax.servlet.http.HttpServletResponse;
04  import javax.servlet.http.HttpSession;
05  import org.springframework.web.servlet.HandlerInterceptor;
06  import org.springframework.web.servlet.ModelAndView;
07  import com.ssm.po.User;
08  public class LoginInterceptor implements HandlerInterceptor{
09      @Override
10      public boolean preHandle(HttpServletRequest request, HttpServletResponse response,
11      Object handler) throws Exception {
12          //获取请求的 URL
13          String url=request.getRequestURI();
14          //允许公开访问"/toLogin"
15          if(url.indexOf("/toLogin")>=0){
16              return true;
17          }
18          //允许公开访问"/login"
19          if(url.indexOf("/login")>=0){
20              return true;
21          }
22          //获取 Session
23          HttpSession session=request.getSession();
```

```
24          User user=(User)session.getAttribute("user_session");
25          //如果 user 不为空，则表示已登录
26          if(user!=null){
27              return true;
28          }
29          //如果 user 为空，则表示未登录
30          request.setAttribute("msg","请先登录！");
31          request.getRequestDispatcher("WEB-INF/jsp/login.jsp").forward(request, response);
32          return false;
33      }
34      @Override
35      public void postHandle(HttpServletRequest arg0, HttpServletResponse arg1, Object
36  arg2, ModelAndView arg3) throws Exception {
37      }
38      @Override
39      public void afterCompletion(HttpServletRequest arg0, HttpServletResponse arg1,
40  Object arg2, Exception arg3) throws Exception {
41      }
42  }
```

在文件 14.8 的 preHandle()方法中，先获取了请求的 URL，然后通过 indexOf()方法判断 URL 中是否有“/toLogin”或“/login”字符串。如果有，就返回 true，即直接放行；如果没有，就继续向下执行拦截处理。接下来获取了 Session 中的用户信息，如果 Session 中包含用户信息，就表示用户已登录，直接放行即可；否则会跳转到登录页面，不再执行本方法中的后续程序。

步骤 04 在配置文件的<mvc: interceptors>元素中配置自定义的登录拦截器信息，代码如下：

```
<mvc:interceptor>
    <mvc:mapping path="/**" />
    <bean class="com.ssm.interceptor.LoginInterceptor" />
</mvc:interceptor>
```

步骤 05 在 WEB-INF 目录下的 jsp 文件夹中创建一个管理主页面文件 main.jsp。在该页面中，使用 EL 表达式获取用户信息，并通过一个超链接来实现退出功能，如文件 14.9 所示。

文件 14.9 main.jsp

```
01  <%@ page language="java" contentType="text/html; charset=UTF-8" pageEncoding="UTF-8"%>
02  <!DOCTYPE HTML>
03  <html>
04  <head>
05    <meta http-equiv="Content-Type" content="text/html; charset=UTF-8">
06    <title>管理主页</title>
07  </head>
08  <body>
09      当前用户信息：${user_session.username}
10      <a href="${pageContext.request.contextPath}/logout">退出</a>
11  </body>
```

```
12 </html>
```

步骤06 在 WEB-INF 目录下的 jsp 文件夹中，创建一个登录页面文件 login.jsp，在页面中编写一个用于实现登录操作的 form 表单，如文件 14.10 所示。

文件 14.10 login.jsp

```
01 <%@ page contentType="text/html;charset=UTF-8" language="java" %>
02 <html>
03 <head>
04 <title>用户登录</title>
05 <script src="https://unpkg.com/axios/dist/axios.min.js"></script>
06 <script src="https://unpkg.com/vue@next"></script>
07 <meta http-equiv="Content-Type" content="text/html; charset=UTF-8">
08 </head>
09 <body>
10 <div id="login">
11 <font style="color:Red;">{{msg}}</font>
12 登录名: <input v-model="username" > <br>
13 密    码: <input v-model="password" /> <br />
14 <button @click="tologin">登录</button>
15 </div>
16 <script type="text/javascript" src="@{/plugins/jquery/jquery.min.js}">
17 const app=Vue.createApp({
18  data(){
19      return{
20          msg:'',
21          username:'',
22          password:'',
23      }
24  }
25 });
26 const vm = app.mount('#login')
27 </script>
28 </body>
29 </html>
```

步骤07 发布并启动项目，在浏览器中访问地址 http://localhost:8080/chapter14/main，程序正确执行后，页面显示效果如图 14.6 所示。

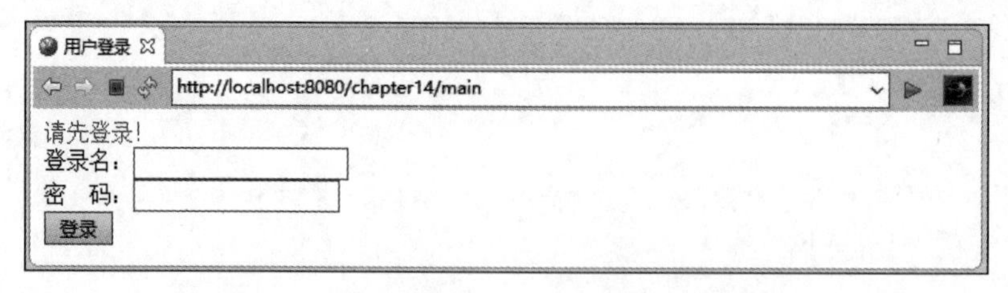

图 14.6 程序执行后 login.jsp 页面的显示结果

从图 14.6 所示的结果中可以看出，当用户未登录而直接访问主页面时，访问请求会被登录拦截器拦截，从而跳转到登录页面，并提示用户未登录信息。如果在“登录名”输入框中输入“zhangsan”，在“密码”框中输入“123456”，当单击“登录”按钮后，浏览器的显示结果如图 14.7 所示。

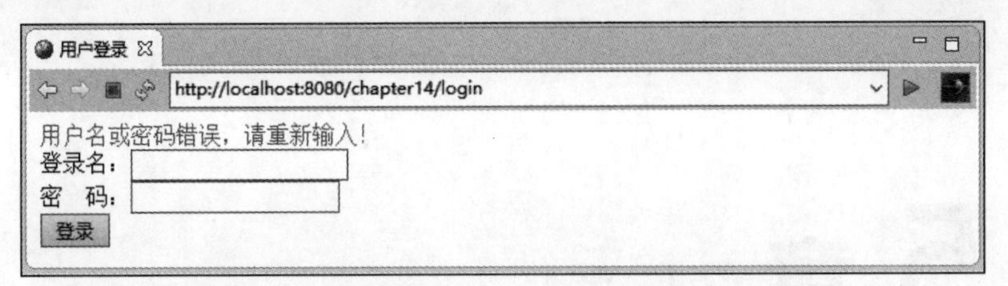

图 14.7　输入错误用户名和密码后页面的显示结果

当输入正确的用户名“wujit”和密码“123456”并单击“登录”按钮后，浏览器会跳转到管理主页面，如图 14.8 所示。当单击图 14.8 所示页面中的“退出”超链接后，用户即可退出当前管理页面，重定向到登录页面。

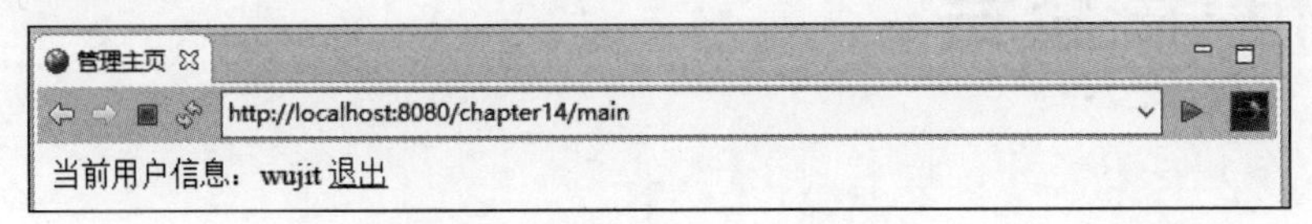

图 14.8　输入正确用户名和密码后

第 15 章

SSM 框架整合

前面已经讲解了 Spring、MyBatis、Spring MVC 三大框架以及 Spring 与 MyBatis 整合的使用。在实际项目开发中经常将这三大框架一起整合使用。本章将对 SSM（Spring、Spring MVC、MyBatis）框架的整合使用进行讲解和应用测试。

本章主要涉及的知识点如下。

- SSM 框架的整合思路。
- SSM 框架整合配置文件的编写。
- SSM 框架整合应用程序的编写。

15.1 整合环境的搭建

Spring MVC 是 Spring 框架中的一个模块，因此 Spring MVC 与 Spring 之间不需要整合，只需要引入相应 JAR 包就可以直接使用。SSM 框架的整合只需要在 Spring 与 MyBatis 之间和 Spring MVC 与 MyBatis 之间进行整合。

15.1.1 整合思路

SSM 框架的整合思路概括如下：

（1）通过 Spring 实例化 Bean，然后调用实例对象中的查询方法来执行 MyBatis 映射文件中的 SQL 语句，如果能够正确查询出数据库中的数据，那么可以认为 Spring 与 MyBatis 框架整合成功。

（2）如果可以通过前台页面来执行查询方法，并且查询出的数据能够在页面中正确显示，那么可以认为 SSM 三大框架整合成功。

15.1.2 准备所需 JAR 包

如果要实现 SSM 框架的整合，需要准备这三个框架的 JAR 包，以及其他整合所需的 JAR 包。在前面讲解 Spring 与 MyBatis 框架整合时，已经介绍了 Spring 与 MyBatis 整合所需要的 JAR 包，这里只需要加入 Spring MVC 的如下两个相关 JAR 包即可：

- spring-web-5.2.9.RELEASE.jar
- spring-webmvc-5.2.9.RELEASE.jar

SSM 整合时所需的全部 JAR 包如图 15.1 所示。

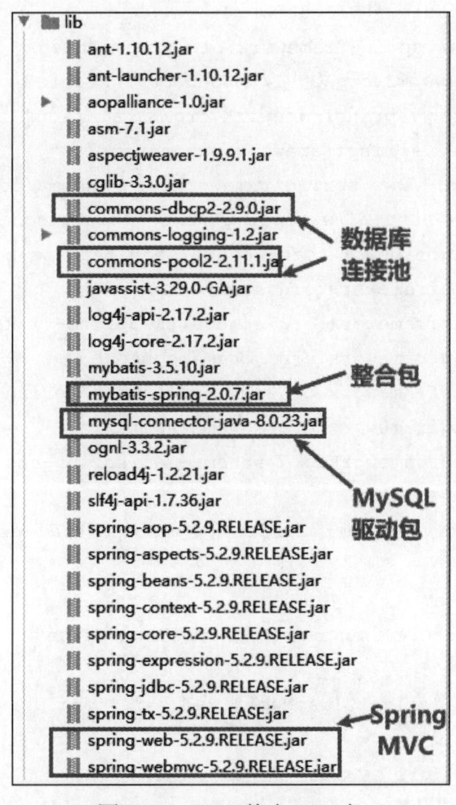

图 15.1 SSM 整合 JAR 包

15.1.3 编写配置文件

编写配置文件的操作步骤如下：

步骤 01 在 IntelliJ IDEA 中创建一个名为 chapter15 的 Web 项目，将整合所需的 JAR 包添加到项目的 lib 目录中，并发布到类路径下。

步骤 02 在项目 src 目录下分别创建数据库常量配置文件 db.properties、Spring 的配置文件 applicationContext.xml 以及 MyBatis 的配置文件 mybatis-config.xml。这 3 个配置文件的实现代码分别如文件 15.1、文件 15.2 和文件 15.3 所示。

文件 15.1 db.properties

```
01 jdbc.driver=com.mysql.cj.jdbc.Driver
02 jdbc.url=jdbc:mysql://localhost:3306/db_mybatis?serverTimezone=UTC
03 jdbc.username=root
04 jdbc.password=root
05 jdbc.maxTotal=30
06 jdbc.maxIdle=10
07 jdbc.initialSize=5
```

文件 15.2 applicationContext.xml

```
01 <?xml version="1.0" encoding="UTF-8"?>
02 <beans xmlns="http://www.springframework.org/schema/beans"
03     xmlns:xsi="http://www.w3.org/2001/XMLSchema-instance"
04     xmlns:aop="http://www.springframework.org/schema/aop"
05     xmlns:tx="http://www.springframework.org/schema/tx"
06     xmlns:context="http://www.springframework.org/schema/context"
07     xsi:schemaLocation="http://www.springframework.org/schema/beans
08       http://www.springframework.org/schema/beans/spring-beans.xsd
09       http://www.springframework.org/schema/tx
10       http://www.springframework.org/schema/tx/spring-tx.xsd
11       http://www.springframework.org/schema/context
12       http://www.springframework.org/schema/context/spring-context.xsd
13       http://www.springframework.org/schema/aop
14       http://www.springframework.org/schema/aop/spring-aop.xsd">
15     <!--读取 db.properties-->
16     <context:property-placeholder location="classpath:db.properties"/>
17     <!--配置数据源 -->
18     <bean id="dataSource"
19         class="org.apache.commons.dbcp2.BasicDataSource">
20         <!--数据库驱动 -->
21         <property name="driverClassName" value="${jdbc.driver}" />
22         <!--连接数据库的 url -->
23         <property name="url" value="${jdbc.url}" />
24         <!--连接数据库的用户名 -->
25         <property name="username" value="${jdbc.username}" />
26         <!--连接数据库的密码-->
27         <property name="password" value="${jdbc.password}" />
28         <!--最大连接数-->
29         <property name="maxTotal" value="${jdbc.maxTotal}" />
30         <!--最大空闲连接-->
31         <property name="maxIdle" value="${jdbc.maxIdle}" />
32         <!--初始化连接数-->
33         <property name="initialSize" value="${jdbc.initialSize}" />
34     </bean>
35     <!--事务管理器，依赖于数据源 -->
36     <bean id="transactionManager"
```

```
37          class="org.springframework.jdbc.datasource.DataSourceTransactionManager">
38          <property name="dataSource" ref="dataSource"/>
39      </bean>
40      <!-- 注册事务管理器驱动，开启事务注解 -->
41      <tx:annotation-driven transaction-manager="transactionManager"/>
42      <!-- 配置 MyBatis 工厂 -->
43      <bean id="sqlSessionFactory" class="org.mybatis.spring.SqlSessionFactoryBean">
44          <!-- 注入数据源 -->
45          <property name="dataSource" ref="dataSource" />
46          <!-- 指定核心配置文件位置 -->
47          <property name="configLocation" value="classpath:mybatis-config.xml" />
48      </bean>
49      <!-- 配置 mapper 扫描器 -->
50      <bean class="org.mybatis.spring.mapper.MapperScannerConfigurer">
51          <property name="basePackage" value="com.ssm.dao"></property>
52      </bean>
53      <!-- 扫描 Service -->
54      <context:component-scan base-package="com.ssm.service"/>
55  </beans>
```

在文件 15.2 中，首先定义了读取 db.properties 文件的配置和数据源配置，然后配置了事务管理器并开启了事务注解。接下来配置了用于整合 MyBatis 框架的 MyBatis 工厂信息，最后定义了 mapper 扫描器来扫描 DAO 层以及 Service 层的配置。

注意：在实际开发时，为了避免 Spring 配置文件中的信息过于臃肿，通常会将 Spring 配置文件中的信息按照不同的功能分散在多个配置文件中。例如可以将事务配置放置在名称为 applicationContext-transaction.xml 的文件中，将数据源等信息放置在名称为 applicationContext-db.xml 的文件中。在 web.xml 中配置加载 Spring 文件信息时，只需通过 applicationContext-*.xml 的方式即可自动加载全部配置文件，如文件 15.3 所示。

文件 15.3 mybatis-config.xml

```
01  <?xml version="1.0" encoding="UTF-8"?>
02  <!DOCTYPE configuration PUBLIC "-//mybatis.org//DTD Config 3.0//EN"
03      "http://mybatis.org/dtd/mybatis-3-config.dtd">
04  <configuration>
05      <!-- 配置别名 -->
06      <typeAliases>
07          <package name="com.ssm.po"/>
08      </typeAliases>
09  </configuration>
```

由于在 Spring 的配置文件中已经配置了数据源信息以及 mapper 接口文件扫描器，因此在 MyBatis 的配置文件中只需要根据 POJO 类路径进行别名配置即可。

步骤 03 在 config 文件夹中创建 Spring MVC 的配置文件 springmvc-config.xml，如文件 15.4 所示。

文件 15.4 springmvc-config.xml

```
01  <?xml version="1.0" encoding="UTF-8"?>
02  <beans xmlns="http://www.springframework.org/schema/beans"
03      xmlns:xsi="http://www.w3.org/2001/XMLSchema-instance"
04      xmlns:mvc="http://www.springframework.org/schema/mvc"
05      xmlns:context="http://www.springframework.org/schema/context"
06      xmlns:tx="http://www.springframework.org/schema/tx"
07      xsi:schemaLocation="http://www.springframework.org/schema/beans
08        http://www.springframework.org/schema/beans/spring-beans.xsd
09        http://www.springframework.org/schema/mvc
10        http://www.springframework.org/schema/mvc/spring-mvc.xsd
11        http://www.springframework.org/schema/context
12        http://www.springframework.org/schema/context/spring-context.xsd">
13      <!-- 指定需要扫描的包 -->
14      <context:component-scan base-package="com.ssm.controller" />
15      <!-- 配置注解驱动 -->
16      <mvc:annotation-driven />
17      <!-- 定义视图解析器 -->
18      <bean id="viewResoler"
19          class="org.springframework.web.servlet.view.InternalResourceViewResolver">
20          <!-- 设置前缀 -->
21          <property name="prefix" value="/WEB-INF/jsp/" />
22          <!-- 设置后缀 -->
23          <property name="suffix" value=".jsp" />
24      </bean>
25  </beans>
```

在文件 15.4 中，主要配置了用于扫描@Controller 注解的包扫描器、注解驱动器以及视图解析器。

步骤 04 在 web.xml 中配置加载 Spring 文件的监听器、编码过滤器以及 Spring MVC 的前端控制器等信息，如文件 15.5 所示。

文件 15.5 web.xml

```
01  <?xml version="1.0" encoding="UTF-8"?>
02  <web-app xmlns:xsi="http://www.w3.org/2001/XMLSchema-instance"
03      xmlns="http://xmlns.jcp.org/xml/ns/javaee"
04      xsi:schemaLocation="http://xmlns.jcp.org/xml/ns/javaee
05      http://xmlns.jcp.org/xml/ns/javaee/web-app_4_0.xsd"
06        id="WebApp_ID" version="4.0">
07      <!-- 配置加载 Spring 文件的监听器 -->
08      <context-param>
09          <param-name>contextConfigLocation</param-name>
10          <param-value>classpath:applicationContext.xml</param-value>
11      </context-param>
12      <listener>
```

```
13            <listener-class>
14                org.springframework.web.context.ContextLoaderListener
15            </listener-class>
16        </listener>
17        <!-- 编码过滤器 -->
18        <filter>
19            <filter-name>encoding</filter-name>
20            <filter-class>org.springframework.web.filter.CharacterEncodingFilter
      </filter-class>
21            <init-param>
22                <param-name>encoding</param-name>
23                <param-value>UTF-8</param-value>
24            </init-param>
25        </filter>
26        <filter-mapping>
27            <filter-name>encoding</filter-name>
28            <url-pattern>*.action</url-pattern>
29        </filter-mapping>
30        <!-- 配置 Spring MVC 的前端控制器 -->
31        <servlet>
32            <!-- 配置前端过滤器 -->
33            <servlet-name>springmvc</servlet-name>
34            <servlet-class>org.springframework.web.servlet.DispatcherServlet
    </servlet-class>
35            <!-- 初始化时加载配置文件 -->
36            <init-param>
37               <param-name>contextConfigLocation</param-name>
38               <param-value>classpath:springmvc-config.xml</param-value>
39            </init-param>
40            <!-- 配置服务器启动时立即加载 Spring MVC 配置文件 -->
41            <load-on-startup>1</load-on-startup>
42        </servlet>
43        <servlet-mapping>
44            <servlet-name>springmvc</servlet-name>
45            <!-- /：拦截所有请求，jsp 除外 -->
46            <url-pattern>/</url-pattern>
47        </servlet-mapping>
48    </web-app>
```

15.2　整合测试

15.1 节完成了 SSM 三大框架的整合工作。接下来，我们以查询用户信息为例讲解 SSM 框架的整合开发，具体实现步骤如下：

步骤 01 在 src 目录下创建一个 com.ssm.po 包，并在包中创建持久化类 User，如文件 15.6 所示。

文件 15.6 User.java

```
01  package com.ssm.po;
02  public class User {
03      private Integer id;
04      private String username;
05      //其他属性省略
06      public Integer getId() {
07          return id;
08      }
09      public void setId(Integer id) {
10          this.id = id;
11      }
12      public String getUsername() {
13          return username;
14      }
15      public void setUsername(String username) {
16          this.username = username;
17      }
18  }
```

在文件 15.6 中编写了一个用于映射数据库表 t_user 的用户持久化类，在类中分别定义了 id、username、password 属性，以及其对应的 getter()方法和 setter()方法。

步骤 02 在 src 目录下创建一个 com.ssm.dao 包，并在包中创建接口文件 UserDao 以及对应的映射文件 UserDao.xml，如文件 15.7 和文件 15.8 所示。

文件 15.7 UserDao.java

```
01  package com.ssm.dao;
02  import com.ssm.po.User;
03  // User 接口文件
04  public interface UserDao {
05      // 根据 id 查询用户信息
06      public User findUserById(Integer id);
07  }
```

从上述代码可以看出，UserDao 中只定义了一个根据 id 查询用户信息的方法。

文件 15.8 UserDao.xml

```
01  <?xml version="1.0" encoding="UTF-8"?>
02  <!DOCTYPE mapper PUBLIC "-//mybatis.org//DTD Mapper 3.0//EN"
03      "http://mybatis.org/dtd/mybatis-3-mapper.dtd">
04  <mapper namespace="com.ssm.dao.UserDao">
05      <!-- 根据用户编号获取用户信息 -->
06      <select id="findUserById" parameterType="Integer" resultType="User">
07          select * from t_user where id=#{id}
```

```
08        </select>
09    </mapper>
```

在文件 15.8 中，根据文件 15.7 中接口文件的方法编写了对应的执行语句。

注意：在前面搭建整合环境时，已经在配置文件 applicationContext.xml 中使用包扫描的形式加入了扫描包 com.ssm.dao 下的所有接口及映射文件，所以在这里完成 DAO 层接口及映射文件开发后，就不必再进行映射文件的扫描配置了。

步骤 03 在 src 目录下创建一个 com.ssm.service 包，然后在包中创建接口文件 UserService，并在 UserService 中定义通过 id 查询用户的方法，如文件 15.9 所示。

文件 15.9　UserService.java

```
01  package com.ssm.service;
02  import com.ssm.po.User;
03  public interface UserService {
04      public User findUserById(Integer id);
05  }
```

步骤 04 在 src 目录下创建一个 com.ssm.service.impl 包，并在包中创建 UserService 接口的实现类 UserServiceImpl，如文件 15.10 所示。

文件 15.10　UserServiceImpl.java

```
01  package com.ssm.service.impl;
02  import org.springframework.beans.factory.annotation.Autowired;
03  import org.springframework.stereotype.Service;
04  import org.springframework.transaction.annotation.Transactional;
05  import com.ssm.dao.UserDao;
06  import com.ssm.po.User;
07  import com.ssm.service.UserService;
08  //使用@Service 注解标识业务层的实现类
09  @Service
10  //使用@Transactional 注解来标识将类中的所有方法都纳入 Spring 的事务管理
11  @Transactional
12  public class UserServiceImpl implements UserService {
13      //注解注入 UserDao
14      @Autowired
15      private UserDao userDao;
16      //查询用户
17      public User findUserById(Integer id) {
18          return this.userDao.findUserById(id);
19      }
20  }
```

在文件 15.10 中，使用@Service 注解来标识业务层的实现类，使用@Transactional 注解来标识将类中的所有方法都纳入 Spring 的事务管理，并使用@Autowired 注解将 UserDao 接口对象注入本类中，然后在本类的查询方法中调用 UserDao 对象的查询用户方法。

注意：在上述代码中，@Transactional 注解主要是针对数据的增加、修改、删除进行事务管理的，上面示例中的查询方法并不需要使用该注解，此处的作用就是告知读者该注解在实际开发中应该如何使用。

步骤 05 在 src 目录下创建一个 com.ssm.controller 包，并在包中创建用于处理页面请求的控制器类 UserController，如文件 15.11 所示。

文件 15.11　UserController.java

```
01  package com.ssm.controller;
02  import org.springframework.beans.factory.annotation.Autowired;
03  import org.springframework.stereotype.Controller;
04  import org.springframework.ui.Model;
05  import org.springframework.web.bind.annotation.RequestMapping;
06  import com.ssm.po.User;
07  import com.ssm.service.UserService;
08  //@Controller 注解控制类
09  @Controller
10  public class UserController {
11  //@Autowired 注解注入
12      @Autowired
13      private UserService userService;
14      // 根据 id 查询用户详情
15      @RequestMapping("/findUserById")
16      public String findUserById(Integer id,Model model){
17          User user=userService.findUserById(id);
18          model.addAttribute("user", user);
19          //返回用户信息展示页面
20          return "user";
21      }
22  }
```

在文件 15.11 中，首先使用 Spring 的注解@Controller 来标识控制器类，然后通过@Autowired 注解将 UserService 接口对象注入本类中，最后编写了一个根据 id 查询用户详情的方法 findUserById()，该方法会将获取的用户详情返回到视图名为 user 的 JSP 页面中。

步骤 06 在 WEB-INF 目录下创建一个 jsp 文件夹，在该文件夹下创建一个用于展示客户详情的页面文件 user.jsp，如文件 15.12 所示。

文件 15.12　user.jsp

```
01  <%@ page contentType="text/html;charset=UTF-8" language="java" %>
02  <html>
03  <head>
04  <title>用户信息</title>
05  <script src="https://unpkg.com/axios/dist/axios.min.js"></script>
```

```
06 <script src="https://unpkg.com/vue@next"></script>
07 <meta http-equiv="Content-Type" content="text/html; charset=UTF-8">
08 </head>
09 <body>
10 <div id="user">
11 <table>
12 <tr>
13 <td>用户 ID: </td>
14 <td>${user.id}</td>
15 </tr>
16 <tr>
17 <td>用户姓名: </td>
18 <td>${user.username}</td>
19 </tr>
20 </table>
21 </div>
22 <script type="text/javascript" src="@{/plugins/jquery/jquery.min.js}">
23 const app=Vue.createApp({
24  data(){
25      return{
26          id:'',
27          username:'',
28      }
29  }
30 });
31 const vm = app.mount('#user')
32 </script>
33 </body>
34 </html>
```

在文件 15.12 中编写了一个用于展示用户信息的表格，表格会通过 EL 表达式来获取后台控制层返回的用户信息。

步骤 07 发布并启动项目，在浏览器中访问地址 http://localhost:8080/chapter15/findUserById?id=1，其显示效果如图 15.2 所示。从图中可以看出，通过浏览器已经成功查询出了 t_user 表中 id 为 1 的用户信息，这说明 SSM 框架整合成功了。

图 15.2 user.jsp 页面

注意：在实际项目开发中不只有查询操作，还有增加、修改和删除等各种复杂业务操作。本节的案例只是测试 SSM 三个框架的整合，即整合和测试三个框架是否能够“协同工作”。在第 17、18 章中，将通过两个综合性的案例来真正体验 SSM 三大框架整合的魅力。

第 16 章

Vue.js 3 入门

在学习了 Spring 框架和 MyBatis 框架的使用及整合后，接下来将讲解 Vue.js 3 框架。

注意：Vue.js 按业内惯例也称为 Vue。

本章主要涉及的知识点如下：

- Vue 的特点。
- Vue 的安装与使用。
- Vue 的应用实例。
- Vue 的模板语法。

16.1 Vue.js 3 概述

Vue（读音/vjuː/，类似于 view）是一套用于构建用户页面的渐进式框架。与其他大型框架不同的是，Vue 被设计为可以自底向上逐层应用。一方面 Vue 的核心库只关注视图层，不仅易于上手，还便于与第三方库或既有项目进行整合。另一方面，当与现代化的工具链以及各种支持类库结合使用时，Vue 也完全能够为复杂的单页应用提供驱动。

Vue 具有如下特点：

- 轻量级，体积小是一个重要指标。Vue.js 压缩后只有 20KB 左右（Angular 压缩后为 56KB+，React 压缩后为 44KB+）。
- 移动优先。更适合移动端，比如移动端的 Touch 事件。
- 易上手，学习曲线平稳，文档齐全。
- 吸取了 Angular（模块化）和 React（虚拟 DOM）的长处，并拥有自己独特的功能，如计算属性。
- 开源，社区活跃度高。

16.2 应用案例——第一个 Vue 应用

了解了什么是 Vue 及其优点后，接下来通过一个简单的入门案例来演示 Vue 的使用。

16.2.1 Vue 的安装与使用

将 Vue.js 添加到项目中主要有 4 种方式：

- 在页面上以 CDN 包的形式导入。
- 下载 JavaScript 文件并自行托管。
- 使用 npm 安装。
- 使用官方的 CLI 来构建一个项目，它为现代前端工作流程提供了功能齐备的构建设置（例如，热重载、保存时的提示等）。

（1）方法一：CDN。对于制作原型或学习，可以使用最新版本：

```
<script src="https://unpkg.com/vue@next"></script>
```

（2）方法二：下载并托管。如果想避免使用构建工具，但又无法在生产环境使用 CDN，那么可以下载相关 JS 文件并自行托管到服务器上，然后通过<script>标签引入，与使用 CDN 的方法类似。

这些文件可以在 unpkg 或者 jsDelivr 这些 CDN 上浏览和下载。各种不同文件将在以后讲解，但通常需要同时下载开发环境构建版本和生产环境构建版本。

（3）方法三：npm 安装。在用 Vue 构建大型应用时推荐使用 npm 安装。npm 能很好地和诸如 webpack 或 Rollup 模块打包器配合使用。

```
# 最新稳定版
$ npm install vue@next
```

（4）方法四：CLI 构建。Vue 提供了一个官方的 CLI，可以为单页面应用（SPA）快速搭建烦琐的脚手架。它为现代前端工作流提供了功能齐备的构建设置，只需要几分钟的时间就可以运行起来并带有热重载、保存时 lint 校验，以及生产环境可用的构建版本。

```
npm install -g @vue/cli
```

16.2.2 Vue 的实例

创建一个 HTML 文件，在 HTML 代码中引入 CDN，如文件 16.1 所示。

文件 16.1 vue1.html

```
01  <!DOCTYPE html>
02  <html>
03  <head>
04  <meta charset="utf-8">
```

```
05  <title>vue</title>
06  <script src="https://unpkg.com/vue@next"></script>
07  </head>
08  <body>
09  <div id="app">
10  {{msg}}
11  </div>
12  </body>
13  <script>
14  const app = Vue.createApp({
15  data() {
16      return {
17          msg: "第一个 Vue 程序"
18      }
19  }
20  });
21  const vm = app.mount('#app')
22  console.log(vm.msg);
23  </script>
24  </html>
```

页面显示效果如图 16.1 所示，从图中可以看到，浏览器页面中已经显示出字符串信息了，这就说明第一个 Vue 程序执行成功。

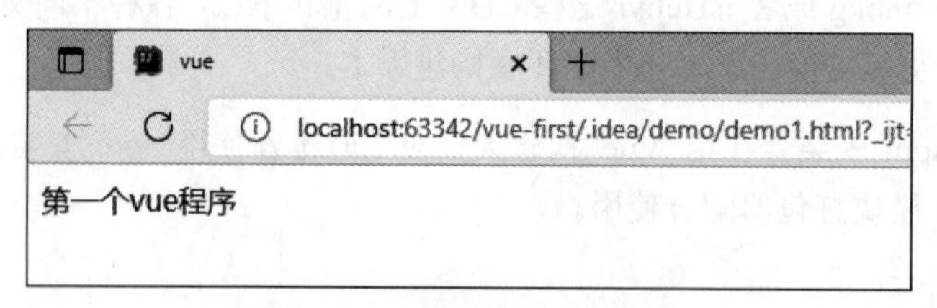

图 16.1　运行成功的页面效果

通过这个入门案例的学习，我们总结一下 Vue 程序的执行流程。

（1）创建一个 HTML 文件。

（2）引入 CDN。

（3）在<body>中建立<div>，将内容写入<div>中。

（4）调用 createApp 函数创建一个新的应用实例，app.mount()方法根据 id 将创建的应用实例与 HTML 模板相关联。

（5）编译代码，启动页面，内容就会显示在浏览器中。

16.3　Vue 的模板语法

在学习了 Vue 入门案例之后，我们再来学习一下 Vue 的模板语法。Vue.js 使用基于 HTML 的模板语法，允许开发者声明式地将 DOM 绑定至底层组件实例的数据。所有 Vue.js 的模板都是合法的 HTML，所以能被遵循规范的浏览器和 HTML 解析器解析。在底层的实现上，Vue 将模板编译成虚拟 DOM 渲染函数。结合响应性系统，Vue 能够智能地计算出最少需要重新渲染多少组件，并把 DOM 操作次数减到最少。

16.3.1 插值

1. 文本

数据绑定中最常见的就是使用 Mustache（双大括号）语法的文本插值，无论何时，绑定的组件实例上的 msg 值发生变化，插值处的内容都会更新。

```
<span>Message: {{ msg }}</span>
```

通过使用 v-once 指令，也能执行一次性地插值，当数据改变时，插值处的内容不会更新。

```
<span v-once>这个将不会改变: {{ msg }}</span>
```

2. 原始 HTML

双大括号会将数据解释为普通文本，而非 HTML 代码。为了输出真正的 HTML，需要使用 v-html 指令，如文件 16.2 所示。

文件 16.2 vue2.html

```
01  <!DOCTYPE html>
02  <html>
03  <head>
04  <meta charset="utf-8">
05  <title>vue</title>
06  <script src="https://unpkg.com/vue@next"></script>
07  </head>
08  <body>
09  <div id="app">
10      <p>{{msg}}</p>
11      <p v-html="msg"></p>
12  </div>
13  </body>
14  <script>
15  const app = Vue.createApp({
16      data() {
17          return {
18              msg: "<b>VUE</b>vue"
19          }
20      }
21  });
22  const vm = app.mount('#app')
23  console.log(vm.msg);
24  </script>
25  </html>
```

打开浏览器，页面显示效果如图 16.2 所示，从图中可以看到，双大括号没有编译标签，而 v-html 对标签进行了编译。

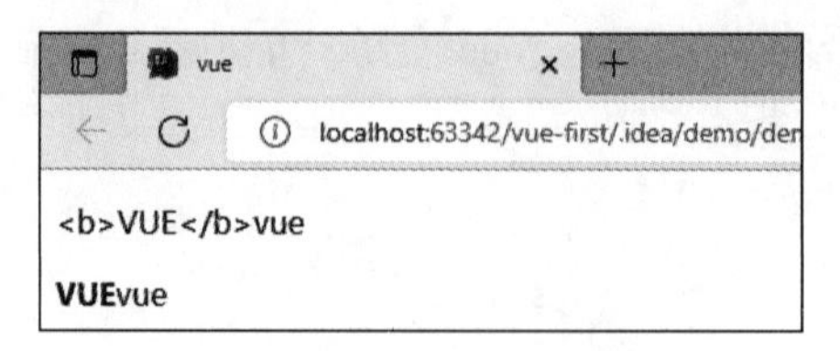

图 16.2 运行结果的页面效果

3. attribute

在 HTML attribute 中不能使用 Mustache 语法，但是可以使用 v-bind 指令，如文件 16.3 所示。

文件 16.3 vue3.html

```
01 <div id="app">
02 <p v-bind:title="msg">hello world</p>
03 <!--v-bind 缩写-->
04 <p :title="msg">hello Vue</p>
05 </div>
06 </body>
07 <script>
08 const app = Vue.createApp({
09  data() {
10      return {
11          msg: "VUE"
12      }
13  }
14 });
15 const vm = app.mount('#app')
16 console.log(vm.msg);
17 </script>
```

Vue 为 v-bind 提供了特定缩写，打开浏览器，页面显示效果如图 16.3 所示，我们可以看见绑定值被渲染了出来。

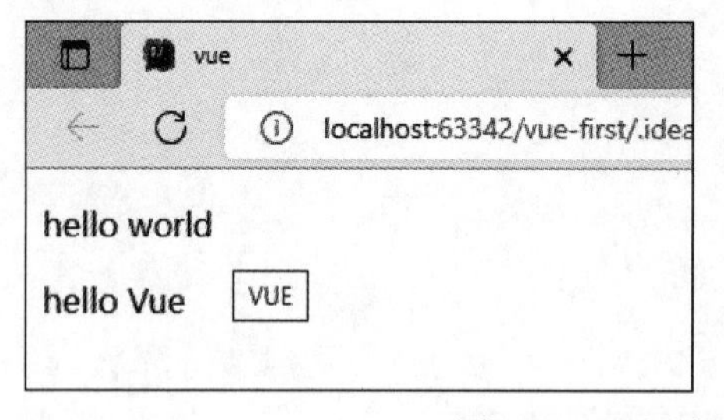

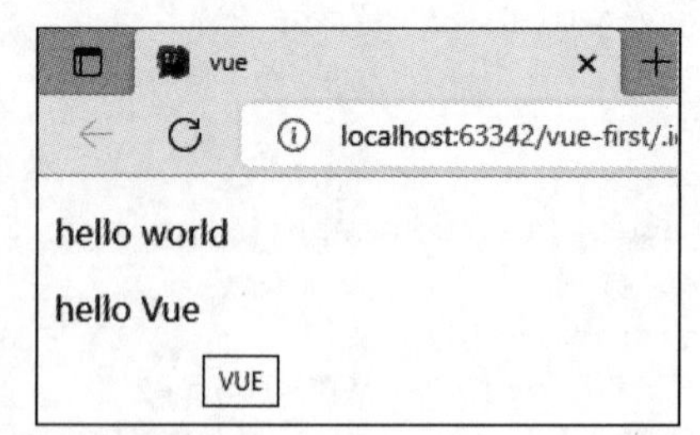

图 16.3 运行结果的页面效果

16.3.2 条件渲染

Vue 中的 v-if 指令用于条件性渲染内容，这部分内容只会在指令的表达式返回 truthy 时被渲染。

v-else-if 作为 v-if 的 else 使用，跟随在 v-if 之后使用。v-else 必须紧跟在带 v-if 或者 v-else-if 的元素的后面。v-show 与 v-if 相似，可以控制元素的显示与隐藏，不同之处在于 v-if 通过移除节点的方式隐藏，而 v-show 通过 CSS 属性进行隐藏与显示。如文件 16.4 所示。

文件 16.4　vue4.html

```
01  <div id="app">
02      <p v-if="islog">欢迎回来！XXX</p>
03      <p v-else>请登录</p>
04      <hr >
05      <p v-if="score>=90">{{score}}优秀</p>
06      <p v-else-if="score>=80">{{score}}良好</p>
07      <p v-else-if="score>=70">{{score}}中等</p>
08      <p v-else-if="score>=60">{{score}}及格</p>
09      <p v-else>{{score}}不及格</p>
10      <hr >
11      <p v-show="show">非常棒</p>
12  </div>
13  </body>
14  <script>
15  const app = Vue.createApp({
16      data() {
17          return {
18              islog:true,
19              score:100,
20              show:true
21          }
22      }
23  });
24  const vm = app.mount('#app')
25  </script>
```

打开浏览器，页面显示效果如图 16.4 所示，从图中可以看到条件渲染运行结果。

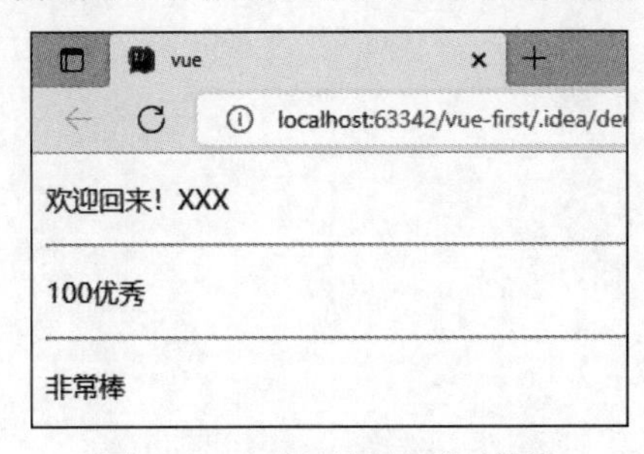

图 16.4　运行结果的页面效果

16.3.3　事件 v-on

我们可以使用 v-on 指令（通常缩写为@符号）来监听 DOM 事件，并在触发事件时执行一些

JavaScript。用法为 v-on:click="methodName"或使用快捷方式@click="methodName"，如文件 16.5 所示。

文件 16.5 vue5.html

```
01 <div id="app">
02 <button type="button" @click="num++">{{num}}</button>
03 </div>
04 </body>
05 <script type="text/javascript">
06 const app=Vue.createApp({
07     data(){
08         return{
09             num:1
10         }
11     }
12 });
13 const vm = app.mount('#app')
14 </script>
```

在单击按钮后数字由 1 开始增加，单击一次数字加 1，运行结果如图 16.5 所示。

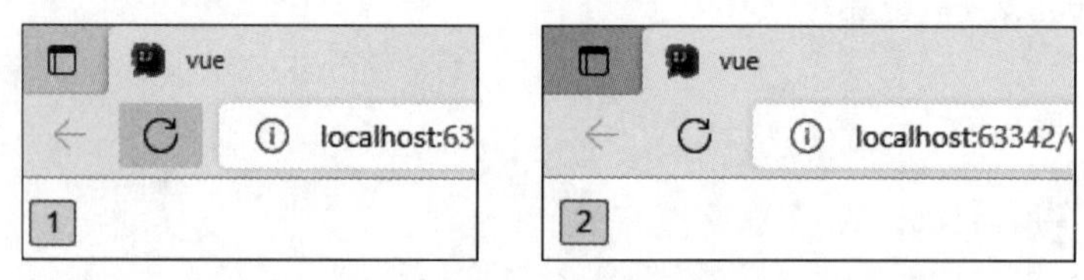

图 16.5 运行结果的页面效果

注意：本章只是介绍了 Vue 框架的一些入门内容，如想深入学习可前往 Vue 官网参看相关学习材料。

第 17 章

SSM+Vue.js 实战：新闻发布管理系统

本章将使用前面介绍的 SSM+Vue.js 框架知识来实现一个新闻发布管理系统，实现系统前台的页面设计，后台的用户管理、用户登录、登录验证、新闻发布管理等功能。

本章主要涉及的知识点如下：

- 系统架构和文件组织结构。
- 数据分析与设计。
- 开发环境和框架搭建。
- 系统功能设计和功能编码实现。

17.1 系统概述

本系统后台使用 SSM 三大框架实现，前台页面使用当前主流的 Vue 框架完成新闻展示。

17.1.1 系统功能需求

系统中主要实现了几大功能模块：用户管理（含角色设置和登录验证）、新闻类别管理、新闻发布管理和前台新闻展示等模块，如图 17.1 所示。

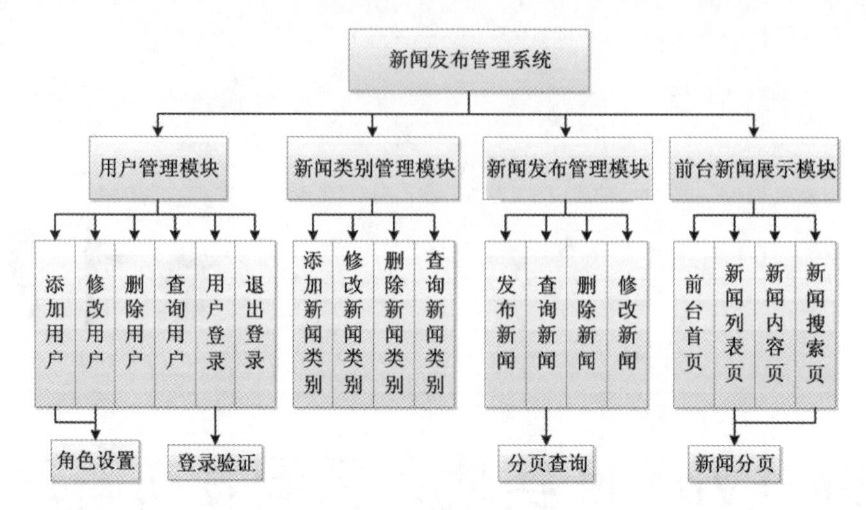

图 17.1 系统功能结构

17.1.2 系统架构设计

本系统根据功能的不同，项目结构可以划分为以下几个层次。

- 持久对象层：由若干持久化类（实体类）组成。
- 数据访问层：由若干 DAO 接口和 MyBatis 映射文件组成。接口的名称统一以 DAO 结尾，且 MyBatis 的映射文件名称要与接口的名称相同。
- 业务逻辑层：该层由若干 Service 接口和实现类组成。在本系统中，业务逻辑层的接口统一使用 Service 结尾，其实现类名称统一在接口名后加 Impl。该层主要用于实现系统的业务逻辑。
- Web 表现层：该层主要包括 Spring MVC 中的 Controller 类。

17.2 数据分析与设计

根据系统的功能需求，本系统的设计与实现中主要涉及角色实体、用户实体、新闻类别实体和新闻实体 4 个实体，其中角色实体与用户实体之间构成一对多的关联关系，新闻类别实体和新闻实体之间构成一对多的关联关系，用户实体和新闻实体之间构成一对多的关联关系。

与对象实体相适应，本系统中涉及角色表、用户表、新闻类别表和新闻表，其中用户表通过 roleId（角色 ID）字段与角色表构成关联关系，新闻表通过 categoryId（类别 ID）字段与新闻类别表构成关联关系，新闻表还通过 userId（用户 ID）字段与用户表构成关联关系。这 4 张表的表结构如表 17.1~表 17.4 所示。

表 17.1 角色表（t_role）

字 段 名	类 型	长 度	字段说明	备 注
roleId	Int	32	角色 id	主键
roleName	VarChar	20	角色名称	

表 17.2 用户表（t_user）

字 段 名	类 型	长 度	字段说明	备 注
userId	Int	32	用户 id	主键
userName	VarChar	20	用户姓名	
loginName	VarChar	20	登录账号	
password	VarChar	20	登录密码	
roleId	Int	32	角色 id	外键
tel	VarChar	50	联系电话	
registerTime	DataTime		注册时间	
status	Char	1	注册状态	'1'：未启用 '2'：已启用 '3'：被禁用

表 17.3 新闻类别表（t_category）

字 段 名	类 型	长 度	字段说明	备 注
categoryId	Int	32	类别 id	主键
categoryName	VarChar	20	类别名称	

表 17.4 新闻表（t_news）

字 段 名	类 型	长 度	字段说明	备 注
newsId	Int	32	类别 id	主键
title	VarChar	60	信息标题	
contentTitle	VarChar	120	信息内容标题	
titlePicUrl	VarChar	120	标题图（路径）	
content	Text		信息内容	
contentAbstract	VarChar	300	内容摘要	
keywords	VarChar	100	关键词	
categoryId	Int	32	信息类别 id	外键
userId	Int	32	发布用户 id	外键
author	VarChar	30	作者（来源）	
publishTime	DataTime		发布时间	
clicks	Int	32	浏览次数	
publishStatus	Char	1	发布状态	'1'：发布 '2'：撤稿

本系统使用 MySQL 数据库。在实现系统前，需要提前准备好系统中的数据库资源，创建数据库 news，并在数据库中创建上述 4 张表，同时添加一些必要的基础数据。创建数据库 news 的 SQL 语句和上述 4 张表的建表及基础数据插入的 SQL 语句如下：

```
# 创建数据库 news
CREATE DATABASE news;
# 使用数据库 news
USE news;
# 创建一张名称为 t_role 的表
CREATE TABLE t_role(
   roleId INT PRIMARY KEY,
   roleName VARCHAR(20)
);
# 插入两条数据
INSERT INTO t_role VALUES(1,'管理员');
INSERT INTO t_role VALUES(2,'信息员');
# 创建一张名称为 t_user 的表
CREATE TABLE t_user(
   userId INT PRIMARY KEY AUTO_INCREMENT,
   userName VARCHAR(20),
   loginName VARCHAR(20),
   password VARCHAR(20),
   tel VARCHAR(50),
   registerTime DATETIME,
   status CHAR(1),
   roleId INT ,
   FOREIGN KEY (roleId) REFERENCES t_role(roleId)
);
# 插入 1 条数据
INSERT INTO t_user(userName, loginName , password ,status,roleId)
    VALUES('无为','admin','123456','2',1);
# 创建一张名称为 t_category 的表
CREATE TABLE t_category(
   categoryId INT PRIMARY KEY,
   categoryName VARCHAR(20)
);
# 插入 4 条数据
INSERT INTO t_category VALUES(1,'今日头条');
INSERT INTO t_category VALUES(2,'综合资讯');
INSERT INTO t_category VALUES(3,'国内新闻');
INSERT INTO t_category VALUES(4,'国际新闻');
# 创建一张名称为 t_news 的表
CREATE TABLE t_news(
   newsId INT PRIMARY KEY AUTO_INCREMENT,
   title VARCHAR(60),
   contentTitle VARCHAR(120),
   titlePicUrl VARCHAR(120),
   content TEXT,
   contentAbstract VARCHAR(300),
   keywords VARCHAR(100),
```

```
    author VARCHAR(30),
    publishTime DATETIME,
    clicks INT,
    publishStatus CHAR(1),
    categoryId INT ,
    userId INT ,
    FOREIGN KEY (categoryId) REFERENCES t_category(categoryId),
    FOREIGN KEY (userId) REFERENCES t_user(userId)
);
```

17.3 系统功能设计与实现

本系统的功能设计与实现分模块进行，主要涉及以下几个方面：

- 开发环境和框架的搭建。
- 用户管理模块。
- 新闻类别管理模块。
- 新闻发布管理模块。
- 前台新闻展示模块。

其中角色管理模块和新闻类别管理模块只创建对应的表格和插入初始化数据，并提供查询方法供用户管理和新闻发布管理等模块调用。前台新闻展示模块不做详细讲解，有兴趣的读者可以查看本章的案例代码。

另外，从安全角度考虑，系统后台中使用拦截器对操作用户进行登录验证。

17.4 开发环境和框架的搭建

系统的开发环境和框架搭建涉及各类文件的创建、引入和编写，其中包括：包文件（内含各类接口和类）、配置文件、页面文件以及相关的 JAR 包文件、资源文件等。下面将对系统开发环境和框架的搭建进行讲解。

17.4.1 创建项目，引入 JAR 包

在 IntelliJ IDEA 中创建一个名称为 news_publish 的 Web 项目，将系统所准备的全部 JAR 包复制到项目的 lib 目录中，并发布到类路径下。

由于本系统使用 SSM 框架开发，因此需要准备三大框架的 JAR 包。另外，系统中还涉及数据库连接、JSTL 标签等，所以还要准备其他包。整个系统中需要准备的 JAR 包共计 34 个，除了第 16 章 SSM 框架整合中用到的 28 个 JAR 包外，还包括以下 6 个 JAR 包：

- JSTL 标签库 JAR 包（2 个）：taglibs-standard-spec-1.2.5.jar、taglibs-standard-impl-1.2.5.jar。
- Jackson 框架所需 JAR 包（3 个）：jackson-annotations-2.13.3.jar、jackson-core-2.13.3.jar、jackson-databind-2.13.3.jar。
- Java 工具类 JAR 包（1 个）：commons-lang3-3.12.0.jar。

17.4.2 编写配置文件

在项目 src 目录下分别创建数据库常量配置文件、Spring 配置文件、MyBatis 配置文件、Spring MVC 配置文件、log4j 配置文件以及资源配置文件。前 4 个文件分别如文件 17.1~文件 17.4 所示。log4j 配置文件请参照前面相关章节的内容进行编写。本系统没有用到资源配置文件。

文件 17.1 db.properties

```
01  jdbc.driver=com.mysql.cj.jdbc.Driver
02  jdbc.url=jdbc:mysql://localhost:3306/news?serverTimezone=UTC
03  jdbc.username=root
04  jdbc.password=root
05  jdbc.maxTotal=30
06  jdbc.maxIdle=10
07  jdbc.initialSize=5
```

文件 17.2 applicationContext.xml

```
01  <?xml version="1.0" encoding="UTF-8"?>
02  <beans xmlns="http://www.springframework.org/schema/beans"
03      xmlns:xsi="http://www.w3.org/2001/XMLSchema-instance"
04      xmlns:mvc="http://www.springframework.org/schema/mvc"
05      xmlns:aop="http://www.springframework.org/schema/aop"
06      xmlns:tx="http://www.springframework.org/schema/tx"
07      xmlns:context="http://www.springframework.org/schema/context"
08      xsi:schemaLocation="http://www.springframework.org/schema/beans
09         http://www.springframework.org/schema/beans/spring-beans.xsd
10         http://www.springframework.org/schema/mvc
11         http://www.springframework.org/schema/mvc/spring-mvc.xsd
12         http://www.springframework.org/schema/tx
13         http://www.springframework.org/schema/tx/spring-tx.xsd
14         http://www.springframework.org/schema/context
15         http://www.springframework.org/schema/context/spring-context.xsd
16         http://www.springframework.org/schema/aop
17         http://www.springframework.org/schema/aop/spring-aop.xsd">
18      <!--读取 db.properties-->
19      <context:property-placeholder location="classpath:db.properties"/>
20      <!--配置数据源 -->
21      <bean id="dataSource"
22          class="org.apache.commons.dbcp2.BasicDataSource">
23          <!--数据库驱动 -->
```

```
24          <property name="driverClassName" value="${jdbc.driver}" />
25          <!--连接数据库的 url -->
26          <property name="url" value="${jdbc.url}" />
27          <!--连接数据库的用户名 -->
28          <property name="username" value="${jdbc.username}" />
29          <!--连接数据库的密码-->
30          <property name="password" value="${jdbc.password}" />
31          <!--最大连接数-->
32          <property name="maxTotal" value="${jdbc.maxTotal}" />
33          <!--最大空闲连接-->
34          <property name="maxIdle" value="${jdbc.maxIdle}" />
35          <!--初始化连接数-->
36          <property name="initialSize" value="${jdbc.initialSize}" />
37      </bean>
38      <!--事务管理器，依赖于数据源 -->
39      <bean id="transactionManager"
40       class="org.springframework.jdbc.datasource.DataSourceTransactionManager">
41          <property name="dataSource" ref="dataSource"/>
42      </bean>
43      <!-- 通知 -->
44      <tx:advice id="txAdvice" transaction-manager="transactionManager">
45          <tx:attributes>
46              <!-- 传播行为 -->
47              <tx:method name="save*" propagation="REQUIRED"/>
48              <tx:method name="insert*" propagation="REQUIRED"/>
49              <tx:method name="add*" propagation="REQUIRED"/>
50              <tx:method name="create*" propagation="REQUIRED"/>
51              <tx:method name="delete*" propagation="REQUIRED"/>
52              <tx:method name="update*" propagation="REQUIRED"/>
53              <tx:method name="find*" propagation="SUPPORTS" read-only="true"/>
54              <tx:method name="select*" propagation="SUPPORTS" read-only="true"/>
55              <tx:method name="get*" propagation="SUPPORTS" read-only="true"/>
56          </tx:attributes>
57      </tx:advice>
58      <!-- 切面 -->
59      <aop:config>
60          <aop:advisor advice-ref="txAdvice" pointcut="execution(* com.ssm.service.*.*(..))" />
61      </aop:config>
62      <!--配置 MyBatis 工厂 -->
63      <bean id="sqlSessionFactory" class="org.mybatis.spring.SqlSessionFactoryBean">
64          <!--注入数据源 -->
65          <property name="dataSource" ref="dataSource" />
66          <!--指定核心配置文件位置 -->
67          <property name="configLocation" value="classpath:mybatis-config.xml" />
68      </bean>
69      <!--配置 mapper 扫描器 -->
```

```
70      <bean class="org.mybatis.spring.mapper.MapperScannerConfigurer">
71          <property name="basePackage" value="com.ssm.dao"></property>
72      </bean>
73      <!--扫描 Service -->
74      <context:component-scan base-package="com.ssm.service"/>
75  </beans>
```

applicationContext.xml 文件与第 16 章讲解 SSM 框架整合时的配置有所不同，这里增加了事务传播行为以及切面的配置。在事务传播行为中，只有查询方法的事务为只读，添加、修改和删除的操作都纳入了事务管理。

文件 17.3　mybatis-config.xml

```
01  <?xml version="1.0" encoding="UTF-8"?>
02  <!DOCTYPE configuration PUBLIC "-//mybatis.org//DTD Config 3.0//EN"
03      "http://mybatis.org/dtd/mybatis-3-config.dtd">
04  <configuration>
05      <!--配置别名 -->
06      <typeAliases>
07          <package name="com.ssm.po"/>
08      </typeAliases>
09  </configuration>
```

文件 17.4　springmvc-config.xml

```
01  <?xml version="1.0" encoding="UTF-8"?>
02  <beans xmlns="http://www.springframework.org/schema/beans"
03      xmlns:xsi="http://www.w3.org/2001/XMLSchema-instance"
04      xmlns:mvc="http://www.springframework.org/schema/mvc"
05      xmlns:context="http://www.springframework.org/schema/context"
06      xmlns:aop="http://www.springframework.org/schema/aop"
07      xmlns:tx="http://www.springframework.org/schema/tx"
08      xsi:schemaLocation="http://www.springframework.org/schema/beans
09        http://www.springframework.org/schema/beans/spring-beans.xsd
10        http://www.springframework.org/schema/mvc
11        http://www.springframework.org/schema/mvc/spring-mvc.xsd
12        http://www.springframework.org/schema/aop
13        http://www.springframework.org/schema/aop/spring-aop.xsd
14        http://www.springframework.org/schema/context
15        http://www.springframework.org/schema/context/spring-context.xsd">
16      <!--指定需要扫描的包 -->
17      <context:component-scan base-package="com.ssm.web.controller" />
18      <!-- 配置注解驱动 -->
19      <mvc:annotation-driven />
20      <!-- 配置静态资源的访问映射，此配置中的文件将不被前端控制器拦截 -->
21      <mvc:resources location="/js/" mapping="/js/**"></mvc:resources>
22      <mvc:resources location="/css/" mapping="/css/**"></mvc:resources>
23      <mvc:resources location="/images/" mapping="/images/**"></mvc:resources>
24      <!-- 定义视图解析器 -->
```

```
25      <bean id="viewResoler"
26          class="org.springframework.web.servlet.view.InternalResourceViewResolver">
27          <!-- 设置前缀 -->
28          <property name="prefix" value="/WEB-INF/jsp/" />
29          <!-- 设置后缀 -->
30          <property name="suffix" value=".jsp" />
31      </bean>
32  </beans>
```

上述代码除了配置需要扫描的包、注解驱动和视图解析器外，还增加了加载属性文件和访问静态资源的配置。

除以上配置文件外，还需要在项目的/WebContent/WEB-INF 目录下编写 web.xml 文件，如文件 17.5 所示。

文件 17.5　web.xml

```
01  <?xml version="1.0" encoding="UTF-8"?>
02  <web-app xmlns:xsi="http://www.w3.org/2001/XMLSchema-instance"
03      xmlns="http://xmlns.jcp.org/xml/ns/javaee"
04      xsi:schemaLocation="http://xmlns.jcp.org/xml/ns/javaee
                           http://xmlns.jcp.org/xml/ns/javaee/web-app_4_0.xsd"
05      id="WebApp_ID" version="4.0">
06      <!-- 配置加载 Spring 文件的监听器 -->
07      <context-param>
08          <param-name>contextConfigLocation</param-name>
09          <param-value>classpath:applicationContext.xml</param-value>
10      </context-param>
11      <listener>
12          <listener-class>
13              org.springframework.web.context.ContextLoaderListener
14          </listener-class>
15      </listener>
16      <!-- 编码过滤器 -->
17      <filter>
18          <filter-name>encoding</filter-name>
19          <filter-class>org.springframework.web.filter.CharacterEncodingFilter </filter-class>
20          <init-param>
21              <param-name>encoding</param-name>
22              <param-value>UTF-8</param-value>
23          </init-param>
24      </filter>
25      <filter-mapping>
26          <filter-name>encoding</filter-name>
27          <url-pattern>*.action</url-pattern>
28      </filter-mapping>
29      <!-- 配置 Spring MVC 前端控制器 -->
30      <servlet>
```

```
31            <!-- 配置前端过滤器 -->
32            <servlet-name>springmvc</servlet-name>
33            <servlet-class>org.springframework.web.servlet.DispatcherServlet </servlet-class>
34            <!-- 初始化时加载配置文件 -->
35            <init-param>
36                <param-name>contextConfigLocation</param-name>
37                <param-value>classpath:springmvc-config.xml</param-value>
38            </init-param>
39            <!-- 配置服务器启动时立即加载 Spring MVC 配置文件 -->
40            <load-on-startup>1</load-on-startup>
41        </servlet>
42        <servlet-mapping>
43            <servlet-name>springmvc</servlet-name>
44            <url-pattern>*.action</url-pattern>
45        </servlet-mapping>
46        <!-- 系统默认页面 -->
47        <welcome-file-list>
48            <welcome-file>index.jsp</welcome-file>
49        </welcome-file-list>
50    </web-app>
```

在 web.xml 文件中配置了 Spring 的监听器、编码过滤器和 Spring MVC 的前端控制器等信息。

17.4.3 创建项目相关目录（包）和文件，并引入相关文件资源

按照如图 17.2 所示的项目文件组织结构，创建项目相关的目录（包）及文件（如相关类和接口的包、Vue 文件对应的文件夹、Vue 文件等），并引入项目开发需要的相关文件资源（如 CSS 样式文件、images 图片文件、JS 文件、标签文件等），为项目开发做好准备工作。在后续项目开发过程中，根据需要可以创建其他文件或引入其他资源文件。

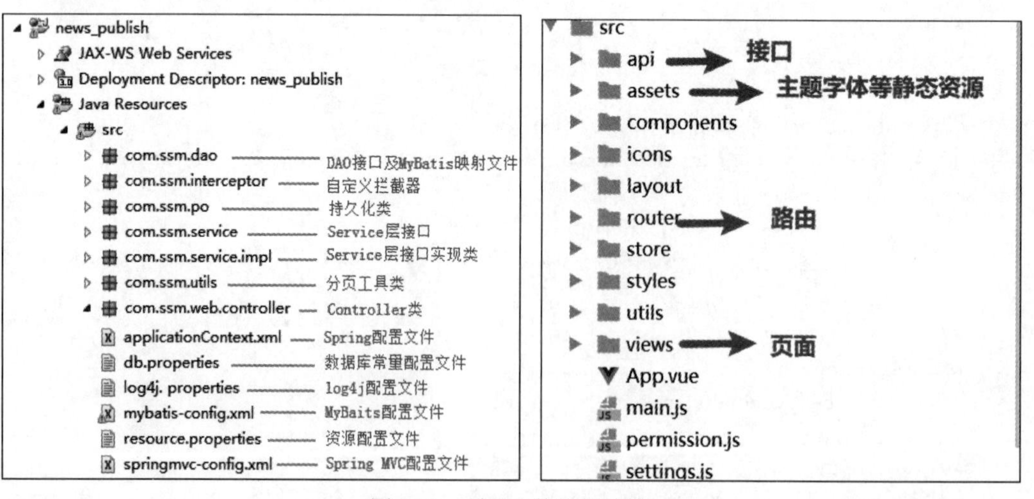

图 17.2　项目文件组织结构

17.5　用户管理模块

用户管理模块涉及用户的添加、修改、查询、删除和设置角色、登录及退出等功能。本节将具体实现这些功能。

17.5.1　创建持久化类

用户管理模块持久化类有角色类 Role 和用户类 User，具体代码分别如文件 17.6 和文件 17.7 所示。

文件 17.6　Role.java

```
01  package com.ssm.po;
02  import java.util.List;
03  //角色类
04  public class Role {
05      private Integer roleId;           //角色 id
06      private String roleName;          //角色名称
07      private List<User> userList;      //对应角色的用户列表
08      public Integer getRoleId() {
09          return roleId;
10      }
11      public void setRoleId(Integer roleId) {
12          this.roleId = roleId;
13      }
14      public String getRoleName() {
15          return roleName;
16      }
17      public void setRoleName(String roleName) {
18          this.roleName = roleName;
19      }
20      public List<User> getUserList() {
21          return userList;
22      }
23      public void setUserList(List<User> userList) {
24          this.userList = userList;
25      }
26      @Override
27      public String toString() {
28          return "Role [roleId=" + roleId + ", roleName="+roleName+", userList="+userList+"]";
29      }
30  }
```

文件 17.7 User.java

```
01  package com.ssm.po;
02  import java.util.Date;
03  //用户类
04  public class User {
05      private Integer userId;       //用户 id
06      private String userName;      //用户姓名
07      private String loginName;     //用户登录名
08      private String password;      //登录密码
09      private String tel;           //联系电话
10      private Date registerTime;    //注册或修改用户时间
11      private String status;        //用户状态('1': 未启用; '2': 已启用; '3': 被禁用)
12      private Integer roleId;       //用户对应的角色 id
13      private String roleName;      //角色名称（为了方便列表页显示角色名，增加此属性）
14      public Integer getUserId() {
15          return userId;
16      }
17      public void setUserId(Integer userId) {
18          this.userId = userId;
19      }
20      public String getUserName() {
21          return userName;
22      }
23      public void setUserName(String userName) {
24          this.userName = userName;
25      }
26      public String getLoginName() {
27          return loginName;
28      }
29      public void setLoginName(String loginName) {
30          this.loginName = loginName;
31      }
32      public String getPassword() {
33          return password;
34      }
35      public void setPassword(String password) {
36          this.password = password;
37      }
38      public String getTel() {
39          return tel;
40      }
41      public void setTel(String tel) {
42          this.tel = tel;
43      }
44      public Date getRegisterTime() {
45          return registerTime;
```

```
46      }
47      public void setRegisterTime(Date registerTime) {
48          this.registerTime = registerTime;
49      }
50      public String getStatus() {
51          return status;
52      }
53      public void setStatus(String status) {
54          this.status = status;
55      }
56      public Integer getRoleId() {
57          return roleId;
58      }
59      public void setRoleId(Integer roleId) {
60          this.roleId = roleId;
61      }
62      public String getRoleName() {
63          return roleName;
64      }
65      public void setRoleName(String roleName) {
66          this.roleName = roleName;
67      }
68      @Override
69      public String toString() {
70          return "User [userId=" + userId + ", userName=" + userName + ", loginName=" +
71           loginName + ", password=" + password + ", tel=" + tel + ", registerTime=" +
72           registerTime + ", status=" + status + ", roleId="         + roleId +
73            ", roleName=" + roleName + "]";
74      }
75  }
```

17.5.2 实现 DAO 层接口

实现 DAO 层接口的操作步骤如下：

步骤 01 创建 DAO 层接口。在 src 目录下的 com.ssm.dao 包中创建一个角色接口 RoleDao 和一个用户接口 UserDao，并在接口中编写增、删、改、查等方法，如文件 17.8 和文件 17.9 所示。

文件 17.8 RoleDao.java

```
01  package com.ssm.dao;
02  import java.util.List;
03  import org.apache.ibatis.annotations.Param;
04  import com.ssm.po.Role;
05  public interface RoleDao {
06      //获取所有角色信息（角色列表）
```

```
07      public List<Role> selectRoleList();
08  }
```

文件 17.9　UserDao.java

```
01  package com.ssm.dao;
02  import java.util.List;
03  import org.apache.ibatis.annotations.Param;
04  import com.ssm.po.User;
05  // 用户 DAO 层接口
06  public interface UserDao {
07      //查询所有用户
08      public List<User> selectUserList(@Param("keywords") String keywords,
09                           @Param("userListRoleId") Integer userListRoleId);
10      //通过账号和密码查询用户
11      public User findUser(@Param("loginName") String loginName,
12                  @Param("password") String password);
13      //通过用户 id 查询用户
14      public User getUserByUserId(Integer userId);
15      //通过用户登录名查询用户（用于判断用户名是否存在）
16      public User getUserByLoginName(String loginName);
17      //添加用户
18      public int addUser(User user);
19      //更新用户
20      public int updateUser(User user);
21      //删除用户
22      public int delUser(Integer userId);
23      //设置用户状态（'1': 未启用; '2': 已启用; '3': 被禁用）
24      public int setUserStatus(User user);
25  }
```

步骤 02 创建映射文件。在 src 目录下的 com.ssm.dao 包中创建 MyBatis 映射文件 RoleDao.xml 和 UserDao.xml，并在映射文件中编写增、删、改、查等方法的执行语句，如文件 17.10 和文件 17.11 所示。

文件 17.10　RoleDao.xml

```
01  <?xml version="1.0" encoding="UTF-8"?>
02  <!DOCTYPE mapper PUBLIC "-//mybatis.org//DTD Mapper 3.0//EN"
03      "http://mybatis.org/dtd/mybatis-3-mapper.dtd">
04  <mapper namespace="com.ssm.dao.RoleDao">
05      <!--查询角色集合列表 -->
06      <select id="selectRoleList" resultType="Role">
07          select roleId,roleName from t_role
08      </select>
09      <!-- 请参看代码-->
10  </mapper>
```

文件 17.11　UserDao.xml

```
01 <?xml version="1.0" encoding="UTF-8"?>
02 <!DOCTYPE mapper PUBLIC "-//mybatis.org//DTD Mapper 3.0//EN"
03     "http://mybatis.org/dtd/mybatis-3-mapper.dtd">
04 <mapper namespace="com.ssm.dao.UserDao">
05     <!--查询所有用户集合的 where 语句 -->
06     <sql id="selectUserListWhere">
07         <where>
08             u.roleId=r.roleId
09             <if test="keywords!=null and keywords!=''">
10                 and (u.username like CONCAT('%',#{keywords},'%') or
11                 u.loginName like CONCAT('%',#{keywords},'%'))
12             </if>
13             <if test="userListRoleId!=null and userListRoleId!=''">
14                 and (u.roleId=#{userListRoleId})
15             </if>
16         </where>
17     </sql>
18     <!--查询所有用户集合列表 -->
19     <select id="selectUserList" parameterType="String" resultType="User">
20       select u.*,r.roleName from t_user as u,t_role as r
21         <include refid="selectUserListWhere" />
22         order by registerTime desc
23     </select>
24     <!--通过账号和密码查询用户 -->
25     <select id="findUser" parameterType="String" resultType="User">
26         select * from t_user where loginName=#{loginName} and password=#{password}
27          limit 0,1
28     </select>
29     <!--通过 userId 查询用户 -->
30     <select id="getUserByUserId" parameterType="Integer" resultType="User">
31         select * from t_user where userId=#{userId}
32     </select>
33     <!--通过登录账号查询用户 -->
34     <select id="getUserByLoginName" parameterType="String" resultType="User">
35         select * from t_user where loginName=#{loginName} limit 0,1
36     </select>
37     <!--添加用户 -->
38     <insert id="addUser" parameterType="User">
39         insert into t_user(
40             userName,
41             loginName,
42             password,
43             tel,
44             registerTime,
45             status,
```

```
46                  roleId
47              )
48              values(
49                  #{userName},
50                  #{loginName},
51                  #{password},
52                  #{tel},
53                  #{registerTime},
54                  #{status},
55                  #{roleId}
56              )
57      </insert>
58      <!-- 更新用户 -->
59      <update id="updateUser" parameterType="User">
60          update t_user
61          <set>
62              registerTime=#{registerTime},
63              status=#{status},
64          <if test="userName!=null and userName!=''">
65              userName=#{userName},
66          </if>
67          <if test="password!=null and password!=''">
68              password=#{password},
69          </if>
70          <if test="tel!=null and tel!=''">
71              tel=#{tel},
72          </if>
73          <if test="roleId!=null and roleId!=''">
74              roleId=#{roleId},
75          </if>
76          </set>
77          where userId=#{userId}
78      </update>
79      <!--删除用户 -->
80      <delete id="delUser" parameterType="Integer">
81          delete from t_user where userId=#{userId}
82      </delete>
83      <!--设置用户状态（status '1':未启用; '2': 已启用; '3': 被禁用）-->
84      <update id="setUserStatus" parameterType="User">
85          update t_user set status=#{status} where userId=#{userId}
86      </update>
87  </mapper>
```

文件 17.11 中的代码通过映射查询、增加、修改和删除等语句来实现用户表中的操作。

17.5.3　实现 Service 层接口

实现 Service 层接口的操作步骤如下：

步骤 01 创建角色和用户的 Service 层接口。在 src 目录下创建一个 com.ssm.service 包，在包中创建 RoleService 接口和 UserService 接口，并在该接口中编写操作用户的相关方法，如文件 17.12 和文件 17.13 所示。

文件 17.12　RoleService.java

```
01  package com.ssm.service;
02  import java.util.List;
03  import com.ssm.po.Role;
04  // 角色 Service 层接口
06  public interface RoleService {
07      public List<Role> findRoleList();
08  }
```

文件 17.13　UserService.java

```
01  package com.ssm.service;
02  import java.util.List;
03  import com.ssm.po.User;
04  // 用户 Service 层接口
06  public interface UserService {
07      public List<User> findUserList(String keywords, Integer userListRoleId);
08      public User findUser(String loginName,String password);
09      public User getUserByUserId(Integer userId);
10      public User getUserByLoginName(String loginName);
11      public int editUser(User user);
12      public int addUser(User user);
13      public int delUser(Integer userId);
14      public int setUserStatus(User user);
15  }
```

步骤 02 创建角色和用户 Service 层接口的实现类。在 src 目录下创建一个 com.ssm.service.impl 包，并在包中创建 RoleService 接口的实现类 RoleServiceImpl 和 UserService 接口的实现类 UserServiceImpl，在类中编辑并实现接口中的方法，如文件 17.14 和文件 17.15 所示。

文件 17.14　RoleServiceImpl.java

```
01  package com.ssm.service.impl;
02  import java.util.List;
03  import org.springframework.beans.factory.annotation.Autowired;
04  import org.springframework.stereotype.Service;
05  import com.ssm.dao.RoleDao;
06  import com.ssm.po.Role;
07  import com.ssm.service.RoleService;
08  @Service("roleService")
```

```
09  // 角色 Service 接口实现类
10  public class RoleServiceImpl implements RoleService {
11      //注入 RoleDao
12      @Autowired
13      private RoleDao roleDao;
14      @Override
15      public List<Role> findRoleList() {
16          List<Role> roleList=roleDao.selectRoleList();
17          return roleList;
18      }
19  }
```

文件 17.15 UserServiceImpl.java

```
01  package com.ssm.service.impl;
02  import java.util.List;
03  import org.springframework.beans.factory.annotation.Autowired;
04  import org.springframework.stereotype.Service;
05  import com.ssm.dao.UserDao;
06  import com.ssm.po.User;
07  import com.ssm.service.UserService;
08  @Service("userService")
09  // 用户 Service 接口实现类
10  public class UserServiceImpl implements UserService{
11      //注入 UserDao
12      @Autowired
13      private UserDao userDao;
14      @Override
15      public List<User> findUserList(String keywords, Integer userListRoleId) {
16          List<User> userList=this.userDao.selectUserList(keywords,userListRoleId);
17          return userList;
18      }
19      @Override
20      public User findUser(String loginName, String password) {
21          User user=this.userDao.findUser(loginName, password);
22          return user;
23      }
24      @Override
25      public User getUserByUserId(Integer userId) {
26          return this.userDao.getUserByUserId(userId);
27      }
28      @Override
29      public User getUserByLoginName(String loginName) {
30          return this.userDao.getUserByLoginName(loginName);
31      }
32      @Override
33      public int addUser(User user) {
```

```
34          return userDao.addUser(user);
35      }
36      @Override
37      public int editUser(User user) {
38          return this.userDao.updateUser(user);
39      }
40      @Override
41      public int delUser(Integer userId) {
42          return userDao.delUser(userId);
43      }
44      @Override
45      public int setUserStatus(User user) {
46          return userDao.setUserStatus(user);
47      }
48  }
```

17.5.4　实现 Controller 类

在 src 目录下创建一个 com.ssm.web.controller 包，在包中创建用户控制器类 UserController，代码如文件 17.16 所示。

文件 17.16　UserController.java

```
01  package com.ssm.web.controller;
02  import java.util.Date;
03  import java.util.List;
04  import javax.servlet.http.HttpSession;
05  import org.springframework.beans.factory.annotation.Autowired;
06  import org.springframework.stereotype.Controller;
07  import org.springframework.ui.Model;
08  import org.springframework.web.bind.annotation.RequestBody;
09  import org.springframework.web.bind.annotation.RequestMapping;
10  import org.springframework.web.bind.annotation.RequestMethod;
11  import org.springframework.web.bind.annotation.ResponseBody;
12  import com.ssm.po.Role;
13  import com.ssm.po.User;
14  import com.ssm.service.RoleService;
15  import com.ssm.service.UserService;
16  // 用户控制类
17  @Controller
18  public class UserController {
19      //依赖注入
20      @Autowired
21      private UserService userService;
22      @Autowired
23      private RoleService roleService;
```

```
24     //查询所有状态的用户集合（用户列表）
25     @RequestMapping(value="/findUserList.action")
26     public String findUserList(String keywords,Integer userListRoleId,Model model){
27         // 获取角色列表
28         List<Role> roleList = roleService.findRoleList();
29         model.addAttribute("roleList", roleList);
30         // 获取用户列表
31         List<User> userList=userService.findUserList(keywords,userListRoleId);
32         model.addAttribute("userList", userList);
33         model.addAttribute("keywords", keywords);
34         model.addAttribute("userListRoleId", userListRoleId);
35         return "user/user_list";
36     }
37     //跳转至添加用户页面
38     @RequestMapping(value="/toAddUser.action")
39     public String toAddUser(Model model){
40         //获取角色列表，用于添加用户页面中的用户角色下拉列表
41         List<Role> roleList=roleService.findRoleList();
42         model.addAttribute("roleList", roleList);
43         return "user/add_user";
44     }
45     //判断登录账号是否已存在
46     @RequestMapping(value = "/checkLoginName.action")
47     @ResponseBody
48     public User checkLoginName(@RequestBody User user, Model model) {
49         User checkUser = userService.getUserByLoginName(user.getLoginName());
50         if (checkUser!=null) {
51             //登录账号已存在
52             return checkUser;
53         }else{
54             checkUser=new User();
55             checkUser.setUserId(0);
56             return checkUser;
57         }
58     }
59     //添加用户
60     @RequestMapping(value = "/addUser.action", method = RequestMethod.POST)
61     public String addUser(User user, Model model) {
62         // 获取角色列表
63         List<Role> roleList = roleService.findRoleList();
64         model.addAttribute("roleList", roleList);
65         model.addAttribute("user", user);
66         //检查登录账号是否已存在
67         User checkUser = userService.getUserByLoginName(user.getLoginName());
68         if (checkUser!=null) {
69             // 登录账号已存在，重新跳转回添加用户页面
```

```
70                model.addAttribute("checkUserLoginNameMsg", "登录账号已存在，请重新输入");
71                return "user/add_user";
72            }else{
73                // 登录账号可用
74                Date date = new Date();
75                user.setRegisterTime(date);
76                //默认设置用户为启用状态"2"
77                user.setStatus("2");
78                //调用 UserService 实例中的添加用户方法
79                int rows = userService.addUser(user);
80                if (rows > 0) {
81                    // 添加成功，跳转回用户列表页面
82                    return "redirect:findUserList.action";
83                } else {
84                    // 添加失败，重新跳转回添加用户页面
85                    return "user/add_user";
86                }
87            }
88        }
89        // 跳转回修改用户页面
90        @RequestMapping(value = "/toEditUser.action")
91        public String toEditUser(Integer userId, Model model) {
92            //通过 userId 获取用户
93            User user = userService.getUserByUserId(userId);
94            if (user != null) {
95                model.addAttribute("user", user);
96                // 获取角色列表
97                List<Role> roleList = roleService.findRoleList();
98                model.addAttribute("roleList", roleList);
99                return "user/edit_user";
100           }else{
101               return "redirect:findUserList.action";
102           }
103       }
104       //修改用户
105       @RequestMapping(value = "/editUser.action", method = RequestMethod.POST)
106       public String editUser(User user, Model model) {
107           // 获取角色列表
108           Date date = new Date();
109           user.setRegisterTime(date);
110           // 默认设置用户为启用状态"2"
111           user.setStatus("2");
112           int rows = userService.editUser(user);
113           if (rows > 0) {
114               // 添加成功，跳转回用户列表页面
115               return "redirect:findUserList.action";
```

```
116         } else {
117             List<Role> roleList = roleService.findRoleList();
118             model.addAttribute("roleList", roleList);
119             model.addAttribute("user", user);
120             // 修改失败，跳转回修改用户页面
121             return "user/edit_user";
122         }
123     }
124     //删除用户（在前台页面中通过 jQuery Ajax 方式调用此方法）
125     @RequestMapping(value = "/delUser.action")
126     @ResponseBody
127     public User delUser(@RequestBody User user, Model model) {
128         int rows = userService.delUser(user.getUserId());
129         if (rows>0) {
130             return user;
131         }else{
132             //此处设置 userId 为 0，只是作为操作失败的标记使用
133             user.setUserId(0);
134             return user;
135         }
136     }
137     //禁用用户（更新 status 字段值为'3'，在前台页面中通过 jQuery Ajax 方式调用此方法）
138     @RequestMapping(value = "/disableUser.action")
139     @ResponseBody
140     public User disableUser(@RequestBody User user, Model model) {
141         int rows = userService.setUserStatus(user);
142         if (rows>0) {
143             return user;
144         }else{
145             //此处设置 userId 为 0，只是作为操作失败的标记使用
146             user.setUserId(0);
147             return user;
148         }
149     }
150     //启用用户（更新 status 字段值为'2'，在前台页面中通过 jQuery Ajax 方式调用此方法）
151     @RequestMapping(value = "/enableUser.action")
152     @ResponseBody
153     public User enableUser(@RequestBody User user, Model model) {
154         int rows = userService.setUserStatus(user);
155         if (rows>0) {
156             return user;
157         }else{
158             //此处设置 userId 为 0，只是作为操作失败的标记使用
159             user.setUserId(0);
160             return user;
161         }
```

```
162     }
163     //用户登录
164     @RequestMapping(value="/login.action",method=RequestMethod.POST)
165     public String login(String loginName,String password,Model model,HttpSession session){
166         //通过用户名和密码查询用户
167         User user=userService.findUser(loginName, password);
168         if(user!=null){
169             if (user.getStatus().equals("2")) {
170                 //用户被启用时，允许登录到后台
171                 session.setAttribute("login_user", user);
172                 return "main";
173             } else {
174                 //账号未启用或被禁用时，不允许登录到后台
175                 model.addAttribute("msg", "账号未启用或被禁用，请联系管理员！");
176                 return "login";
177             }
178         }
179         //账号或密码错误时，不允许登录到后台
180         model.addAttribute("msg","账号或密码错误，请重新登录！");
181         return "login";
182     }
183     //退出登录
184     @RequestMapping(value="/logout.action")
185     public String logout(HttpSession session){
186         //清空 Session
187         session.invalidate();
188         return "login";
189     }
190 }
```

在文件 17.16 中，首先通过@Autowired 注解将 RoleService 对象和 UserService 对象注入本类中，然后创建对用户进行增、删、改、查、登录、退出等的方法。

注意：UserController 类的一些方法中涉及一些业务逻辑，请读者仔细阅读，并将 UserController 类与系统中的用户管理相关页面联系起来推敲。

① UserController 类中每个方法前通过注解@RequestMapping 设置了方法的访问路径，例如@RequestMapping(value = "/addUser.action", method = RequestMethod.POST)。

② addUser()方法用于添加用户。添加用户前，先调用 toAddUser()方法获取角色集合列表并传递到添加用户页面，再通过 c 标签循环添加到角色选择的下拉列表框中。在添加用户页面输入用户信息后，单击“添加”按钮，调用 addUser()方法，addUser()方法先调用 UserService 接口实例的 getUserByLoginName()方法判断输入的登录账号是否已经存在。如果已存在，就返回添加用户页面，并给出提示信息；如果不存在，就继续添加用户，添加成功的话，跳转回用户列表页面，添加不成功则返回添加用户页面，让用户重新添加用户。

③ login()方法用于用户登录。由于在用户登录时，表单会以 POST 方式提交，因此将

RequestMapping 注解的 method 属性值设置为 RequestMethod.POST。在 login()方法中，首先通过页面中传递过来的账号和密码查询用户，然后通过 if 语句判断是否存在该用户。如果不存在，就提示错误信息，并返回登录页面；如果存在，但用户账号未启用或被禁用，就提示相关信息，并返回登录页面；如果存在且已启用，就将用户信息存储在 session 中，并跳转到系统后台主页面。

④ 删除用户的 delUser()方法、禁用用户的 disableUser()方法和启用用户的 enableUser()方法，系统页面采用 jQuery Ajax 方式进行调用。相关交互方式在“13.1 JSON 数据交互”中进行了介绍，在此不再赘述。

⑤ 系统页面与 UserController 类中方法的交互过程的参数传递，在前面章节中有相关介绍，请读者参考前面相关章节中的内容。

17.5.5 实现页面功能

页面功能使用 Vue.js 3 进行开发，实现页面功能的操作步骤如下：

步骤 01 在项目 src/views/login 目录下创建 login.vue 页面文件。login.vue 主要实现了用户登录到后台的功能，代码如文件 17.17 所示。

文件 17.17 login.vue

```
01  <template>
02    <div class="login-container">
03      <el-form ref="loginForm" :model="loginForm" :rules="loginRules" class="login-form"
04  auto-complete="on" label-position="left">
05        <div class="title-container">
06          <h3 class="title">登录</h3>
07        </div>
08
09        <el-form-item prop="username">
10          <span class="svg-container">
11            <svg-icon icon-class="user" />
12          </span>
13          <el-input
14            ref="username"
15            v-model="loginForm.username"
16            placeholder="Username"
17            name="username"
18            type="text"
19            tabindex="1"
20            auto-complete="on"
21          />
22        </el-form-item>
23
24        <el-form-item prop="password">
25          <span class="svg-container">
26            <svg-icon icon-class="password" />
```

```
27          </span>
28          <el-input
29            :key="passwordType"
30            ref="password"
31            v-model="loginForm.password"
32            :type="passwordType"
33            placeholder="Password"
34            name="password"
35            tabindex="2"
36            auto-complete="on"
37            @keyup.enter.native="handleLogin"
38          />
39          <span class="show-pwd" @click="showPwd">
40            <svg-icon :icon-class="passwordType === 'password' ? 'eye' : 'eye-open'" />
41          </span>
42        </el-form-item>
43        <el-button :loading="loading" type="primary"  tyle="width:100%;margin-bottom:
44  30px;"  @click.native.prevent="handleLogin">登录</el-button>
45
46      </el-form>
47    </div>
48  </template>
49
50  <script>
51  import { validUsername } from '@/utils/validate'
52
53  export default {
54    name: 'Login',
55    data() {
56      const validateUsername = (rule, value, callback) => {
57        if (!validUsername(value)) {
58          callback(new Error('Please enter the correct user name'))
59        } else {
60          callback()
61        }
62      }
63      const validatePassword = (rule, value, callback) => {
64        if (value.length < 6) {
65          callback(new Error('The password can not be less than 6 digits'))
66        } else {
67          callback()
68        }
69      }
70      return {
71        loginForm: {
72          username: '',
```

```
73          password: ''
74        },
75        loginRules: {
76          username: [{ required: true, trigger: 'blur', validator: validateUsername }],
77          password: [{ required: true, trigger: 'blur', validator: validatePassword }]
78        },
79        loading: false,
80        passwordType: 'password',
81        redirect: undefined
82      }
83    },
84    watch: {
85      $route: {
86        handler: function(route) {
87          this.redirect = route.query && route.query.redirect
88        },
89        immediate: true
90      }
91    },
92    methods: {
93      showPwd() {
94        if (this.passwordType === 'password') {
95          this.passwordType = ''
96        } else {
97          this.passwordType = 'password'
98        }
99        this.$nextTick(() => {
100         this.$refs.password.focus()
101       })
102     },
103     handleLogin() {
104       this.$refs.loginForm.validate(valid => {
105         if (valid) {
106           this.loading = true
107           this.$store.dispatch('user/login', this.loginForm).then(() => {
108             this.$router.push({ path: this.redirect || '/' })
109             this.loading = false
110           }).catch(() => {
111             this.loading = false
112           })
113         } else {
114           console.log('error submit!!')
115           return false
116         }
117       })
118     }
```

```
119     }
120   }
121   </script><!-- 此处省略部分 CSS 代码 -->
```

在文件 17.17 中，核心代码是用户登录操作的 form 表单，该表单在提交时会通过 checkValue()方法检查账户或密码是否为空，如果为空，就弹出消息框提示用户；如果账号和密码都已填写，就将表单提交到以“/login.action”结尾的请求中。

步骤 02 在 src/router 目录下的 index.js 文件夹的 views 文件下创建 main.vue、top.vue、left.vue、right.vue 页面。这几个页面主要实现后台框架页面功能，以及根据登录后保存在 session 中的用户角色信息显示与其权限相适应的操作页面和菜单选项的功能，具体代码读者可参考本章案例代码（在本书配套的下载资源中获取）。

步骤 03 在 src/views 目录下创建 user 文件夹，并在该文件夹下创建用户列表页面 user_list.vue、添加用户页面 add_user.vue 和更新用户页面 edit_user.vue，具体实现代码如文件 17.18~文件 17.20 所示。

文件 17.18　user_list.vue

```
01   <template>
02   <div class="app-container">
03   <el-table
04   :data="tableData"
05   style="width: 100%"
06   >
07   <el-table-column
08   label="注册/修改日期"
09   width="150"
10   >
11   <template slot-scope="scope">
12   <i class="el-icon-time" />
13   <span style="margin-left: 10px">{{ scope.row.registerTime }}</span>
14   </template>
15   </el-table-column>
16   <el-table-column
17   label="用户姓名"
18   width="150"
19   >
20   <template slot-scope="scope">
21   <el-popover trigger="hover" placement="top">
22   <p>姓名: {{ scope.row.userName }}</p>
23   <div slot="reference" class="name-wrapper">
24   <el-tag size="medium">{{ scope.row.userName}}</el-tag>
25   </div>
26   </el-popover>
27   </template>
28   </el-table-column>
29   <el-table-column
30   label="登录账号"
31   width="150"
32   >
```

```
33  <template slot-scope="scope">
34  <span style="margin-left: 10px">{{ scope.row.loginName }}</span>
35  </template>
36  </el-table-column>
37  <el-table-column
38  label="联系电话"
39  width="120"
40  >
41  <template slot-scope="scope">
42  <span style="margin-left: 10px">{{ scope.row.tel }}</span>
43  </template>
44  </el-table-column>
45  <el-table-column
46  label="用户角色"
47  width="150"
48  >
49  <template slot-scope="scope">
50  <span style="margin-left: 10px">{{ scope.row.roleName }}</span>
51  </template>
52  </el-table-column>
53  <el-table-column
54  label="审核状态"
55  width="150"
56  >
57  <template slot-scope="scope">
58  <span style="margin-left: 10px">{{ scope.row.status }}</span>
59  </template>
60  </el-table-column>
61  <el-table-column label="操作">
62  <template slot-scope="scope">
63  <el-button
64  size="mini"
65  @click="handleEdit(scope.$index, scope.row)"
66  >编辑</el-button>
67  <el-button
68  size="mini"
69  type="danger"
70  @click="handleDelete(scope.$index, scope.row)"
71  >删除</el-button>
72  </template>
73  </el-table-column>
74  </el-table>
75  </div>
76  </template>
77  <script>
78  export default {
79      filters: {
80          statusFilter(status) {
81              const statusMap = {
82                  published: 'success',
```

```
83                  draft: 'gray',
84                  deleted: 'danger'
85              }
86              return statusMap[status]
87          }
88      },
89      data() {
90          return {
91              tableData: [{
92                  loginName: '',
93                  registerTime: '',
94                  userName: '',
95                  tel: '',
96                  roleName: '',
97                  status: ''
98              }
99          ]}
100 }
101 </script>
102 <!-- 此处省略部分代码 -->
```

用户列表页面 user_list.vue 文件主要用来展示用户列表，通过访问 UserController 类的 findUserList()方法，可以跳转到 user_list.vue 页面，并在页面中迭代显示用户列表信息。在此基础上可以通过链接对用户进行修改、删除、禁用/启用等相关操作。

文件 17.19　add_user.vue

```
01  <template>
02  <div class="app-container">
03  <el-form
04  ref="detailData"
05  style="margin-left: 10%"
06  :model="detailData"
07  label-width="80px"
08  label-position="right"
09  :rules="essayRules"
10  >
11  <el-form-item label="登录账号" style="width: 300px" prop="loginName">
12  <el-input v-model="detailData.loginName" placeholder="请输入登录账号"/>
13  </el-form-item>
14  <el-form-item label="密码:" style="width: 300px" prop="password">
15  <el-input v-model="detailData.password" placeholder="请输入密码"/>
16  </el-form-item>
17  <el-form-item label="姓名:" style="width: 300px" prop="userName">
18  <el-input v-model="detailData.userName" placeholder="请输入姓名"/>
19  </el-form-item>
20  <el-form-item label="手机号:" style="width: 300px" prop="tel">
21  <el-input v-model="detailData.tel" placeholder="请输入手机号"/>
22  </el-form-item>
23  <el-form-item label="角色:" style="width: 300px" prop="roleName">
```

```
24  <el-select v-model="detailData.roleName" placeholder="请选择角色">
25          <el-option
26            v-for="item in options"
27            :key="item.roleId"
28            :label="item.roleList"
29            :value="item.roleId">
30          </el-option>
31        </el-select>
32  </el-form-item>
33  <el-form-item>
34  <el-button style="margin-left: 10%" type="primary" @click="addclick('detailData')">添
    加</el-button>
35  </el-form-item>
36  </el-form>
37  </div>
38  </template>
    <!-- 此处省略部分代码 -->
39  <script>
40  export default {
41      data() {
42          return {
43              detailData: {
44                  loginName: '',
45                  password: '',
46                  userName: '',
47                  tel: '',
48                  roleName: ''
49              },
50              essayRules: {
51                  loginName: [{ required: true, message: '姓名不能为空', trigger: 'blur' }],
52                  password: [{ required: true, message: '密码不能为空', trigger: 'blur' }],
53                  userName: [{ required: true, message: '用户名不能为空', trigger: 'blur' }],
54                  tel: [{ required: true, message: '手机号不能为空', trigger: 'blur' }],
55                  roleName: [{ required: true, message: '请选择角色', trigger: 'blur' }]
56              }
57          }
58      },
59      methods: {
60          addclick(param) {
61              this.$refs[param].validate((valid) => {
62                  if (valid) {
63                      this.$message({
64                          type: 'success',
65                          message: '添加成功'
66                      })
67                  } else {
68                      this.$message({
69                          type: 'error',
70                          message: '添加失败'
71                      })
```

```
72                }
73            })
74        }
75    }
76 }
77 </script>
78 <!-- 此处省略部分代码 -->
```

文件 17.20　edit_user.vue

```
01 <template>
02 <div class="app-container">
03 <el-form
04 ref="detailData"
05 style="margin-left: 10%"
06 :model="detailData"
07 label-width="80px"
08 label-position="right"
09 :rules="essayRules"
10 >
11 <el-form-item label="登录账号" style="width: 300px" prop="loginName">
12 <el-input v-model="detailData.loginName" placeholder="请输入登录账号"/>
13 </el-form-item>
14 <el-form-item label="密码:" style="width: 300px" prop="password">
15 <el-input v-model="detailData.password" placeholder="请输入密码"/>
16 </el-form-item>
17 <el-form-item label="姓名:" style="width: 300px" prop="userName">
18 <el-input v-model="detailData.userName" placeholder="请输入姓名"/>
19 </el-form-item>
20 <el-form-item label="手机号:" style="width: 300px" prop="tel">
21 <el-input v-model="detailData.tel" placeholder="请输入手机号"/>
22 </el-form-item>
23 <el-form-item label="角色:" style="width: 300px" prop="roleName">
24 <el-select v-model="detailData.roleName" placeholder="请选择角色">
25 <el-option
26 v-for="item in options"
27 :key="item.roleId"
28 :label="item.roleList"
29 :value="item.roleId">
30 </el-option>
31 </el-select>
32 </el-form-item>
33 <el-form-item>
34 <el-button style="margin-left: 10%" type="primary" @click="addclick('detailData')">确
   定修改</el-button>
35 </el-form-item>
36 </el-form>
37 </div>
38 </template>
   <!-- 此处省略部分代码 -->
39 <script>
```

```
40  export default {
41  data() {
42      return {
43          detailData: {
44              loginName: '',
45              password: '',
46              userName: '',
47              tel: '',
48              roleName: ''
49          },
50          essayRules: {
51              loginName: [{ required: true, message: '姓名不能为空', trigger: 'blur' }],
52              password: [{ required: true, message: '密码不能为空', trigger: 'blur' }],
53              userName: [{ required: true, message: '用户名不能为空', trigger: 'blur' }],
54              tel: [{ required: true, message: '手机号不能为空', trigger: 'blur' }],
55              roleName: [{ required: true, message: '请选择角色', trigger: 'blur' }]
56          }
57      }
58  },
59  methods: {
60      addclick(param) {
61          this.$refs[param].validate((valid) => {
62              if (valid) {
63                  this.$message({
64                      type: 'success',
65                      message: '修改成功'
66                  })
67              } else {
68                  this.$message({
69                      type: 'error',
70                      message: '修改失败'
71                  })
72              }
73          })
74      }
75  }
76  }
77  <!-- 此处省略部分代码 -->
78  </script>
```

步骤04 启动项目，测试登录和用户管理功能。在浏览器中访问地址 http://localhost:8080/news_publish/login，进入登录页面输入账号和密码登录系统，如图 17.3 所示。登录时，存在 3 种情况：

- 第 1 种情况是，如果账号或密码错误，则不允许登录，如图 17.4 所示。
- 第 2 种情况是，如果账号未启用或被禁用，则不允许登录，如图 17.5 所示。

图 17.3　系统后台登录页面 1

图 17.4　系统后台登录页面 2

图 17.5　系统后台登录页面 3

- 第 3 种情况是，若账号和密码正确，则允许登录系统后台，登录后根据用户角色的不同，后面操作页面和功能菜单选项会有所不同。如图 17.6 所示为管理员后台页面，如图 17.7 所示为信息员后台页面。

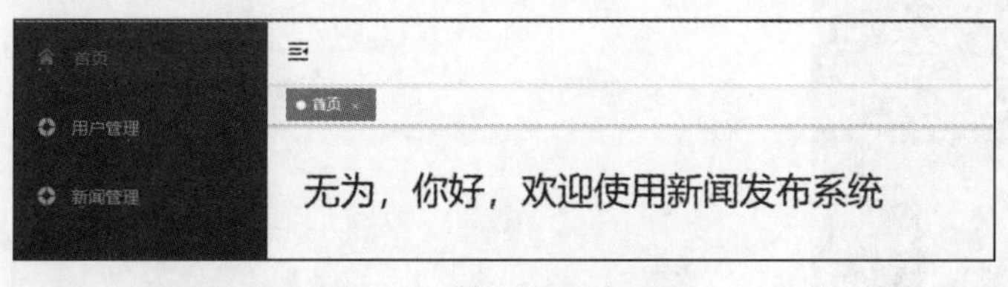

图 17.6　管理员后台页面

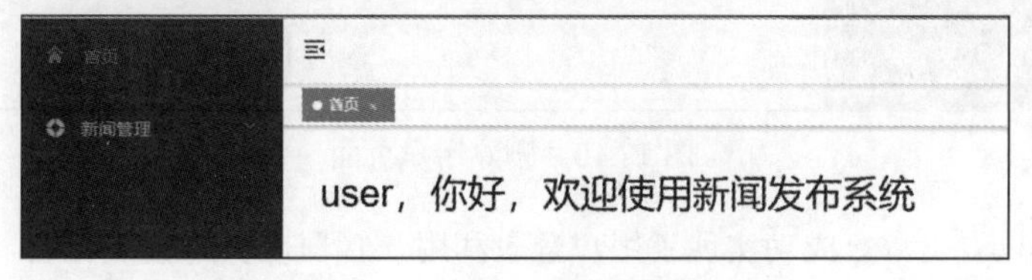

图 17.7　信息员后台页面

管理员登录系统后台页面后，可以对系统用户进行相关操作。在如图 17.8~图 17.10 所示的页面中，管理员可以进行查询用户、删除用户、启用/禁用用户、修改用户和添加用户操作。为了保证管理员 admin 不被误删除和禁用，页面代码中进行了判断和限制操作，即只可以进行密码修改等操作。

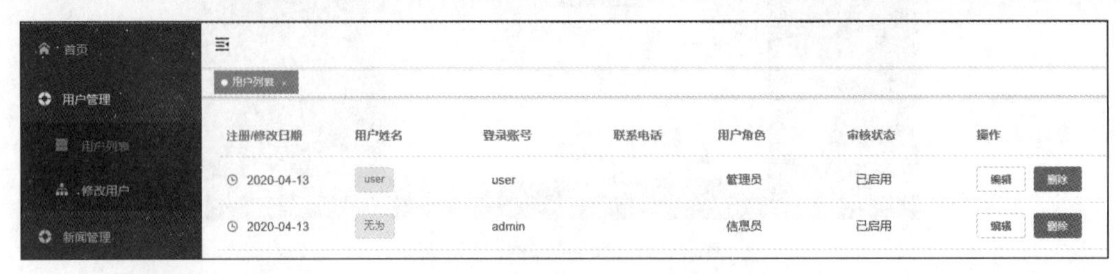

图 17.8　查询用户页面

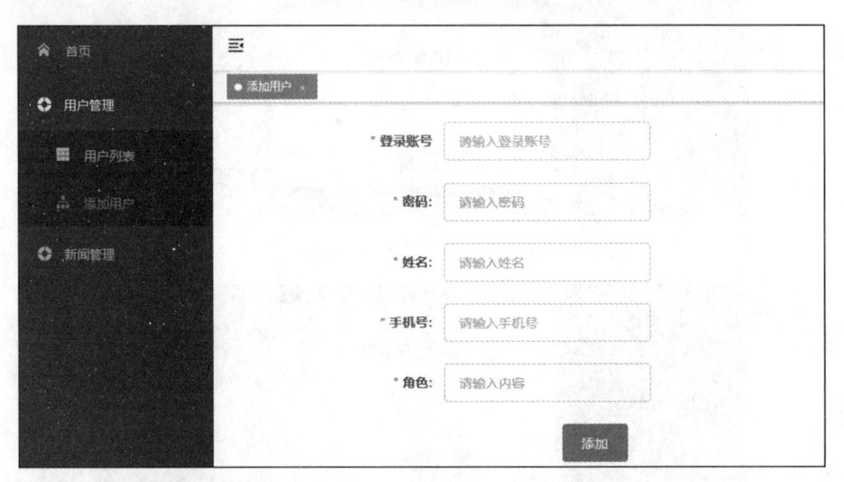

图 17.9　添加用户页面

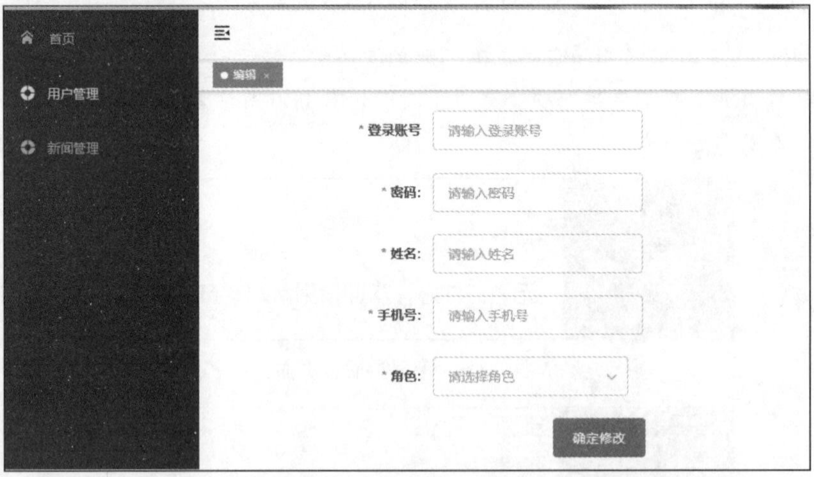

图 17.10　删除用户页面

从以上测试可以看出，已成功实现了用户登录和用户管理模块的相关功能。

17.6　新闻管理模块

新闻管理模块涉及新闻发布、修改、查询、删除、撤稿等功能。本节将详细讲解这些功能的具体实现。

17.6.1　创建持久化类

用户管理模块持久化类有新闻类别类 Category 和新闻类 News，具体代码如文件 17.21 和文件 17.22 所示。

文件 17.21　Category.java

```
01  package com.ssm.po;
02  //新闻类别实体类
03  public class Category {
04      private Integer categoryId;    //新闻类别 id
05      private String categoryName;   //新闻类别名称
06      public Integer getCategoryId() {
07          return categoryId;
08      }
09      public void setCategoryId(Integer categoryId) {
10          this.categoryId = categoryId;
11      }
12      public String getCategoryName() {
13          return categoryName;
14      }
15      public void setCategoryName(String categoryName) {
16          this.categoryName = categoryName;
17      }
18      @Override
19      public String toString() {
20          return "Category [categoryId=" + categoryId + ", categoryName=" + categoryName + "]";
21      }
22  }
```

文件 17.22　News.java

```
01  package com.ssm.po;
02  import java.util.Date;
03  //新闻实体类
04  public class News {
05      private Integer newsId;              //新闻 id
06      private String title;                //新闻标题（用于新闻列表页）
07      private String contentTitle;         //新闻内容页标题（用于新闻内容页）
08      private String content;              //新闻内容
```

```
09      private String contentAbstract;   //内容摘要
10      private String keywords;          //新闻关键词
11      private String author;            //作者或来源
12      private Date publishTime;         //发布时间
13      private Integer clicks;           //点击率（量）
14      private String publishStatus;     //发布状态（'1'：发布；'2'：撤稿）
15      private Integer categoryId;       //新闻类别 id
16      private String categoryName;      //新闻类别名称（为了方便新闻列表页显示，特添加此属性）
17      private Integer userId;
18      public Integer getNewsId() {
19          return newsId;
20      }
21      public void setNewsId(Integer newsId) {
22          this.newsId = newsId;
23      }
24      public String getTitle() {
25          return title;
26      }
27      public void setTitle(String title) {
28          this.title = title;
29      }
30      public String getContentTitle() {
31          return contentTitle;
32      }
33      public void setContentTitle(String contentTitle) {
34          this.contentTitle = contentTitle;
35      }
36      public String getContent() {
37          return content;
38      }
39      public void setContent(String content) {
40          this.content = content;
41      }
42      public String getContentAbstract() {
43          return contentAbstract;
44      }
45      public void setContentAbstract(String contentAbstract) {
46          this.contentAbstract = contentAbstract;
47      }
48      public String getKeywords() {
49          return keywords;
50      }
51      public void setKeywords(String keywords) {
52          this.keywords = keywords;
53      }
54      public String getAuthor() {
```

```
55            return author;
56        }
57        public void setAuthor(String author) {
58            this.author = author;
59        }
60        public Date getPublishTime() {
61            return publishTime;
62        }
63        public void setPublishTime(Date publishTime) {
64            this.publishTime = publishTime;
65        }
66        public Integer getClicks() {
67            return clicks;
68        }
69        public void setClicks(Integer clicks) {
70            this.clicks = clicks;
71        }
72        public String getPublishStatus() {
73            return publishStatus;
74        }
75        public void setPublishStatus(String publishStatus) {
76            this.publishStatus = publishStatus;
77        }
78        public Integer getCategoryId() {
79            return categoryId;
80        }
81        public void setCategoryId(Integer categoryId) {
82            this.categoryId = categoryId;
83        }
84        public String getCategoryName() {
85            return categoryName;
86        }
87        public void setCategoryName(String categoryName) {
88            this.categoryName = categoryName;
89        }
90        public Integer getUserId() {
91            return userId;
92        }
93        public void setUserId(Integer userId) {
94            this.userId = userId;
95        }
96        @Override
97        public String toString() {
98            return "News [newsId=" + newsId + ", title="+title+",contentTitle="+contentTitle
99             + ", content=" + content + ", contentAbstract=" + contentAbstract + ", keywords="
100            + keywords + ", author=" + author+ ", publishTime=" + publishTime +",clicks=" +
```

```
101             clicks + ", publishStatus=" + publishStatus+ ", categoryId=" + categoryId +
102             ", categoryName=" + categoryName + ", userId=" + userId + "]";
103     }
104 }
```

后台新闻管理列表需要进行分页显示，在项目 src 目录下创建包 com.ssm.utils，并定义分页类 PageBean，如文件 17.23 所示。

文件 17.23 PageBean.java

```
01  package com.ssm.utils;
02  import java.util.List;
03  public class PageBean<T> {
04      private int currentPage;        //当前页码
05      private int pageSize;           //每页显示新闻条数
06      private int count;              //新闻数量
07      private int totalPage;          //总页数
08      private List<T> list;           //当前页的新闻集合列表
09      public int getCurrentPage() {
10          return currentPage;
11      }
12      public void setCurrentPage(int currentPage) {
13          this.currentPage = currentPage;
14      }
15      public int getPageSize() {
16          return pageSize;
17      }
18      public void setPageSize(int pageSize) {
19          this.pageSize = pageSize;
20      }
21      public int getCount() {
22          return count;
23      }
24      public void setCount(int count) {
25          this.count = count;
26      }
27      public int getTotalPage() {
28          return totalPage;
29      }
30      public void setTotalPage(int totalPage) {
31          this.totalPage = totalPage;
32      }
33      public List<T> getList() {
34          return list;
35      }
36      public void setList(List<T> list) {
37          this.list = list;
38      }
```

```
39  }
```

17.6.2　实现 DAO 层接口

实现 DAO 层接口的操作步骤如下：

步骤 01 创建 DAO 层接口。在 src 目录下的 com.ssm.dao 包中创建一个类别接口 CategoryDao 和一个新闻接口 NewsDao，并在接口中编写增、删、改、查等方法，如文件 17.24 和文件 17.25 所示。

文件 17.24　CategoryDao.java

```
01  package com.ssm.dao;
02  import java.util.List;
03  import com.ssm.po.Category;
04  //新闻类别 DAO 层接口
05  public interface CategoryDao {
06      //查询所有新闻类别
07      public List<Category> selectCategoryList();
08      //根据新闻类别 id 查询新闻类别
09      public Category getCategoryById(Integer categoryId);
10  }
```

文件 17.25　NewsDao.java

```
01  package com.ssm.dao;
02  import java.util.List;
03  import org.apache.ibatis.annotations.Param;
04  import com.ssm.po.News;
05  //新闻类别 DAO 层接口
06  public interface NewsDao {
07      //获取当前类别新闻数量
08      int getNewsCount(@Param("keywords") String keywords,
09                @Param("newsListCategoryId") Integer newsListCategoryId);
10      //获取当前类别新闻列表
11      List<News> findNewsList(@Param("keywords") String keywords,
12                    @Param("newsListCategoryId") Integer newsListCategoryId,
13                    @Param("startRows") Integer startRows,
14                    @Param("pageSize") Integer pageSize);
15      //根据新闻 id 获取新闻
16      News getNewsByNewsId(Integer newsId);
17      //添加新闻
18      int addNews(News news);
19      //更新新闻
20      int updateNews(News news);
21      //设置新闻状态（'1'：发布；'2'：撤稿）
22      int setNewsPublishStatus(News news);
23      //删除新闻
```

```
24      int delNews(Integer newsId);
25  }
```

步骤 02 创建映射文件。在 src 目录下的 com.ssm.dao 包中创建 MyBatis 映射文件 CategoryDao.xml 和 NewsDao.xml，并在映射文件中编写增、删、改、查等方法的执行语句，如文件 17.26 和文件 17.27 所示。

文件 17.26 CategoryDao.xml

```
01  <?xml version="1.0" encoding="UTF-8"?>
02  <!DOCTYPE mapper PUBLIC "-//mybatis.org//DTD Mapper 3.0//EN"
03      "http://mybatis.org/dtd/mybatis-3-mapper.dtd">
04  <mapper namespace="com.ssm.dao.CategoryDao">
05      <!--查询新闻类别集合列表 -->
06      <select id="selectCategoryList" resultType="Category">
07          select * from t_category
08      </select>
09      <!--通过 categoryId 查询新闻类别 -->
10      <select id="getCategoryById" parameterType="Integer" resultType="Category">
11          select * from t_category where categoryId=#{categoryId}
12      </select>
13  </mapper>
```

文件 17.27 NewsDao.xml

```
01  <?xml version="1.0" encoding="UTF-8"?>
02  <!DOCTYPE mapper PUBLIC "-//mybatis.org//DTD Mapper 3.0//EN"
03      "http://mybatis.org/dtd/mybatis-3-mapper.dtd">
04  <mapper namespace="com.ssm.dao.NewsDao">
05  <!--查询新闻集合的 Where 语句 -->
06      <sql id="selectNewsListWhere">
07        <where>
08           n.categoryId=c.categoryId
09          <if test="keywords!=null and keywords!=''" >
10             and (n.title like CONCAT('%',#{keywords},'%') or
11                  n.keywords like CONCAT('%',#{keywords},'%'))
12          </if>
13          <if test="newsListCategoryId!=null and newsListCategoryId!=''" >
14             and (n.categoryId=#{newsListCategoryId})
15          </if>
16        </where>
17      </sql>
18      <!--查询新闻集合 -->
19      <select id="findNewsList" parameterType="String" resultType="News">
20         select n.*,c.categoryName from t_news as n,t_category as c
21          <include refid="selectNewsListWhere" />
22          order by publishTime desc
23          limit #{startRows},#{pageSize}
```

```
24      </select>
25      <sql id="getNewsCountWhere">
26        <where>
27          <if test="keywords!=null and keywords!=''" >
28            and (n.title like CONCAT('%',#{keywords},'%') or
29              n.keywords like CONCAT('%',#{keywords},'%'))
30          </if>
31          <if test="newsListCategoryId!=null and newsListCategoryId!=''" >
32            and (n.categoryId=#{newsListCategoryId})
33          </if>
34        </where>
35      </sql>
36      <!--查询新闻数量 -->
37      <select id="getNewsCount" parameterType="String" resultType="Integer">
38         select count(*) from t_news as n
39          <include refid="getNewsCountWhere" />
40      </select>
41      <!--通过 newsId 查询新闻 -->
42      <select id="getNewsByNewsId" parameterType="Integer" resultType="News">
43          select *,categoryName from t_news as n,t_category as c
44              where newsId=#{newsId} and n.categoryId=c.categoryId
45      </select>
46      <!--添加新闻 -->
47      <insert id="addNews" parameterType="News">
48          insert into t_news(
49              title,
50              contentTitle,
51              content,
52              contentAbstract,
53              keywords,
54              author,
55              publishTime,
56              publishStatus,
57              categoryId
58          )
59          values(
60              #{title},
61              #{contentTitle},
62              #{content},
63              #{contentAbstract},
64              #{keywords},
65              #{author},
66              #{publishTime},
67              #{publishStatus},
68              #{categoryId}
69          )
```

```
70      </insert>
71      <!-- 更新新闻 -->
72      <update id="editNews" parameterType="News">
73          update t_news
74          <set>
75              publishTime=#{publishTime},
76              publishStatus=#{publishStatus},
77              title=#{title},
78              contentTitle=#{contentTitle},
79              content=#{content},
80              contentAbstract=#{contentAbstract},
81              keywords=#{keywords},
82              author=#{author},
83              categoryId=#{categoryId}
84          </set>
85          where newsId=#{newsId}
86      </update>
87      <!--设置新闻发布状态（status '1':已发布；'2':被撤稿）-->
88      <update id="setNewsPublishStatus" parameterType="News">
89          update t_news set publishStatus=#{publishStatus} where newsId=#{newsId}
90      </update>
91      <!--删除新闻 -->
92      <delete id="delNews" parameterType="Integer">
93          delete from t_news where newsId=#{newsId}
94      </delete>
95  </mapper>
```

文件 17.27 中的代码通过映射查询、增加、修改和删除等语句来实现新闻表中的相关操作。

17.6.3 实现 Service 层接口

实现 Service 层接口的操作步骤如下：

步骤 01 创建新闻类别和新闻的 Service 层接口。在 src 目录下创建一个 com.ssm.service 包，在包中创建 CategoryService 接口和 NewsService 接口，并在该接口中编写操作新闻的相关方法，如文件 17.28 和文件 17.29 所示。

文件 17.28 CategoryService.java

```
01  package com.ssm.service;
02  import java.util.List;
03  import com.ssm.po.Category;
04  // 新闻类别 Service 层接口
05  public interface CategoryService {
06      public List<Category> findCategoryList();
07      public Category findCategoryById(Integer categoryId);
08  }
```

文件 17.29　NewsService.java

```
01  package com.ssm.service;
02  import com.ssm.po.News;
03  import com.ssm.utils.PageBean;
04  // 新闻 Service 层接口
05  public interface NewsService {
06      PageBean<News> findNewsByPage(String keywords, Integer newsListCategoryId,
07                              Integer currentPage, Integer pageSize);
08      News getNewsByNewsId(Integer newsId);
09      int setNewsPublishStatus(News news);
10      int addNews(News news);
11      int editNews(News news);
12      int delNews(Integer newsId);
13  }
```

步骤 02 创建新闻类别和新闻 Service 层接口的实现类。在 src 目录下创建一个 com.ssm.service.impl 包，并在包中创建 CategoryService 接口的实现类 CategoryServiceImpl 和 NewsService 接口的实现类 NewsServiceImpl，在类中编辑并实现接口中的方法，如文件 17.30 和文件 17.31 所示。

文件 17.30　CategoryServiceImpl.java

```
01  package com.ssm.service.impl;
02  import java.util.List;
03  import org.springframework.beans.factory.annotation.Autowired;
04  import org.springframework.stereotype.Service;
05  import com.ssm.dao.CategoryDao;
06  import com.ssm.po.Category;
07  import com.ssm.service.CategoryService;
08  // 新闻类别 Service 接口实现类
09  @Service("categoryService")
10  public class CategoryServiceImpl implements CategoryService {
11      //注入 RoleDao
12      @Autowired
13      private CategoryDao categoryDao;
14      @Override
15      public List<Category> findCategoryList() {
16          return this.categoryDao.selectCategoryList();
17      }
18      @Override
19      public Category findCategoryById(Integer categoryId) {
20          return this.categoryDao.getCategoryById(categoryId);
21      }
22  }
```

文件 17.31　NewsServiceImpl.java

```
01  package com.ssm.service.impl;
```

```
02  import java.util.List;
03  import org.springframework.beans.factory.annotation.Autowired;
04  import org.springframework.stereotype.Service;
05  import com.ssm.dao.NewsDao;
06  import com.ssm.po.News;
07  import com.ssm.service.NewsService;
08  import com.ssm.utils.PageBean;
09  // 新闻类别 Service 接口实现类
10  @Service("newsService")
11  public class NewsServiceImpl implements NewsService {
12      //注入 NewsDao
13      @Autowired
14      private NewsDao newsDao;
15      @Override
16      public PageBean<News> findNewsByPage(String keywords, Integer newsListCategoryId,
17                                     Integer currentPage,Integer pageSize) {
18          //获取当前类别信息数量
19          int count=newsDao.getNewsCount(keywords,newsListCategoryId);
20          //求总页数
21          int totalPage = (int) Math.ceil(count*1.0/pageSize);
22          //获取新闻列表
23          List<News> newsList= newsDao.findNewsList(keywords,
24    newsListCategoryId,(currentPage-1)*pageSize, pageSize);
25          //创建 PageBean 实例 pb
26          PageBean<News> pb = new PageBean<>();
27          //对 PageBean 实例 pb 设置值
28          pb.setCount(count);
29          if(currentPage==0) currentPage=1;
30          pb.setCurrentPage(currentPage);
31          pb.setList(newsList);
32          pb.setPageSize(pageSize);
33          pb.setTotalPage(totalPage);
34          return pb;
35      }
36      @Override
37      public News getNewsByNewsId(Integer newsId) {
38          return newsDao.getNewsByNewsId(newsId);
39      }
40      @Override
41      public int setNewsPublishStatus(News news) {
42          return newsDao.setNewsPublishStatus(news);
43      }
44      @Override
45      public int addNews(News news) {
46          return newsDao.addNews(news);
47      }
```

```
48      @Override
49      public int editNews(News news) {
50          return newsDao.updateNews(news);
51      }
52      @Override
53      public int delNews(Integer newsId) {
54          return newsDao.delNews(newsId);
55      }
56  }
```

17.6.4　实现 Controller 类

在 src 目录下创建一个 com.ssm.web.controller 包，在包中创建新闻控制器类 NewsController，代码如文件 17.32 所示。

文件 17.32　NewsController.java

```
01  package com.ssm.web.controller;
02  import org.springframework.stereotype.Controller;
03  import org.springframework.ui.Model;
04  //此处省略其他导入包语句
05  import com.ssm.utils.PageBean;
06  // 新闻控制类
07  @Controller
08  public class NewsController {
09      //依赖注入
10      @Autowired
11      private NewsService newsService;
12      @Autowired
13      private CategoryService categoryService;
14      //查询新闻分页
15      @RequestMapping(value="/findNewsByPage.action")
16      public String findNewsByPage(String keywords,Integer newsListCategoryId,
17                  @RequestParam(defaultValue="1")Integer currentPage,
18                  @RequestParam(defaultValue="10")Integer pageSize,Model model){
19          // 获取类别列表
20          List<Category> categoryList = categoryService.findCategoryList();
21          model.addAttribute("categoryList", categoryList);
22          // 获取新闻 PageBean 实例
23          PageBean<News> pb=
24  newsService.findNewsByPage(keywords,newsListCategoryId,currentPage,pageSize);
25          //向 model 添加对象属性，用于页面显示
26          model.addAttribute("pb", pb);
27          model.addAttribute("keywords", keywords);
28          model.addAttribute("newsListCategoryId", newsListCategoryId);
29          model.addAttribute("currentPage", currentPage);
```

```
30          model.addAttribute("pageSize", pageSize);
31          return "news/news_list";
32      }
33      //设置新闻的状态（publishStatus：'1'：发布；'2'：撤稿）
34      @RequestMapping(value = "/setNewsPublishStatus.action")
35      @ResponseBody
36      public News setNewsPublishStatus(@RequestBody News news, Model model) {
37          int rows = newsService.setNewsPublishStatus(news);
38          if (rows>0) {
39              return news;
40          }else{
41              news.setNewsId(0);
42              return news;
43          }
44      }
45      //跳转至添加新闻页面
46      @RequestMapping(value = "/toAddNews.action")
47      public String toAddNews(Model model) {
48          // 获取新闻类别列表
49          List<Category> categoryList = categoryService.findCategoryList();
50          model.addAttribute("categoryList", categoryList);
51          return "news/add_news";
52      }
53      // 添加新闻
54      @RequestMapping(value = "/addNews.action", method = RequestMethod.POST)
55      public String addNews(News news, Model model) {
56          Date date = new Date();
57          news.setPublishTime(date);
58          news.setPublishStatus("1");// 默认设置新闻为已发布状态
59          int rows = newsService.addNews(news);
60          // 添加成功，跳转至新闻列表页面
61          if (rows > 0) {
62              return "redirect:findNewsByPage.action";
63          } else {
64              // 获取新闻类别列表
65              List<Category> categoryList = categoryService.findCategoryList();
66              model.addAttribute("categoryList", categoryList);
67              model.addAttribute("news", news);
68              // 添加失败，跳转至添加新闻页面
69              return "news/add_news";
70          }
71      }
72      // 修改新闻
73      @RequestMapping(value = "/editNews.action", method = RequestMethod.POST)
74      public String editNews(News news, Model model) {
75          Date date = new Date();
```

```
76          news.setPublishTime(date);
77          news.setPublishStatus("1");// 默认设置新闻为已发布状态
78          int rows = newsService.editNews(news);
79          // 添加成功，转向新闻列表页面
80          if (rows > 0) {
81              return "redirect:findNewsByPage.action";
82          } else {
83              // 获取新闻类别列表
84              List<Category> categoryList = categoryService.findCategoryList();
85              model.addAttribute("categoryList", categoryList);
86              model.addAttribute("news", news);
87              // 添加失败，转回添加新闻页面
88              return "news/edit_news";
89          }
90      }
91      // 跳转至修改新闻页面
92      @RequestMapping(value = "/toEditNews.action")
93      public String toEditNews(Integer newsId, Model model) {
94          News news = newsService.getNewsByNewsId(newsId);
95          if (news != null) {
96              model.addAttribute("news", news);
97              // 获取新闻类别列表
98              List<Category> categoryList = categoryService.findCategoryList();
99              model.addAttribute("categoryList", categoryList);
100         }
101         return "news/edit_news";
102     }
103     //删除新闻
104     @RequestMapping(value = "/delNews.action")
105     @ResponseBody
106     public News delNews(@RequestBody News news, Model model) {
107         int rows = newsService.delNews(news.getNewsId());
108         if (rows>0) {
109             return news;
110         }else{
111             news.setNewsId(0);
112             return news;
113         }
114     }
115     //根据新闻类别 id 查询新闻分页（用于前台首页）
116     @RequestMapping(value = "/index.action")
117     public String index(HttpServletRequest request, HttpServletResponse response,
118                 String keywords, Integer newsListCategoryId,
119                 @RequestParam(defaultValue = "1") Integer currentPage,
120                  @RequestParam(defaultValue = "10") Integer pageSize,
121                         Model model) throws ServletException, IOException {
```

```
122         // 获取新闻 PageBean 实例（参数新闻类别"今日头条"id 为 1）
123         PageBean<News> pb1 =
124                 newsService.findNewsByPage(keywords, 1, currentPage, pageSize);
125         model.addAttribute("pb1", pb1);
126         // 获取新闻 PageBean 实例（参数新闻类别"综合资讯"id 为 2）
127         PageBean<News> pb2 =
128                 newsService.findNewsByPage(keywords, 2, currentPage, pageSize);
129         model.addAttribute("pb2", pb2);
130         return "../../first";
131     }
132     //根据新闻类别 id 查询新闻分页（用于前台新闻列表页）
133     @RequestMapping(value = "/findNewsByCategoryIdPage.action")
134     public String findNewsByCategoryIdPage(HttpServletRequest request,
135             HttpServletResponse response, String keywords,
136             Integer newsListCategoryId,
137             @RequestParam(defaultValue = "1") Integer currentPage,
138             @RequestParam(defaultValue = "1") Integer pageSize,   Model model){
139         // 获取类别列表
140         Category category = categoryService.findCategoryById(newsListCategoryId);
141         model.addAttribute("category", category);
142         // 获取新闻 PageBean 实例
143         PageBean<News> pb =
144     newsService.findNewsByPage(keywords, newsListCategoryId, currentPage, pageSize);
145         model.addAttribute("pb", pb);
146         model.addAttribute("newsListCategoryId", newsListCategoryId);
147         model.addAttribute("currentPage", currentPage);
148         model.addAttribute("pageSize", pageSize);
149         return "../../list";
150     }
151     //查询新闻（用于前台新闻内容页）
152     @RequestMapping(value = "/findFrontNewsByNewsId.action")
153     public String findFrontNewsByNewsId(Integer newsId,Model model) {
154         News news = newsService.getNewsByNewsId(newsId);
155         if (news != null) {
156             model.addAttribute("news", news);
157         }
158         return "../../detail";
159     }
160 }
```

文件 17.32 中，首先通过@Autowired 注解将 CategoryService 对象和 NewsService 对象注入本类中，然后创建对新闻进行操作的方法。

注意：NewsController 类与 UserController 类一样，在一些方法中涉及业务逻辑，请读者仔细阅读，并将 NewsController 类与用户管理相关页面联系起来推敲。NewsController 类与 news_list.vue 页面在交互过程中采用 PageBean 对新闻进行了封装，并实现了分页功能。

17.6.5　实现页面功能

实现页面功能的操作步骤如下：

步骤 01 在 src/views 目录下创建 news 文件夹，并在该文件夹下创建新闻列表页面 news_list.vue、添加新闻页面 add_news.vue 和更新新闻页面 edit_news.vue，具体实现代码如文件 17.33~文件 17.35 所示。

文件 17.33　news_list.jsp

```
01  <template>
02  <div class="app-container">
03  <el-table
04  :data="tableData"
05  style="width: 100%"
06  >
07  <el-table-column
08  label="发布/更新时间"
09  width="150"
10  >
11  <template slot-scope="scope">
12  <i class="el-icon-time" />
13  <span style="margin-left: 10px">{{ scope.row.publishTime }}</span>
14  </template>
15  </el-table-column>
16  <el-table-column
17  label="新闻标题"
18  width="250"
19  >
20  <template slot-scope="scope">
21  <el-popover trigger="hover" placement="top">
22  <p>标题: {{ scope.row.title }}</p>
23  <div slot="reference" class="name-wrapper">
24  <el-tag size="medium">{{ scope.row.title }}</el-tag>
25  </div>
26  </el-popover>
27  </template>
28  </el-table-column>
29  <el-table-column
30  label="新闻类别"
31  width="150"
32  >
33  <template slot-scope="scope">
34  <span style="margin-left: 10px">{{ scope.row.categoryName }}</span>
35  </template>
36  </el-table-column>
```

```
37 <el-table-column
38 label="发布状态"
39 width="150"
40 >
41 <template slot-scope="scope">
42 <span style="margin-left: 10px">{{ scope.row.publishStatus }}</span>
43 </template>
44 </el-table-column>
45 <el-table-column label="操作">
46 <template slot-scope="scope">
47 <el-button
48 size="mini"
49 @click="handleEdit(scope.$index, scope.row)"
50 >编辑</el-button>
51 <el-button
52 size="mini"
53 type="danger"
54 @click="handleDelete(scope.$index, scope.row)"
55 >删除</el-button>
56 </template>
57 </el-table-column>
58 </el-table>
59 </div>
60 </template>
61 <!-- 此处省略部分代码 -->
62 <script>
63 export default {
64 filters: {
65 statusFilter(status) {
66 const statusMap = {
67 published: 'success',
68 draft: 'gray',
69 deleted: 'danger'
70 }
71 return statusMap[status]
72 }
73 },
74 data() {
75 return {
76 tableData: [{
77 categoryName: '',
78 publishTime: '',
79 title: '',
80 publishStatus: ''
81 }]
82 }
```

```
83  <!-- 此处省略部分代码 -->
84  </script>
```

新闻列表页面 news_list.vue 主要用来展示新闻列表，通过访问 NewsController 类的 findNewsByPage()可以跳转到 news_list.vue 页面，并在页面中迭代显示新闻列表信息，在此基础上还可以通过链接对用户进行修改、删除、发布/撤稿等相关操作。

文件 17.34　add_news.vue

```
01  <template>
02  <div class="app-container">
03  <el-form
04  ref="detailData"
05  style="margin-left: 10%"
06  :model="detailData"
07  label-width="130px"
08  label-position="right"
09  :rules="essayRules"
10  >
11  <el-form-item label="新闻标题:" style="width: 500px" prop="title">
12  <el-input v-model="detailData.title" placeholder="请输入标题" />
13  </el-form-item>
14  <el-form-item label="新闻类别:" style="width: 300px" prop="categoryId">
15  <el-select v-model="categoryId" placeholder="请选择">
16  <el-option
17  v-for="item in options"
18  :key="item.c.categoryId"
19  :label="item.categoryList"
20  :value="item.c.categoryId">
21  </el-option>
22  </el-select>
23  </el-form-item>
24  <el-form-item label="新闻内容页标题:" style="width: 500px" prop="contentTitle">
25  <el-input  v-model="detailData.contentTitle"  placeholder=" 请 输 入 新 闻 内 容 页 标 题 "
    /></el-form-item>
26  <el-form-item label="内容摘要:" style="width: 800px" prop="contentAbstract">
27  <el-input
28  placeholder="请输入内容摘要"
29  v-model="detailData.contentAbstract"
30  type="textarea"
31  :rows="3"
32  />
33  </el-form-item>
34  <el-form-item label="关键词:" style="width: 300px" prop="keywords">
35  <el-input v-model="detailData.keywords" placeholder="请输入关键词" />
36  </el-form-item>
37  <el-form-item label="作者/来源:" style="width: 300px" prop="author">
```

```
38  <el-input v-model="detailData.author" placeholder="请输入作者/来源" />
39  </el-form-item>
40  <el-form-item label="内容:" style="width: 800px" prop="content" >
41  <el-input
42  placeholder="请输入内容"
43  v-model="detailData.content"
44  type="textarea"
45  :rows="10"
46  />
47  </el-form-item>
48  <el-form-item>
49  <el-button style="margin-left: 20%" type="primary" @click="addclick('detailData')">确
    定添加</el-button>
50  </el-form-item>
51  </el-form>
52  </div>
53  </template>
    <!-- 此处省略部分代码 -->
54  <script>
55  export default {
56  data() {
57  return {
58  detailData: {
59  title: '',
60  categoryId: '',
61  contentTitle: '',
62  contentAbstract: '',
63  content: '',
64  keywords: '',
65  author:'',
66  },
67  essayRules: {
68  title: [{ required: true, message: '新闻标题不能为空', trigger: 'blur' }],
69  categoryId: [{ required: true, message: '新闻类别不能为空', trigger: 'blur' }],
70  contentTitle: [{ required: true, message: '新闻内容页标题不能为空', trigger: 'blur' }],
71  content: [{ required: true, message: '内容不能为空', trigger: 'blur' }],
72  }
73  }
74  },
75  methods: {
76  addclick(param) {
77  this.$refs[param].validate((valid) => {
78  if (valid) {
79  this.$message({
80  type: 'success',
81  message: '添加成功'
```

```
82  })
83  } else {
84  this.$message({
85  type: 'error',
86  message: '添加失败'
87  })
88  }
89  })
90  }
91  }
92  }
93  </script>
94  <!-- 此处省略部分代码 -->
```

文件 17.35　edit_news.vue

```
01  <template>
02  <div class="app-container">
03  <el-form
04  ref="detailData"
05  style="margin-left: 10%"
06  :model="detailData"
07  label-width="130px"
08  label-position="right"
09  :rules="essayRules"
10  >
11  <el-form-item label="新闻标题:" style="width: 500px" prop="title">
12  <el-input v-model="detailData.title" placeholder="请输入标题" />
13  </el-form-item>
14  <el-form-item label="新闻类别:" style="width: 300px" prop="categoryId">
15  <el-select v-model="category.categoryId" placeholder="请选择">
16  <el-option
17  v-for="item in options"
18  :key="item.c.categoryId"
19  :label="item.categoryList"
20  :value="item.c.categoryId">
21  </el-option>
22  </el-select>
23  </el-form-item>
24  <el-form-item label="新闻内容页标题:" style="width: 500px" prop="contentTitle">
25  <el-input v-model="detailData.contentTitle" placeholder="请输入新闻内容页标题"
    /></el-form-item>
26  <el-form-item label="内容摘要:" style="width: 800px" prop="contentAbstract">
27  <el-input
28  placeholder="请输入内容摘要"
29  v-model="detailData.contentAbstract"
30  type="textarea"
```

```
31  :rows="3"
32  />
33  </el-form-item>
34  <el-form-item label="关键词:" style="width: 300px" prop="keywords">
35  <el-input v-model="detailData.keywords" placeholder="请输入关键词" />
36  </el-form-item>
37  <el-form-item label="作者/来源:" style="width: 300px" prop="author">
38  <el-input v-model="detailData.author" placeholder="请输入作者/来源" />
39  </el-form-item>
40  <el-form-item label="内容:" style="width: 800px" prop="content" >
41  <el-input
42  placeholder="请输入内容"
43  v-model="detailData.content"
44  type="textarea"
45  :rows="10"
46  />
47  </el-form-item>
48  <el-form-item>
49  <el-button style="margin-left: 20%" type="primary" @click="addclick('detailData')">确
    定修改</el-button>
50  </el-form-item>
51  </el-form>
52  </div>
53  </template>
54  <script>
55  export default {
56  data() {
57  return {
58  detailData: {
59  title: '',
60  options: [],
61  contentTitle: '',
62  contentAbstract: '',
63  content: '',
64  keywords: '',
65  author: '',
66  },
67  category: [{
68  categoryId: '',
69  categoryList: ''
70  }],
71  essayRules: {
72  title: [{ required: true, message: '新闻标题不能为空', trigger: 'blur' }],
73  categoryId: [{ required: true, message: '新闻类别不能为空', trigger: 'blur' }],
74  contentTitle: [{ required: true, message: '新闻内容页标题不能为空', trigger: 'blur' }],
75  content: [{ required: true, message: '内容不能为空', trigger: 'blur' }],
```

```
76  }
77  }
78  }
79  <!-- 此处省略部分代码 -->
```

步骤 02 启动项目，测试新闻管理功能，在浏览器中访问地址 http://localhost:8080/news_publish/login，进入登录页面，以信息员身份登录后进入主页面，如图 17.11 所示。

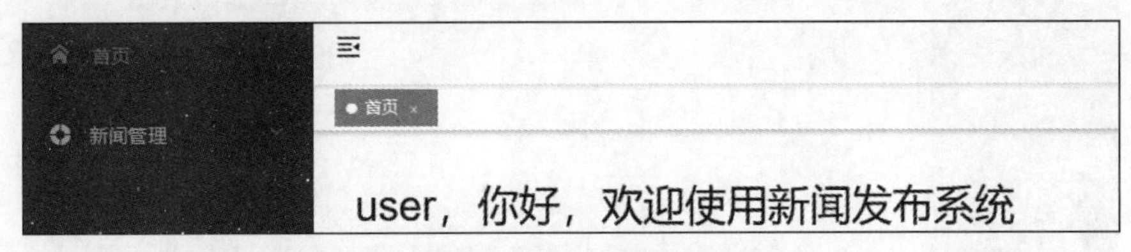

图 17.11　信息员后台页面

登录系统后台页面后，可以对新闻进行相关操作。在如图 17.12~图 17.14 所示的页面中，可以进行查询新闻、删除新闻、发布/撤稿新闻、修改新闻等操作。其中，查询新闻实现了分页效果。另外，在发布新闻和修改新闻页面中，新闻内容的输入使用的是多行文本框，在实际项目开发中可以用“在线编辑器”代替，以便使用富媒体的形式编辑和发布新闻内容。

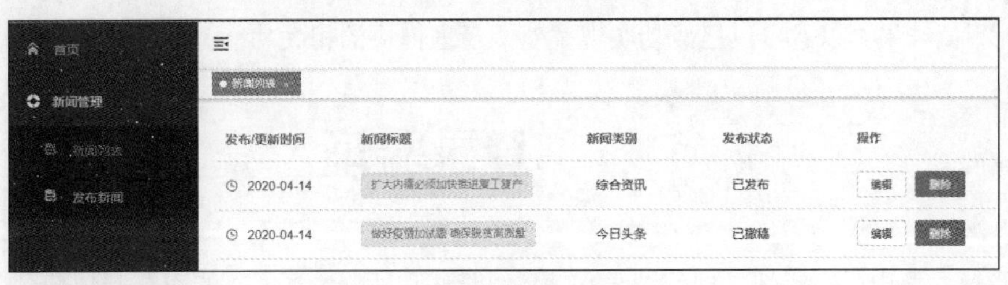

图 17.12　查询新闻页面

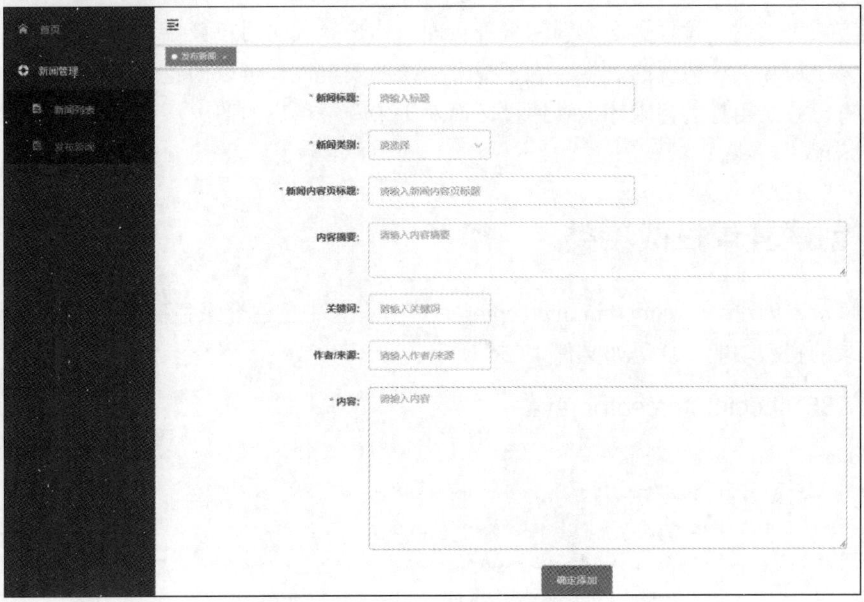

图 17.13　发布新闻页面

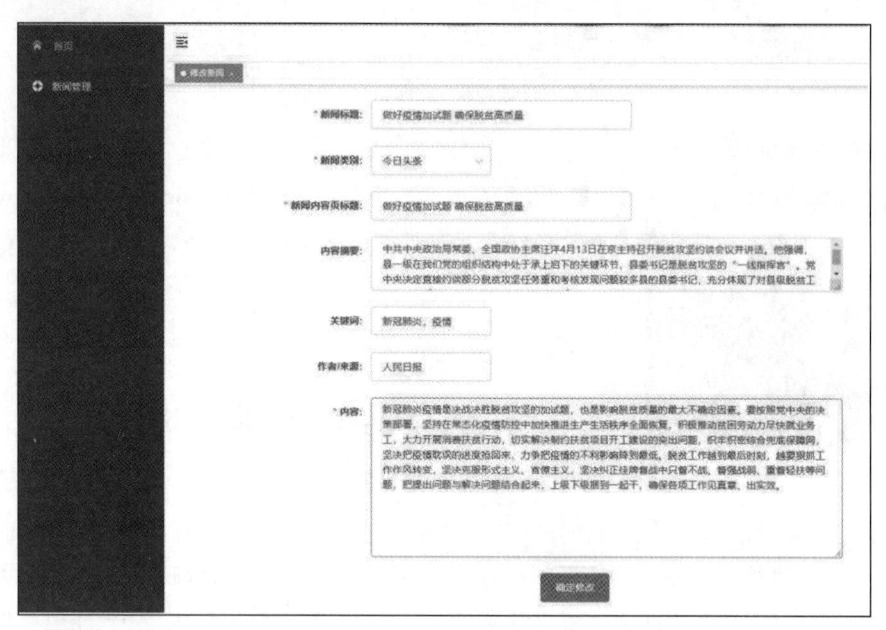

图 17.14　修改新闻页面

从以上测试结果可以看出，已成功实现了新闻管理模块的相关功能。

17.7　登录验证

虽然在 17.5 节中已经实现了用户登录功能，但是此功能并不完善。假设在其他控制器类中也包含一个访问用户管理页面或新闻管理页面的方法，那么用户完全可以绕过登录步骤，而直接通过访问该方法的方式进入用户管理页面或新闻管理页面。让未登录的用户直接访问后台管理页面是十分不安全的。为了避免这种情况的发生，为了提升系统的安全性，可以创建一个登录拦截器来拦截所有请求。只有已登录用户的请求才能够通过，而对于未登录用户的请求，系统会将请求转发到登录页面，并提示用户登录。

17.7.1　创建登录拦截器类

在 src 目录下创建一个 com.ssm.interceptor 包，并在包中创建登录拦截器类 LoginInterceptor 来实现用户登录的拦截功能，代码如文件 17.36 所示。

文件 17.36　LoginInterceptor.java

```
01  package com.ssm.interceptor;
02  import javax.servlet.http.HttpServletRequest;
03  import javax.servlet.http.HttpServletResponse;
04  import javax.servlet.http.HttpSession;
05  import org.springframework.web.servlet.HandlerInterceptor;
06  import org.springframework.web.servlet.ModelAndView;
```

```
07  import com.ssm.po.User;
08  public class LoginInterceptor implements HandlerInterceptor{
09      @Override
10      public boolean preHandle(HttpServletRequest request, HttpServletResponse response,
11                          Object handler) throws Exception {
12          //获取请求的 URL
13          String url=request.getRequestURI();
14          //除了登录请求外，其他的 URL 都进行拦截控制
15          if(url.indexOf("/login.action")>=0){
16              return true;
17          }
18          //获取 Session
19          HttpSession session=request.getSession();
20          User user=(User)session.getAttribute("login_user");
21          //判断 Session 中是否有用户数据
22          //如果有，就返回 true，表示用户已登录，继续向下执行
23          if(user!=null){
24              return true;
25          }
26          //不符合条件，给出提示信息，并转发到登录页面
27          request.setAttribute("msg", "您还没有登录，请先登录！");
28          request.getRequestDispatcher("/WEB-INF/jsp/skip.jsp").forward(request,
    response);
29          return false;
30      }
31      @Override
32      public void afterCompletion(HttpServletRequest arg0, HttpServletResponse arg1,
33                          Object arg2, Exception arg3) throws Exception { }
34      @Overrid
35      public void postHandle(HttpServletRequest arg0, HttpServletResponse arg1,
36                      Object arg2, ModelAndView arg3) throws Exception { }
37  }
```

在文件 17.36 的 preHandle()方法中，首先获取了用户 URL 请求，然后通过请求来判断是否为用户登录操作，只有用户登录的请求才不需要拦截。接下来获取了 Session 对象，并获取 Session 中的用户信息。如果 Session 中的用户信息不为空，那就表示用户已经登录，拦截器将放行；如果 Session 中的用户信息为空，那就表示用户未登录，系统会转发到登录页面，并提示用户登录。

17.7.2 配置拦截器

在 springmvc-config.xml 文件中配置登录拦截器信息，配置代码如下：

```
<!-- 配置拦截器 -->
<mvc:interceptors>
<!-- 使用 bean 直接定义在<mvc:interceptors>下面的 Interceptor 将拦截所有请求 -->
```

```
<mvc:interceptor>
    <!-- 配置拦截器作用的路径 -->
    <mvc:mapping path="/**" />
    <!-- 配置不需要拦截器作用的路径（前台页面需要访问控制器的路径） -->
    <mvc:exclude-mapping path="/index.action"/>
    <mvc:exclude-mapping path="/findNewsByCategoryIdPage.action"/>
    <mvc:exclude-mapping path="/findFrontNewsByNewsId.action"/>
    <bean class="com.ssm.interceptor.LoginInterceptor" />
</mvc:interceptor>
</mvc:interceptors>
```

上述配置代码会将所有的用户请求都交由登录拦截器来处理。当然，也可以通过<mvc:exclude-mapping path="">配置不需要拦截器作用的路径，例如<mvc:exclude-mapping path="/index.action"/>是前台首页需要访问控制器 NewsController 的路径。至此，登录拦截器的实现工作就已经完成。未登录的用户执行访问后台管理页面的方法，会被拦截器转发到系统登录页面，并在登录窗口中给出提示信息，如图 17.15 所示。

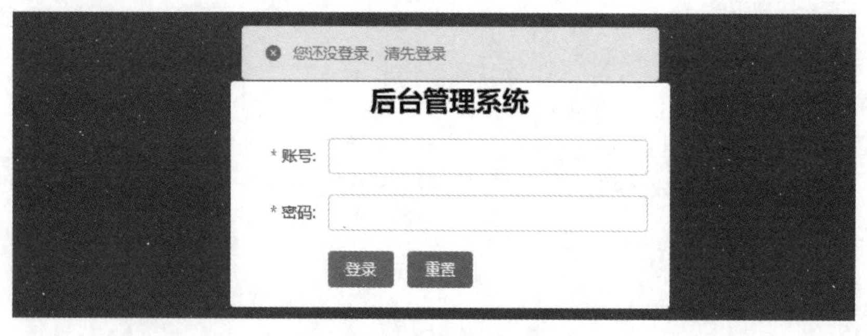

图 17.15　转发到登录页面

注意： 作为一个新闻管理系统，除了后台以外，还要实现前台新闻的展示（新闻首页、新闻列表页和新闻详细内容显示页等），可以使用 SSM 框架和 Bootstrap、jQuery 框架的相关技术进行实现，因为前端知识不是本书的重点内容，所以本章不再做详细介绍，读者可以参考本章的案例代码。

17.8 项目小结

本章主要通过一个新闻管理系统讲解 SSM+Vue.js 框架的整合使用。首先对系统的功能、结构等进行简单的介绍；然后对系统数据库表进行分析和设计；接着详细地讲解了系统的开发环境搭建工作；最后详细地讲解了系统用户管理模块、新闻管理模块和登录验证的实现。

通过本章的学习，读者可以熟练地掌握 SSM+Vue.js 框架的整合使用，并能熟练地使用 SSM+Vue.js 框架实现系统功能模块的开发工作。本系统是 SSM+Vue.js 框架综合使用的案例，读者要多加练习，并熟练编写各个功能模块的实现代码，这样才能将前面所学的知识融会贯通。

第 18 章

SSM+Vue.js 实战：图书管理系统

本章使用 SSM 框架构建后台、Vue 编写前台的方式来实现一个图书管理系统，做到信息化的图书管理，管理员可以通过该系统实施更加容易的管理操作，读者也可以有一个比较好的图书服务体验。

本章主要涉及的知识点如下：

- 系统需求及功能分析。
- 数据库设计。
- 开发环境框架搭建。
- 系统功能设计与实现。

18.1 系统概述

在本章的图书管理系统实现过程中，后台使用 SSM 三大框架实现，前台页面使用当前主流的 Vue.js 框架完成信息展示。SSM+Vue.js 相结合的方式总体来说比较简洁、易上手，非常适合系统的前后端开发。

18.1.1 系统功能需求

图书管理系统的功能首先是用户登录管理功能，然后通过管理员和读者这两种登录类型再来细分相应的功能：读者有借阅记录管理、个人信息管理和公告管理；管理员有读者管理、图书分类管理、借阅管理、图书管理、管理员管理以及公告管理。

其中，用户登录管理通过给定的登录类型去登录图书馆管理系统，在系统的登录页面输入对应用户名、密码及验证码；借阅管理通过管理员来管理所有读者的借书、还书；图书管理用于图书信息的增、删、改、查等操作；图书分类管理则包含了该系统所有图书的类别；读者管理让管理员可以看到所有读者的信息，还可以添加新读者信息，也可以注销没有借书记录的读者；公告管理让管理员可以管理所有发布的公告信息。

18.1.2 功能模块设计

图书管理系统通过登录模块的用户登录类型来区分相应的功能模块，大致的功能模块有：图书管理、借阅管理、图书分类管理、读者管理、管理员管理、公告管理、个人信息管理、借阅记录管理，当然还有退出登录操作。其中的公告管理是管理员和读者共有的功能，但是读者只能查看公告信息不能修改。系统管理员以管理员登录方式来管理读者和图书的信息，他们需要管理读者，同时管理书籍的增、删、改、查以及借出/归还。用户则以读者登录方式来编辑自己的信息，查看借书信息以及相关公告。系统所有的功能模块如图 18.1 所示。

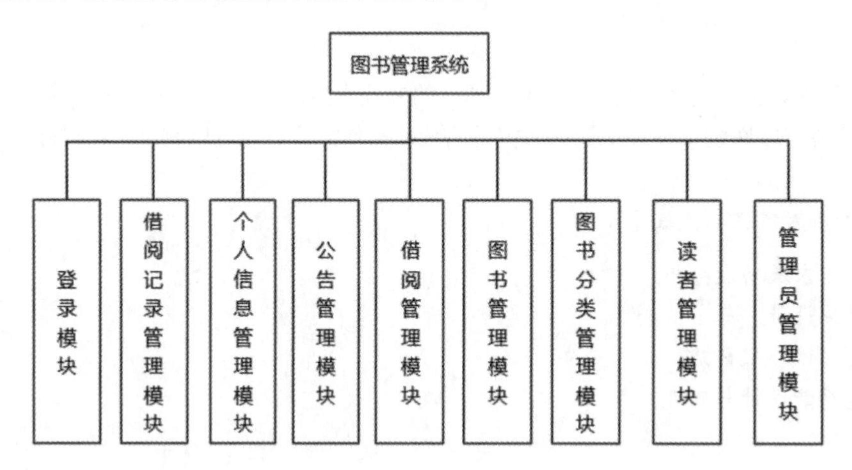

图 18.1 功能模块设计

部分功能模块介绍如下：

- 借阅管理模块：使用管理员的身份登录，可以对读者所有的借阅信息进行编辑。要添加读者的一些借书信息，只需输入图书名称、勾选图书再填写读者借阅卡卡号即可。
- 图书管理模块：使用管理员的身份登录，可以实现上架、下架图书，修改图书的信息（如作者名、图书名、图书的编号、更新的日期、出版社、语言、图书类型、价格、图书相关介绍等）等操作。
- 图书分类管理模块：使用管理员的身份登录，可以对图书类型进行增加、删除以及修改，还可以根据需要打印、导出、导入图书类型信息。
- 读者管理模块：使用管理员身份登录，可以新增或注销读者、修改读者信息，并支持导出或导入读者信息。
- 管理员管理模块：使用管理员身份登录，可以添加新的管理员、修改管理员用户名及密码。
- 公告管理模块：使用管理员身份登录，可以发送新的消息。

18.2　数据分析与设计

根据系统的功能需求，该图书管理系统的设计与实现主要涉及管理员实体、读者实体、图书实体和借阅记录实体 4 个实体。其中，管理员实体与图书、读者实体之间构成一对多的关联关系，读者实体和借阅记录实体之间构成一对一的关联关系，如图 18.2 所示。

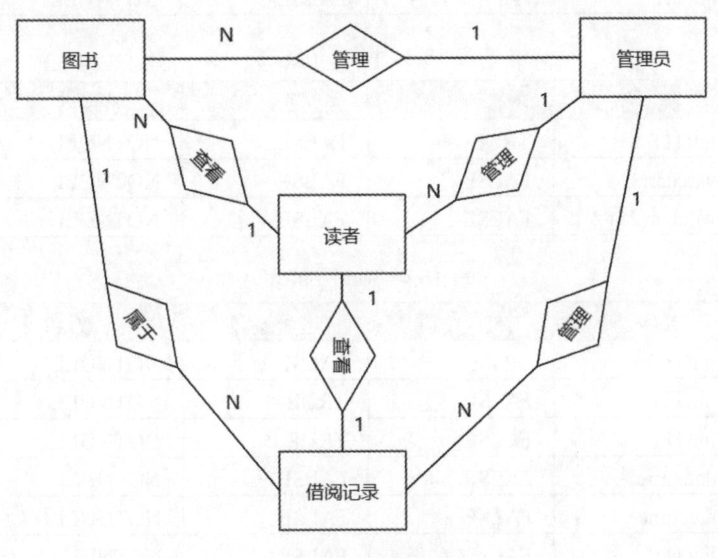

图 18.2　数据设计

与上述实体相对应，该图书管理系统中涉及数据库管理员表 admin、图书的信息表 book_info、图书的分类表 type_info、借阅信息表 lend_list、公告信息表 notice、读者借阅卡信息表 reader_card 等 6 个基本表，如表 18.1~表 18.6 所示。

表 18.1　admin 表

属 性 名	类　型	是否主键	是否外键	是否为空	说　明
id	int(11)	TRUE	FALSE	NOTNULL	管理员 id
username	varchar(20)	FALSE	FALSE	NOTNULL	用户名
password	varchar(20)	FALSE	FALSE	NOTNULL	密码
type	int(11)	FALSE	FALSE	NULL	类型

表 18.2　book_info 表

属 性 名	类　型	是否主键	是否外键	是否为空	说　明
id	int(11)	TRUE	FALSE	NOTNULL	图书 id
name	varchar(15)	FALSE	FALSE	NOTNULL	书名
author	varchar(25)	FALSE	FALSE	NOTNULL	作者
publish	varchar(25)	FALSE	FALSE	NOTNULL	出版社
ISBN	varchar(25)	FALSE	FALSE	NOTNULL	图书编号
introduction	text	FALSE	FALSE	NOTNULL	介绍

（续表）

属 性 名	类 型	是否主键	是否外键	是否为空	说 明
language	varchar(20)	FALSE	FALSE	NOTNULL	语言
price	double	FALSE	FALSE	NOTNULL	价格
pub_date	date	FALSE	FALSE	NOTNULL	出版日期
type_id	int(11)	FALSE	FALSE	NOTNULL	类型 id
status	int(11)	FALSE	FALSE	NOTNULL	借出状态

表 18.3 type_info 表

属 性 名	类 型	是否主键	是否外键	是否为空	说 明
id	int(11)	TRUE	FALSE	NOTNULL	图书 id
name	varchar(15)	FALSE	FALSE	NOTNULL	书名
remarks	varchar(25)	FALSE	FALSE	NOTNULL	备注

表 18.4 lend_list 表

属 性 名	类 型	是否主键	是否外键	是否为空	说 明
id	int(11)	TRUE	FALSE	NOTNULL	借阅 id
book_id	int(11)	FALSE	TRUE	NOTNULL	书的编号
reader_id	int(11)	FALSE	TRUE	NOTNULL	读者 id
lend_date	datetime	FALSE	FALSE	NOTNULL	借出时间
back_date	datetime	FALSE	FALSE	NOTNULL	归还时间
type	int(11)	FALSE	FALSE	NOTNULL	归还类型
remarks	varchar(255)	FALSE	FALSE	NOTNULL	备注

表 18.5 notice 表

属 性 名	类 型	是否主键	是否外键	是否为空	说 明
id	int(11)	TRUE	FALSE	NOTNULL	公告 id
author	varchar(25)	FALSE	TRUE	NOTNULL	发布人
create_date	date	FALSE	FALSE	NOTNULL	发布日期
content	text	FALSE	FALSE	NOTNULL	内容信息

表 18.6 reader_card 表

属 性 名	类 型	是否主键	是否外键	是否为空	说 明
id	int(11)	TRUE	FALSE	NOTNULL	读者 id
username	varchar(25)	FALSE	FALSE	NOTNULL	用户名
password	varchar(25)	FALSE	FALSE	NOTNULL	密码
number	int(11)	FALSE	FALSE	NOTNULL	借书时间
name	varchar(25)	FALSE	FALSE	NOTNULL	真实姓名
sex	varchar(5)	FALSE	FALSE	NOTNULL	性别
birthday	date	FALSE	FALSE	NOTNULL	生日
address	varchar(150)	FALSE	FALSE	NOTNULL	地址
tel	varchar(20)	FALSE	FALSE	NOTNULL	电话

（续表）

属 性 名	类　型	是否主键	是否外键	是否为空	说　明
email	varchar(50)	FALSE	FALSE	NOTNULL	邮箱
creat_date	datetime	FALSE	FALSE	NOTNULL	创建日期
cjr	int(11)	FALSE	FALSE	NOTNULL	无
cardnumber	varchar(25)	FALSE	FALSE	NOTNULL	借阅卡号

从本项目的实际需求出发，结合 MySQL 的优点，选用 MySQL 数据库作为图书管理系统后的台数据库最为合适。MySQl 优点非常多，比如代码开发方便、需要的资源成本非常低、可以存储大量的数据记录等。在实现系统前，需要提前准备好系统中的数据库资源，创建好数据库，并在数据库中创建上述 6 张表，同时添加一些必要的基础数据。创建数据库 news 的 SQL 语句和上述 6 张表的建表及基础数据插入的 SQL 语句如下：

```
01  CREATE TABLE `admin`  (
02  `id` int(11) NOT NULL AUTO_INCREMENT COMMENT 'id',
03  `username` varchar(20) CHARACTER SET utf8 COLLATE utf8_general_ci NULL DEFAULT NULL COMMENT
    '用户名',
04  `password` varchar(20) CHARACTER SET utf8 COLLATE utf8_general_ci NULL DEFAULT NULL COMMENT
    '密码',
05  `adminType` int(11) NULL DEFAULT NULL COMMENT '管理员类型',
06  PRIMARY KEY (`id`) USING BTREE
07  ) ENGINE = InnoDB AUTO_INCREMENT = 5 CHARACTER SET = utf8 COLLATE = utf8_general_ci COMMENT
    = '管理员' ROW_FORMAT = Dynamic;
08  DROP TABLE IF EXISTS `book_info`;
09  CREATE TABLE `book_info`  (
10  `id` int(11) NOT NULL AUTO_INCREMENT COMMENT 'id',
11  `name` varchar(20) CHARACTER SET utf8 COLLATE utf8_general_ci NULL DEFAULT NULL COMMENT
    '图书名称',
12  `author` varchar(30) CHARACTER SET utf8 COLLATE utf8_general_ci NULL DEFAULT NULL COMMENT
    '作者',
13  `publish` varchar(30) CHARACTER SET utf8 COLLATE utf8_general_ci NULL DEFAULT NULL COMMENT
    '出版社',
14  `isbn` varchar(30) CHARACTER SET utf8 COLLATE utf8_general_ci NULL DEFAULT NULL COMMENT
    '书籍编号',
15  `introduction` varchar(50) CHARACTER SET utf8 COLLATE utf8_general_ci NULL DEFAULT NULL
    COMMENT '简介',
16  `language` varchar(20) CHARACTER SET utf8 COLLATE utf8_general_ci NULL DEFAULT NULL COMMENT
    '语言',
17  `price` double NULL DEFAULT NULL COMMENT '价格',
18  `publish_date` date NULL DEFAULT NULL COMMENT '出版时间',
19  `type_id` int(11) NULL DEFAULT NULL COMMENT '书籍类型',
20  `status` int(11) NULL DEFAULT NULL COMMENT '状态：0 未借出，1 已借出',
21  PRIMARY KEY (`id`) USING BTREE
22  ) ENGINE = InnoDB AUTO_INCREMENT = 7 CHARACTER SET = utf8 COLLATE = utf8_general_ci COMMENT
    = '图书信息' ROW_FORMAT = Dynamic;
```

```
23  DROP TABLE IF EXISTS `lend_list`;
24  CREATE TABLE `lend_list`  (
25  `id` int(11) NOT NULL AUTO_INCREMENT COMMENT 'id',
26  `bookId` int(11) NULL DEFAULT NULL COMMENT '图书id',
27  `readerId` int(11) NULL DEFAULT NULL COMMENT '读者id',
28  `lendDate` datetime(0) NULL DEFAULT NULL COMMENT '借书时间',
29  `backDate` datetime(0) NULL DEFAULT NULL COMMENT '还书时间',
30  `backType` int(11) NULL DEFAULT NULL,
31  `exceptRemarks` varchar(255) CHARACTER SET utf8 COLLATE utf8_general_ci NULL DEFAULT NULL
    COMMENT '备注信息',
32  PRIMARY KEY (`id`) USING BTREE
33  ) ENGINE = InnoDB AUTO_INCREMENT = 40 CHARACTER SET = utf8 COLLATE = utf8_general_ci COMMENT
    = '借阅记录（谁在何时借走了什么书，并且有没有归还，归还时间）' ROW_FORMAT = Dynamic;
34  DROP TABLE IF EXISTS `notice`;
35  CREATE TABLE `notice`  (
36  `id` int(11) NOT NULL AUTO_INCREMENT COMMENT 'id',
37  `topic` varchar(50) CHARACTER SET utf8 COLLATE utf8_general_ci NULL DEFAULT NULL,
38  `content` varchar(255) CHARACTER SET utf8 COLLATE utf8_general_ci NULL DEFAULT NULL COMMENT
    '公告内容',
39  `author` varchar(20) CHARACTER SET utf8 COLLATE utf8_general_ci NULL DEFAULT NULL COMMENT
    '发布人',
40  `createDate` datetime(0) NULL DEFAULT NULL COMMENT '公告发布时间',
41  PRIMARY KEY (`id`) USING BTREE
42  ) ENGINE = InnoDB AUTO_INCREMENT = 7 CHARACTER SET = utf8 COLLATE = utf8_general_ci COMMENT
    = '公告' ROW_FORMAT = Dynamic;
43  DROP TABLE IF EXISTS `reader_info`;
44  CREATE TABLE `reader_info`  (
45  `id` int(11) NOT NULL AUTO_INCREMENT COMMENT 'id',
46  `username` varchar(20) CHARACTER SET utf8 COLLATE utf8_general_ci NULL DEFAULT NULL COMMENT
    '用户名',
47  `password` varchar(20) CHARACTER SET utf8 COLLATE utf8_general_ci NULL DEFAULT NULL COMMENT
    '密码',
48  `realName` varchar(20) CHARACTER SET utf8 COLLATE utf8_general_ci NULL DEFAULT NULL COMMENT
    '真实姓名',
49  `sex` varchar(5) CHARACTER SET utf8 COLLATE utf8_general_ci NULL DEFAULT NULL COMMENT '
    性别',
50  `birthday` date NULL DEFAULT NULL COMMENT '出生日期',
51  `address` varchar(30) CHARACTER SET utf8 COLLATE utf8_general_ci NULL DEFAULT NULL COMMENT
    '籍贯',
52  `tel` varchar(11) CHARACTER SET utf8 COLLATE utf8_general_ci NULL DEFAULT NULL COMMENT
    '电话',
53  `email` varchar(15) CHARACTER SET utf8 COLLATE utf8_general_ci NULL DEFAULT NULL COMMENT
    '邮箱',
54  `registerDate` datetime(0) NULL DEFAULT NULL COMMENT '注册日期',
55  `readerNumber` varchar(20) CHARACTER SET utf8 COLLATE utf8_general_ci NULL DEFAULT NULL
    COMMENT '读者编号',
```

```
56  PRIMARY KEY (`id`) USING BTREE
57  ) ENGINE = InnoDB AUTO_INCREMENT = 4 CHARACTER SET = utf8 COLLATE = utf8_general_ci COMMENT
    = '读者信息（包括登录账号、密码等）' ROW_FORMAT = Dynamic;
58  DROP TABLE IF EXISTS `type_info`;
59  CREATE TABLE `type_info`  (
60  `id` int(11) NOT NULL AUTO_INCREMENT COMMENT 'id',
61  `name` varchar(20) CHARACTER SET utf8 COLLATE utf8_general_ci NULL DEFAULT NULL COMMENT
    '图书分类名称',
62  `remarks` char(10) CHARACTER SET utf8 COLLATE utf8_general_ci NULL DEFAULT NULL COMMENT
    '备注',
63  PRIMARY KEY (`id`) USING BTREE
64  ) ENGINE = InnoDB AUTO_INCREMENT = 7 CHARACTER SET = utf8 COLLATE = utf8_general_ci COMMENT
    = '图书类型表' ROW_FORMAT = Dynamic;
```

18.3 开发环境和框架的搭建

图书管理系统基于 SSM+Vue.js 框架进行需求分析和系统设计，结合数据库设计，选择该平台的开发环境为 IntelliJ IDEA +MySQL+WebStorm，系统的开发环境和框架的搭建涉及各类文件的创建、引入和编写，包括包文件（内含各类接口和类）、配置文件、页面文件以及相关的 JAR 包文件、资源文件等。本节将详细介绍系统开发环境和框架的搭建。

18.3.1 创建项目

在 IntelliJ IDEA 中创建一个名称为 library 的 Web 项目，将系统所准备的全部 JAR 包复制到项目的 lib 目录中，并发布到类路径下。

由于本系统使用 SSM 框架开发，因此需要准备三大框架的 JAR 包。另外，系统中还涉及数据库连接、JSTL 标签等，所以还要准备其他包。整个系统中需要准备的 JAR 包共计 35 个，在第 17 章已经详细介绍过了，此处不再赘述。

18.3.2 编写配置文件

在项目 src 目录下分别创建数据库常量配置文件、Spring 配置文件、MyBatis 配置文件、Spring MVC 配置文件、log4j 配置文件以及资源配置文件。前 4 个文件分别如文件 18.1~文件 18.4 所示。log4j 配置文件请参照前面相关章节的内容进行编写。本系统没有用到资源配置文件。

文件 18.1　db.properties

```
01  jdbc.driver = com.mysql.jdbc.Driver
02  jdbc.url = jdbc:mysql://localhost:3306/library?useUnicode=true&characterEncoding=utf-8
03  jdbc.username = root
04  jdbc.password = 123456
```

文件 18.2 spring.xml

```
01  <?xml version="1.0" encoding="UTF-8"?>
02  <beans xmlns="http://www.springframework.org/schema/beans"
03  xmlns:tx="http://www.springframework.org/schema/tx"
04  xmlns:xsi="http://www.w3.org/2001/XMLSchema-instance"
05  xmlns:context="http://www.springframework.org/schema/context"
06  xsi:schemaLocation="http://www.springframework.org/schema/beans
07  http://www.springframework.org/schema/beans/spring-beans.xsd
08  http://www.springframework.org/schema/tx
09  http://www.springframework.org/schema/tx/spring-tx.xsd
10  http://www.springframework.org/schema/context
11  http://www.springframework.org/schema/context/spring-context.xsd">
12  <!--Spring 接管 MyBatis 内容-->
13  <!--DataSource Druid 连接池-->
14  <context:property-placeholder location="classpath:db.properties"/>
15  <bean id="dataSource" class="com.alibaba.druid.pool.DruidDataSource" init-method="init"
    destroy-method="close">
16      <!--基本配置-->
17      <property name="driverClassName" value="${jdbc.driver}"/>
18      <property name="url" value="${jdbc.url}"/>
19      <property name="username" value="${jdbc.username}"/>
20      <property name="password" value="${jdbc.password}"/>
21  </bean>
22  <!--SqlSessionFactory 要 DataSource 支持-->
23  <bean id="sqlSessionFactory" class="org.mybatis.spring.SqlSessionFactoryBean">
24      <!--注入连接池-->
25      <property name="dataSource" ref="dataSource"/>
26      <!--在实体类中用类名作为别名-->
27      <property name="typeAliasesPackage" value="com.library.po"></property>
28      <!--分页-->
29      <property name="plugins">
30      <array>
31      <bean class="com.github.pagehelper.PageInterceptor">
32          <property name="properties">
33          <props>
34              <!--页号的合理值 0 - max-->
35              <prop key="reasonable">true</prop>
36          </props>
37          </property>
38      </bean>
39      </array>
40      </property>
41  </bean>
42  <!--Dao 需要 MapperScannerConfigurer 支持-->
43  <bean id="mapperScannerConfigurer"
    class="org.mybatis.spring.mapper.MapperScannerConfigurer">
```

```
44        <property name="basePackage" value="com.library.dao"/>
45        <property name="sqlSessionFactoryBeanName" value="sqlSessionFactory"/>
46    </bean>
47    <!--告知 spring 注解位置，保证注解有效性，下面注解扫描表示不扫描 Controller，扫描 service 和 dao-->
48    <context:component-scan base-package="com.library">
49    <context:exclude-filter type="annotation" expression="org.springframework.stereotype.
      Controller"></context:exclude-filter>
50    </context:component-scan>
51    <!--引入一个事务管理器-->
52    <bean id="tx"
      class="org.springframework.jdbc.datasource.DataSourceTransactionManager">
53        <property name="dataSource" ref="dataSource"/>
54    </bean>
55    <!--@Transactional 告诉 Spring，定制事务是基于 DataSourceTransactionManager-->
56    <tx:annotation-driven transaction-manager="tx"></tx:annotation-driven>
57    </beans>
```

文件 18.3　mybatis-config.xml

```
01    <?xml version="1.0" encoding="UTF-8"?>
02    <!DOCTYPE generatorConfiguration PUBLIC "-//mybatis.org//DTD MyBatis Generator
      Configuration 1.0//EN" "http://mybatis.org/dtd/mybatis-generator-config_1_0.dtd">
03    <generatorConfiguration>
04    <classPathEntry
      location="D:\Maven\apache-maven-3.6.0\mvn\repository\mysql\mysql-connector-java\8.0.2
      3\mysql-connector-java-8.0.23.jar" />
05    <context id="msqlTables" targetRuntime="MyBatis3">
06    <plugin type="org.mybatis.generator.plugins.SerializablePlugin"></plugin>
07    <jdbcConnection
      connectionURL="jdbc:mysql://localhost:3306/library?useUnicode=true&amp;characterEncod
      ing=utf-8"
08    driverClass="com.mysql.jdbc.Driver" password="root" userId="123456" >
09        <property name="nullCatalogMeansCurrent" value="true"/>
10    </jdbcConnection>
11    <javaTypeResolver>
12        <property name="forceBigDecimals" value="false" />
13    </javaTypeResolver>
14    <!-- 生成的实体类和数据库表一一对应 -->
15    <javaModelGenerator targetPackage="com.library.po"
      targetProject="D:\IDEAProject\libraryProject\src\main\java">
16        <property name="enableSubPackages" value="true"/>
17        <!–清理从数据库返回的值的前后空格  -->
18        <property name="trimStrings" value="true" />
19    </javaModelGenerator>
20    <!--映射 XML 文件及 DAO 接口-->
21    <sqlMapGenerator targetPackage="com.library.dao" targetProject="D:\IDEAProject\
      libraryProject\src\main\resources">
```

```
22      <property name="enableSubPackages" value="true"/>
23  </sqlMapGenerator>
24  <javaClientGenerator type="XMLMAPPER" targetPackage="com.library.dao"
    targetProject="D:\IDEAProject\libraryProject\src\main\java">
25      <property name="enableSubPackages" value="true"/>
26  </javaClientGenerator>
27  <!--数据库表-->
28  <table tableName="admin" domainObjectName="Admin"
29  enableCountByExample="false" enableUpdateByExample="false"
    enableDeleteByExample="false"
30  enableSelectByExample="false" selectByExampleQueryId="false" >
31      <property name="useActualColumnNames" value="false"/>
32  </table>
33  <table tableName="book_info" domainObjectName="BookInfo"
34  enableCountByExample="false" enableUpdateByExample="false"
    enableDeleteByExample="false"
35  enableSelectByExample="false" selectByExampleQueryId="false" >
36      <property name="useActualColumnNames" value="false"/>
37  </table>
38  <table tableName="lend_list" domainObjectName="LendList"
39  enableCountByExample="false" enableUpdateByExample="false"
    enableDeleteByExample="false"
40  enableSelectByExample="false" selectByExampleQueryId="false" >
41      <property name="useActualColumnNames" value="false"/>
42  </table>
43  <table tableName="notice" domainObjectName="Notice"
44  enableCountByExample="false" enableUpdateByExample="false"
    enableDeleteByExample="false"
45  enableSelectByExample="false" selectByExampleQueryId="false" >
46      <property name="useActualColumnNames" value="false"/>
47  </table>
48  <table tableName="reader_info" domainObjectName="ReaderInfo"
49  enableCountByExample="false" enableUpdateByExample="false"
    enableDeleteByExample="false"
50  enableSelectByExample="false" selectByExampleQueryId="false" >
51      <property name="useActualColumnNames" value="false"/>
52  </table>
53  <table tableName="type_info" domainObjectName="TypeInfo"
54  enableCountByExample="false" enableUpdateByExample="false"
    enableDeleteByExample="false"
55  enableSelectByExample="false" selectByExampleQueryId="false" >
56      <property name="useActualColumnNames" value="false"/>
57  </table>
58  </context>
59  </generatorConfiguration>
```

文件 18.4 springmvc-config.xml

```
01  <?xml version="1.0" encoding="UTF-8"?>
02  <beans xmlns="http://www.springframework.org/schema/beans"
03  xmlns:mvc="http://www.springframework.org/schema/mvc"
    xmlns:context="http://www.springframework.org/schema/context"
04  xmlns:xsi="http://www.w3.org/2001/XMLSchema-instance"
05  xsi:schemaLocation="
06  http://www.springframework.org/schema/beans
07  http://www.springframework.org/schema/beans/spring-beans.xsd
08  http://www.springframework.org/schema/mvc
09  http://www.springframework.org/schema/mvc/spring-mvc.xsd
10  http://www.springframework.org/schema/context
11  http://www.springframework.org/schema/context/spring-context.xsd">
12  <!--开启注解扫描，只扫描 Controller 注解-->
13  <context:component-scan
    base-package="com.library.controller"></context:component-scan>
14  <!--注册注解驱动（开启 SpringMVC 注解的支持）-->
15  <mvc:annotation-driven>
16  <mvc:message-converters>
17      <bean class="com.alibaba.fastjson.support.spring.FastJsonHttpMessageConverter">
18      <!--声明类型转换，若返回的是 JSON 就需要加这个进行转换-->
19      <property name="supportedMediaTypes">
20          <list>
21              <value>application/json</value>
22          </list>
23      </property>
24      </bean>
25  </mvc:message-converters>
26  </mvc:annotation-driven>
27  <!--配置的视图解析器对象-->
28  <bean id="internalResourceViewResolver"
    class="org.springframework.web.servlet.view.InternalResourceViewResolver">
29      <property name="prefix" value="/WEB-INF/pages/"/>
30      <property name="suffix" value=".jsp"/>
31  </bean>
32  <!--配置拦截器-->
33  <mvc:interceptors>
34  <mvc:interceptor>
35      <mvc:mapping path="/**"/>
36      <mvc:exclude-mapping path="/login"/>
37      <mvc:exclude-mapping path="/loginIn"/>
38      <mvc:exclude-mapping path="/verifyCode"/>
39      <bean class="com.library.interceptor.LoginInterceptor"></bean>
40  </mvc:interceptor>
41  </mvc:interceptors>
42  </beans>
```

以上配置文件中，除了配置需要扫描的包、注解驱动和视图解析器外，还增加了加载属性文件和访问静态资源的配置。

除以上配置文件外，还需要在项目的/WebContent/WEB-INF 目录下编写 web.xml 文件，内容如文件 18.5 所示。

文件 18.5 web.xml

```
01  <?xml version="1.0" encoding="UTF-8"?>
02  <web-app xmlns="http://xmlns.jcp.org/xml/ns/javaee"
03  xmlns:xsi="http://www.w3.org/2001/XMLSchema-instance"
04  xsi:schemaLocation="http://xmlns.jcp.org/xml/ns/javaee
    http://xmlns.jcp.org/xml/ns/javaee/web-app_4_0.xsd"
05  version="4.0">
06  <!--SpringMVC 前端（核心）控制器，启动 Controller 组件-->
07  <servlet>
08      <servlet-name>dispatcherServlet</servlet-name>
09      <servlet-class>org.springframework.web.servlet.DispatcherServlet</servlet-class>
10      <!--声明配置文件的位置-->
11      <init-param>
12          <param-name>contextConfigLocation</param-name>
13          <param-value>classpath:springmvc.xml</param-value>
14      </init-param>
15      <!--创建 servlet-->
16      <load-on-startup>1</load-on-startup>
17  </servlet>
18  <servlet-mapping>
19      <servlet-name>dispatcherServlet</servlet-name>
20      <url-pattern>/</url-pattern>
21  </servlet-mapping>
22  <!--防乱码-->
23  <filter>
24  <filter-name>encoding</filter-name>
25  <filter-class>org.springframework.web.filter.CharacterEncodingFilter</filter-class>
26      <init-param>
27          <param-name>encoding</param-name>
28          <param-value>utf-8</param-value>
29      </init-param>
30  </filter>
31  <filter-mapping>
32      <filter-name>encoding</filter-name>
33      <url-pattern>/*</url-pattern>
34  </filter-mapping>
35  <!--配置监听器，在项目启动时，同时启动 Spring 工厂，使得 DAO、Service 以及 Druid 连接池也启动-->
36  <listener>
37      <listener-class>org.springframework.web.context.ContextLoaderListener
    </listener-class>
```

```
38  </listener>
39  <!--配置静态资源-->
40  <servlet-mapping>
41      <servlet-name>default</servlet-name>
42      <url-pattern>*.js</url-pattern>
43      <url-pattern>*.css</url-pattern>
44      <url-pattern>*.png</url-pattern>
45      <url-pattern>*.jpg</url-pattern>
46      <url-pattern>/api/**</url-pattern>
47      <url-pattern>/css/**</url-pattern>
48      <url-pattern>/images/**</url-pattern>
49      <url-pattern>/js/*</url-pattern>
50      <url-pattern>/lib/*</url-pattern>
51      <url-pattern>/page/*</url-pattern>
52  </servlet-mapping>
53  <!--配置 Spring 文件路径-->
54  <context-param>
55      <param-name>contextConfigLocation</param-name>
56      <param-value>classpath:spring.xml</param-value>
57  </context-param>
58  </web-app>
```

18.3.3　创建相关包和文件

按照如图 18.3 所示的项目文件组织结构创建项目相关的目录（包）及文件（如相关类和接口的包、Vue 文件对应的文件夹、Vue 文件等），并引入项目开发需要的相关文件资源（如 CSS 样式文件、images 图片文件、JS 文件、标签文件）等，为项目开发做好准备。在后续项目开发过程中，根据需要可以创建其他文件或引入其他资源文件。

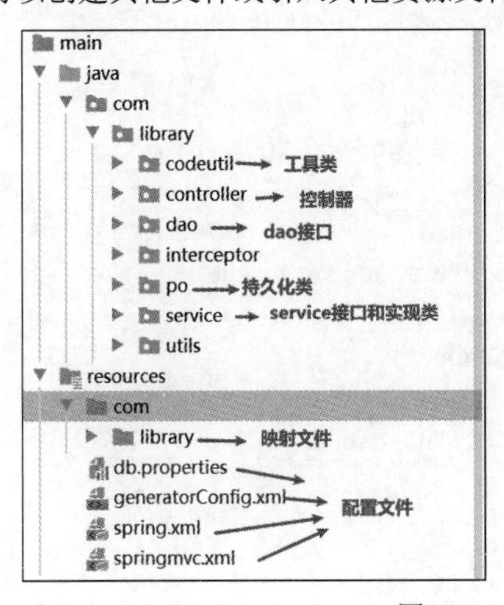

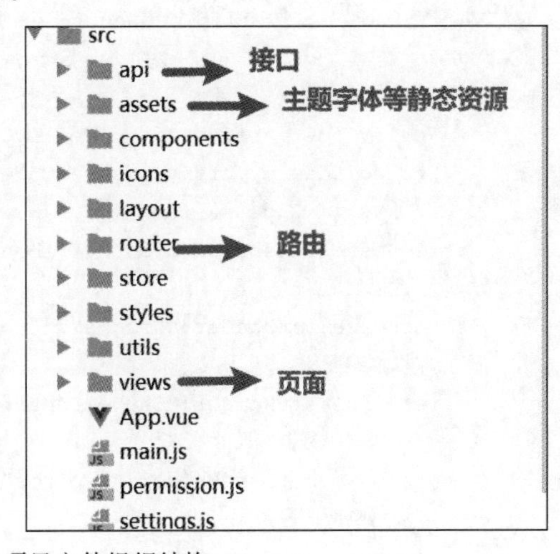

图 18.3　项目文件组织结构

18.4 系统功能设计与实现

图书管理系统的设计与实现需要分模块进行，主要涉及用户登录、图书（分类）管理、借阅管理、读者（管理员）管理、公告管理等几个功能模块。其中，图书管理包含图书分类管理，读者管理包含管理员管理。本节将介绍上述功能模块的设计与实现。

18.4.1 用户登录模块

在 src 目录下创建一个 com.library.controller 包，在包中创建控制器类 LoginController，代码如文件 18.6 所示。

文件 18.6 LoginController.java

```
01  @Controller
02  public class LoginController {
03      @Autowired
04      private AdminService adminService;
05      @Autowired
06      private ReaderInfoService readerService;
07      // 登录页面的转发
08      @GetMapping("/login")
09      public String login(){
10          return "login";
11      }
12      // 获取验证码方法
13      @RequestMapping("/verifyCode")
14      public void verifyCode(HttpServletRequest request, HttpServletResponse response) {
15          IVerifyCodeGen iVerifyCodeGen = new SimpleCharVerifyCodeGenImpl();
16          try {
17              //设置长和宽
18              VerifyCode verifyCode = iVerifyCodeGen.generate(80, 28);
19              String code = verifyCode.getCode();
20              //将 VerifyCode 绑定 Session
21              request.getSession().setAttribute("VerifyCode", code);
22              //设置响应头
23              response.setHeader("Pragma", "no-cache");
24              //设置响应头
25              response.setHeader("Cache-Control", "no-cache");
26              //在代理服务器端防止缓冲
27              response.setDateHeader("Expires", 0);
28              //设置响应内容类型
29              response.setContentType("image/jpeg");
30              response.getOutputStream().write(verifyCode.getImgBytes());
31              response.getOutputStream().flush();
```

```
32            } catch (IOException e) {
33                System.out.println("异常处理");
34            }
35        }
36        // 登录验证
37        @RequestMapping("/loginIn")
38        public String loginIn(HttpServletRequest request, Model model){
39            //获取用户名与密码
40            String username = request.getParameter("username");
41            String password = request.getParameter("password");
42            String code=request.getParameter("captcha");
43            String type=request.getParameter("type");
44            //判断验证码是否正确（验证码已经放入 Session）
45            HttpSession session = request.getSession();
46            String realCode = (String)session.getAttribute("VerifyCode");
47            if (!realCode.toLowerCase().equals(code.toLowerCase())){
48                model.addAttribute("msg","验证码错误");
49                return "login";
50            }else{
51                //验证码正确则判断用户名和密码
52                if(type.equals("1")){//管理员信息
53                    //用户名和密码是否正确
54                    Admin admin
    =adminService.queryUserByNameAndPassword(username,password);
55                    if(admin==null){//该用户不存在
56                        model.addAttribute("msg","用户名或密码错误");
57                        return "login";
58                    }
59                    session.setAttribute("user",admin);
60                    session.setAttribute("type","admin");
61                }else{//来自读者信息表
62                    ReaderInfo readerInfo=readerService.queryUserInfoByNameAndPassword
63    (username,password);
64                    if(readerInfo==null){
65                        model.addAttribute("msg","用户名或密码错误");
66                        return "login";
67                    }
68                    session.setAttribute("user",readerInfo);
69                    session.setAttribute("type","reader");
70                }
71                return "index";
72            }
73        }
74        // 退出功能
75        @GetMapping("loginOut")
76        public String loginOut(HttpServletRequest request){
```

```
77          HttpSession session = request.getSession();
78          session.invalidate();//注销
79          return "/login";
80      }
81  }
```

在项目 src/views/login 目录下创建 login.vue 页面文件。login.vue 主要实现用户登录到后台的功能，代码如文件 18.7 所示。

文件 18.7 login.vue

```
01  <template>
02  <div class="login-container">
03  <div class="login-box">
04  <!-- 头像区域 -->
05  <div class="avatar-box">
06  <img src="../assets/logo.png" alt="" />
07  </div>
08  <!-- 登录表单区域 -->
09  <el-form ref="loginFormRef" label-position="left"
    label-width="55px" :model="loginForm" :rules="loginFormRules"
10  class="login-form" size="medium" hide-required-asterisk>
11  <!-- 账号 -->
12  <el-form-item prop="username" label="账号">
13  <el-input prefix-icon="iconfont icon-ziyuan" v-model="loginForm.username">
14  </el-input>
15  </el-form-item>
16  <!-- 密码 -->
17  <el-form-item prop="password" label="密码">
18  <el-input prefix-icon="iconfont icon-mima" v-model="loginForm.password"
    type="password">
19  </el-input>
20  </el-form-item>
21  <!-- 验证码 -->
22  <el-form-item prop="inputCode" label="验证码">
23  <el-input prefix-icon="iconfont icon-yanzhengma" v-model="loginForm.inputCode">
24  <i slot="suffix" class="verificationCode" @click="makePassCode">{{passCode}}</i>
25  </el-input>
26  </el-form-item>
27  <!-- 用户角色 -->
28  <el-form-item prop="adminType" label="身份">
29  <el-radio-group v-model="loginForm.adminType">
30  <el-radio label="0" border>读者</el-radio>
31  <el-radio label="1" border>管理员</el-radio>
32  </el-radio-group>
33  </el-form-item>
34  <el-form-item class="btns">
```

```
35 <el-button type="primary" @click="login">登录</el-button>
36 <el-button type="info" @click="resetLoginForm">重置</el-button>
37 </el-form-item>
38 </el-form>
39 </div>
40 </div>
41 </template>
42 <script>
43 export default {
44     data() {
45     // 判断验证码输入是否正确
46     var validatePass = (rule, value, callback) => {
47     if (value === '') {
48         callback(new Error('请输入验证码'));
49     } else {
50         if (value !== this.passCode) {
51             callback(new Error('验证码错误'));
52             this.makePassCode();
53         }
54         callback();
55     }
56 };
57 return {
58     passCode: '',
59     loginForm: {
60     username: '',
61     password: '',
62     inputCode: '',
63     role: '0'
64     },
65 loginFormRules: {
66     username: [{
67             required: true,
68             message: '请输入账号',
69             trigger: 'blur'
70         }, {
71             min: 2,
72             max: 10,
73             message: '长度为 2 到 10 个字符',
74             trigger: 'blur'
75         }],
76         password: [{
77             required: true,
78             message: '请输入登录密码',
79             trigger: 'blur'
80         }, {
```

```
81                 min: 3,
82                 max: 8,
83                 message: '长度为 3 到 8 个字符',
84                 trigger: 'blur'
85             }],
86             inputCode: [{
87                 validator: validatePass,
88                 trigger: 'blur'
89             }],
90             adminType: [{
91                 required: true,
92                 message: '请选择用户角色',
93                 trigger: 'change'
94             }],
95             }
96         }
97     },
98     created() {
99         this.makePassCode();
100    },
101    methods: {
102        //生成验证码
103        makePassCode() {
104            var code = "";
105            const codeLength = 4; //验证码的长度
106            const random = new Array(
107    0, 1, 2, 3, 4, 5, 6, 7, 8, 9,
108    "A", "B", "C", "D", "E", "F", "G", "H", "I", "J", "K", "L", "M", "N", "O", "P", "Q", "R",
       "S", "T", "U",
109    "V", "W", "X", "Y", "Z",
110    "a", "b", "c", "d", "e", "f", "g", "h", "i", "j", "k", "l", "m", "n", "o", "p", "q", "r",
       "s", "t", "u",
111    "v", "w", "x", "y", "z",
112        ); //随机数数组
113        for (let i = 0; i < codeLength; i++) {
114            //循环操作
115            let index = Math.floor(Math.random() * 62);
116            code += random[index];
117        }
118        this.passCode = code;
119    },
120    <!--省略部分代码-->
```

系统登录页面如图 18.4 所示。

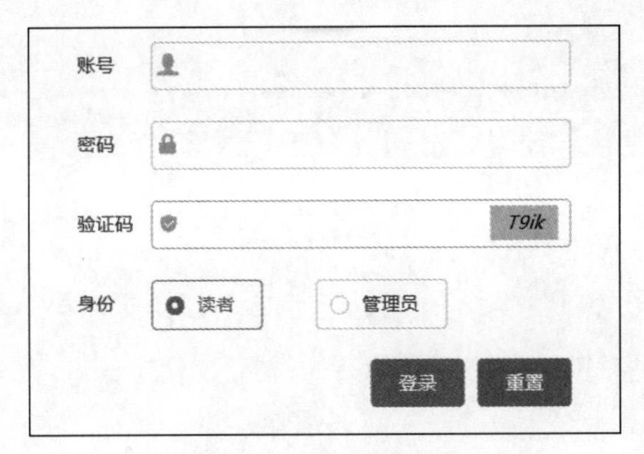

图 18.4　系统登录页面

18.4.2　图书（分类）管理模块

图书（分类）管理模块涉及图书和图书类别的添加、修改、查询、删除等功能。接下来我们将具体实现这些功能。

1. 创建持久化类

图书（分类）管理模块持久化类有图书详情类 BookInfo 和图书类型类 TypeInfo，具体代码分别如文件 18.8 和文件 18.9 所示。

文件 18.8　BookInfo.java

```
01  public class BookInfo implements Serializable {
02      private Integer id;
03      private String name;
04      private String author;
05      private String publish;
06      private String isbn;
07      private String introduction;
08      private String language;
09      private Double price;
10      private Date publishDate;
11      private Integer typeId;
12      private Integer status;
13      private TypeInfo typeInfo;//图书类型，在图书馆中显示什么类型的书
14      private Integer counts;
15      public Integer getCounts() {
16          return counts;
17      }
18      public void setCounts(Integer counts) {
19          this.counts = counts;
20      }
21      public TypeInfo getTypeInfo() {
22          return typeInfo;
```

```
23      }
24      public void setTypeInfo(TypeInfo typeInfo) {
25          this.typeInfo = typeInfo;
26      }
27      private static final long serialVersionUID = 1L;
28      public Integer getId() {
29          return id;
30          }
31      public void setId(Integer id) {
32          this.id = id;
33      }
34      public String getName() {
35          return name;
36      }
37      public void setName(String name) {
38          this.name = name == null ? null : name.trim();
39      }
40      public String getAuthor() {
41          return author;
42      }
43      public void setAuthor(String author) {
44          this.author = author == null ? null : author.trim();
45      }
46      public String getPublish() {
47          return publish;
48      }
49      public void setPublish(String publish) {
50          this.publish = publish == null ? null : publish.trim();
51      }
52      public String getIsbn() {
53          return isbn;
54      }
55      public void setIsbn(String isbn) {
56          this.isbn = isbn == null ? null : isbn.trim();
57      }
58      public String getIntroduction() {
59          return introduction;
60      }
61      public void setIntroduction(String introduction) {
62          this.introduction = introduction == null ? null : introduction.trim();
63      }
64      public String getLanguage() {
65          return language;
66      }
67      public void setLanguage(String language) {
68          this.language = language == null ? null : language.trim();
```

```
69      }
70      public Double getPrice() {
71          return price;
72      }
73      public void setPrice(Double price) {
74          this.price = price;
75      }
76      public Date getPublishDate() {
77          return publishDate;
78      }
79      public void setPublishDate(Date publishDate) {
80          this.publishDate = publishDate;
81      }
82      public Integer getTypeId() {
83          return typeId;
84      }
85      public void setTypeId(Integer typeId) {
86          this.typeId = typeId;
87      }
88      public Integer getStatus() {
89          return status;
90      }
91      public void setStatus(Integer status) {
92          this.status = status;
93      }
94  }
```

文件 18.9　TypeInfo.java

```
01  public class TypeInfo implements Serializable {
02      private Integer id;
03      private String name;
04      private String remarks;
05      public String getRemarks() {
06          return remarks;
07      }
08      public void setRemarks(String remarks) {
09          this.remarks = remarks;
10      }
11      private static final long serialVersionUID = 1L;
12      public Integer getId() {
13          return id;
14      }
15      public void setId(Integer id) {
16          this.id = id;
17      }
18      public String getName() {
```

```
19          return name;
20      }
21      public void setName(String name) {
22          this.name = name == null ? null : name.trim();
23      }
24  }
```

2. 实现 DAO 层接口

实现 DAO 层接口的操作步骤如下：

步骤 01 创建 DAO 层接口。在 src 目录下的 com.library.dao 包中创建接口 BookInfoMapper 和 TypeInfoMapper，并在接口中编写增、删、改、查等方法，如文件 18.10 和文件 18.11 所示。

文件 18.10　BookInfoMapper.java

```
01  public interface BookInfoMapper {
02      int deleteByPrimaryKey(Integer id);
03      int insert(BookInfo record);
04      int insertSelective(BookInfo record);
05      BookInfo selectByPrimaryKey(Integer id);
06      int updateByPrimaryKeySelective(BookInfo record);
07      int updateByPrimaryKey(BookInfo record);
08      List<BookInfo> queryBookInfoAll(BookInfo bookInfo);
09      // 根据类型获取图书数量
10      List<BookInfo> getBookCountByType();
11  }
```

文件 18.11　TypeInfoMapper.java

```
01  public interface TypeInfoMapper {
02      // 查询所有的记录信息
03      List<TypeInfo> queryTypeInfoAll(@Param(value = "name") String name);
04      // 添加图书类型
05      void addTypeSubmit(TypeInfo info);
06      // 修改，根据 id 查询记录信息
07      TypeInfo queryTypeInfoById(Integer id);
08      // 修改提交
09      void updateTypeSubmit(TypeInfo info);
10      // 根据 id 删除记录信息
11      void deleteTypeByIds(List<Integer> id);
12  }
```

步骤 02 创建映射文件。在 resources 目录下的 com.library.dao 包中创建 MyBatis 映射文件 BookInfoMapper.xml 和 TypeInfoMapper.xml，并在映射文件中编写增、删、改、查等方法的执行语句，如文件 18.12 和文件 18.13 所示。

文件 18.12　BookInfoMapper.xml

```
01  <?xml version="1.0" encoding="UTF-8"?>
02  <!DOCTYPE mapper PUBLIC "-//mybatis.org//DTD Mapper 3.0//EN"
    "http://mybatis.org/dtd/mybatis-3-mapper.dtd">
03  <mapper namespace="com.library.dao.BookInfoMapper">
04  <resultMap id="BaseResultMap" type="com.library.po.BookInfo">
05  <id column="id" jdbcType="INTEGER" property="id" />
06  <result column="name" jdbcType="VARCHAR" property="name" />
07  <result column="author" jdbcType="VARCHAR" property="author" />
08  <result column="publish" jdbcType="VARCHAR" property="publish" />
09  <result column="isbn" jdbcType="VARCHAR" property="isbn" />
10  <result column="introduction" jdbcType="VARCHAR" property="introduction" />
11  <result column="language" jdbcType="VARCHAR" property="language" />
12  <result column="price" jdbcType="DOUBLE" property="price" />
13  <result column="publish_date" jdbcType="DATE" property="publishDate" />
14  <result column="type_id" jdbcType="INTEGER" property="typeId" />
15  <result column="status" jdbcType="INTEGER" property="status" />
16  </resultMap>
17  <sql id="Base_Column_List">
18      id, name, author, publish, isbn, introduction, language, price, publish_date, type_id,
19      status
20  </sql>
21  <select id="selectByPrimaryKey" parameterType="java.lang.Integer"
        resultMap="BaseResultMap">
22      select
23      <include refid="Base_Column_List" />
24      from book_info
25      where id = #{id,jdbcType=INTEGER}
26  </select>
27  <delete id="deleteByPrimaryKey" parameterType="java.lang.Integer">
28      delete from book_info
29      where id = #{id,jdbcType=INTEGER}
30  </delete>
31  <insert id="insert" parameterType="com.library.po.BookInfo">
32      insert into book_info (id, name, author,
33      publish, isbn, introduction,
34      language, price, publish_date,
35      type_id, status)
36  values (#{id,jdbcType=INTEGER}, #{name,jdbcType=VARCHAR}, author,jdbcType=VARCHAR},
37  #{publish,jdbcType=VARCHAR}, #{isbn,jdbcType=VARCHAR}, introduction,jdbcType=VARCHAR},
38  #{language,jdbcType=VARCHAR}, #{price,jdbcType=DOUBLE}, #{publishDate,jdbcType=DATE},
39  #{typeId,jdbcType=INTEGER}, #{status,jdbcType=INTEGER})
40  </insert>
41  <insert id="insertSelective" parameterType="com.library.po.BookInfo">
42      insert into book_info
43      <trim prefix="(" suffix=")" suffixOverrides=",">
```

```
44          <if test="id != null">
45              id,
46          </if>
47          <if test="name != null">
48              name,
49          </if>
50          <if test="author != null">
51              author,
52          </if>
53          <if test="publish != null">
54              publish,
55          </if>
56          <if test="isbn != null">
57              isbn,
58          </if>
59          <if test="introduction != null">
60              introduction,
61          </if>
62          <if test="language != null">
63              language,
64          </if>
65          <if test="price != null">
66              price,
67          </if>
68          <if test="publishDate != null">
69              publish_date,
70          </if>
71          <if test="typeId != null">
72              type_id,
73          </if>
74          <if test="status != null">
75              status,
76          </if>
77      </trim>
78      <trim prefix="values (" suffix=")" suffixOverrides=",">
79          <if test="id != null">
80              #{id,jdbcType=INTEGER},
81          </if>
82          <if test="name != null">
83              #{name,jdbcType=VARCHAR},
84          </if>
85          <if test="author != null">
86              #{author,jdbcType=VARCHAR},
87          </if>
88          <if test="publish != null">
89              #{publish,jdbcType=VARCHAR},
```

```
90          </if>
91          <if test="isbn != null">
92              #{isbn,jdbcType=VARCHAR},
93          </if>
94          <if test="introduction != null">
95              #{introduction,jdbcType=VARCHAR},
96          </if>
97          <if test="language != null">
98              #{language,jdbcType=VARCHAR},
99          </if>
100         <if test="price != null">
101             #{price,jdbcType=DOUBLE},
102         </if>
103         <if test="publishDate != null">
104             #{publishDate,jdbcType=DATE},
105         </if>
106         <if test="typeId != null">
107             #{typeId,jdbcType=INTEGER},
108         </if>
109         <if test="status != null">
110             #{status,jdbcType=INTEGER},
111         </if>
112     </trim>
113 </insert>
114 <update id="updateByPrimaryKeySelective" parameterType="com.library.po.BookInfo">
115     update book_info
116     <set>
117         <if test="name != null">
118             name = #{name,jdbcType=VARCHAR},
119         </if>
120         <if test="author != null">
121             author = #{author,jdbcType=VARCHAR},
122         </if>
123         <if test="publish != null">
124             publish = #{publish,jdbcType=VARCHAR},
125         </if>
126         <if test="isbn != null">
127             isbn = #{isbn,jdbcType=VARCHAR},
128         </if>
129         <if test="introduction != null">
130             introduction = #{introduction,jdbcType=VARCHAR},
131         </if>
132         <if test="language != null">
133             language = #{language,jdbcType=VARCHAR},
134         </if>
135         <if test="price != null">
```

```
136             price = #{price,jdbcType=DOUBLE},
137         </if>
138         <if test="publishDate != null">
139             publish_date = #{publishDate,jdbcType=DATE},
140         </if>
141         <if test="typeId != null">
142             type_id = #{typeId,jdbcType=INTEGER},
143         </if>
144         <if test="status != null">
145             status = #{status,jdbcType=INTEGER},
146         </if>
147     </set>
148     where id = #{id,jdbcType=INTEGER}
149 </update>
150 <update id="updateByPrimaryKey" parameterType="com.library.po.BookInfo">
151     update book_info
152     set name = #{name,jdbcType=VARCHAR},
153     author = #{author,jdbcType=VARCHAR},
154     publish = #{publish,jdbcType=VARCHAR},
155     isbn = #{isbn,jdbcType=VARCHAR},
156     introduction = #{introduction,jdbcType=VARCHAR},
157     language = #{language,jdbcType=VARCHAR},
158     price = #{price,jdbcType=DOUBLE},
159     publish_date = #{publishDate,jdbcType=DATE},
160     type_id = #{typeId,jdbcType=INTEGER},
161     status = #{status,jdbcType=INTEGER}
162     where id = #{id,jdbcType=INTEGER}
163 </update>
164 <resultMap id="queryBookAllMap" type="com.library.po.BookInfo" extends="BaseResultMap">
165     <association property="typeInfo" javaType="com.library.po.TypeInfo">
166     <id column="id" property="id"></id>
167     <result column="type_name" property="name"></result>
168     </association>
169 </resultMap>
170 <select id="queryBookInfoAll" parameterType="com.library.po.BookInfo"
    resultMap="queryBookAllMap">
171     select book_info.*,type_info.name as type_name
172     from book_info,type_info
173     where type_info.id = book_info.type_id
174     <if test="name!=null">
175         and book_info.name like '%${name}%'
176     </if>
177     <if test="isbn!=null">
178         and book_info.isbn like '%${isbn}%'
179     </if>
180     <if test="typeId!=null">
```

```
181             and  book_info.type_id like '%${typeId}%'
182         </if>
183     </select>
184     <select id="getBookCountByType" resultType="com.library.po.BookInfo">
185         SELECT
186         count( book.id ) AS counts,
187         type.NAME
188         FROM
189         book_info book
190         LEFT JOIN type_info type ON type.id = book.type_id
191         GROUP BY
192         book.type_id
193     </select>
194     </mapper>
```

文件 18.13　TypeInfoMapper.xml

```
01  <?xml version="1.0" encoding="UTF-8" ?>
02  <!DOCTYPE mapper PUBLIC "-//mybatis.org//DTD Mapper 3.0//EN"
    "http://mybatis.org/dtd/mybatis-3-mapper.dtd" >
03  <mapper namespace="com.library.dao.TypeInfoMapper" >
04  <!--查询全部类型信息-->
05  <select id="queryTypeInfoAll" resultType="com.library.po.TypeInfo">
06      select * from type_info
07      <where>
08          <if test="name!=null">
09              and name like '%${name}%'
10          </if>
11      </where>
12  </select>
13  <!--类型的添加-->
14  <insert id="addTypeSubmit">
15      insert into type_info (name,remarks)values(#{name},#{remarks})
16  </insert>
17  <!--根据 id 查询类型信息-->
18  <select id="queryTypeInfoById" resultType="com.library.po.TypeInfo">
19      select * from type_info where id=#{id}
20  </select>
21  <!--修改图书类型-->
22  <update id="updateTypeSubmit">
23      update type_info set name=#{name},remarks=#{remarks} where id=#{id}
24  </update>
25  <!--删除类型-->
26  <delete id="deleteTypeByIds" parameterType="List">
27      delete from type_info where id in
28      <foreach collection="list" item="id" open="(" separator="," close=")">
29          #{id}
```

```
30      </foreach>
31  </delete>
32  <select id="queryTypeName" resultType="com.library.po.TypeInfo">
33      select type.name
34      from type_info type
35  </select>
36  </mapper>
```

3. 实现 Service 层接口

实现 Service 层接口的操作步骤如下：

步骤01 创建 Service 层接口。在 src 目录下创建一个 com.library.service 包，在包中创建 BookInfoService 和 TypeInfoService 接口，并在该接口中编写相关方法，如文件 18.14 和文件 18.15 所示。

文件 18.14　BookInfoService.java

```
01  public interface BookInfoService {
02      // 查询所有记录
03      PageInfo<BookInfo> queryBookInfoAll(BookInfo bookInfo,Integer
    pageNum,Integerlimit);
04      // 添加图书记录
05      void addBookSubmit(BookInfo bookInfo);
06      // 修改，根据 id 查询记录信息
07      BookInfo queryBookInfoById(Integer id);
08      // 修改提交
09      void updateBookSubmit(BookInfo info);
10      // 根据 id 删除记录信息
11      void deleteBookByIds(List<String> ids);
12      // 根据类型获取图书数量
13      List<BookInfo> getBookCountByType();
14  }
```

文件 18.15　TypeInfoService.java

```
01  public interface TypeInfoService {
02      // 查询所有记录
03      PageInfo<TypeInfo> queryTypeInfoAll(String name, Integer pageNum, Integer limit);
04      // 添加图书类型
05      void addTypeSubmit(TypeInfo info);
06      // 修改，根据 id 查询记录信息
07      TypeInfo queryTypeInfoById(Integer id);
08      // 修改提交
09      void updateTypeSubmit(TypeInfo info);
10      // 根据 id 删除记录信息
11      void deleteTypeByIds(List<String> id);
12  }
```

步骤 02 创建 Service 层接口的实现类。在 src 目录下创建一个 com.library.service.impl 包，并在包中创建 BookInfoService 接口的实现类 BookInfoServiceImpl 和 TypeInfoService 接口的实现类 TypeInfoServiceImpl，在类中编辑并实现接口中的方法，如文件 18.16 和文件 18.17 所示。

文件 18.16 BookInfoServiceImpl.java

```
01  @Service("bookInfoService")
02  public class BookInfoServiceImpl implements BookInfoService {
03      @Autowired
04      private BookInfoMapper bookInfoMapper;
05      @Override
06      public PageInfo<BookInfo> queryBookInfoAll(BookInfo bookInfo, Integer pageNum,
    Integer limit) {
07          PageHelper.startPage(pageNum,limit);
08          List<BookInfo> bookInfoList = bookInfoMapper.queryBookInfoAll(bookInfo);
09          return new PageInfo<>(bookInfoList);
10      }
11      @Override
12      public void addBookSubmit(BookInfo bookInfo) {
13          bookInfoMapper.insert(bookInfo);
14      }
15      @Override
16      public BookInfo queryBookInfoById(Integer id) {
17          return bookInfoMapper.selectByPrimaryKey(id);
18      }
19      @Override
20      public void updateBookSubmit(BookInfo info) {
21          bookInfoMapper.updateByPrimaryKeySelective(info);
22      }
23      @Override
24      public void deleteBookByIds(List<String> ids) {
25          for (String id : ids){
26              bookInfoMapper.deleteByPrimaryKey(Integer.parseInt(id));
27          }
28      }
29      @Override
30      public List<BookInfo> getBookCountByType() {
31          return bookInfoMapper.getBookCountByType();
32      }
33  }
```

文件 18.17 TypeInfoServiceImpl.java

```
01  @Service("typeInfoService")
02  public class TypeInfoServiceImpl implements TypeInfoService {
03      @Autowired
04      private TypeInfoMapper typeInfoMapper;
05      @Override
```

```
06      public PageInfo<TypeInfo> queryTypeInfoAll(String name,Integer pageNum, Integer
    limit) {
07          PageHelper.startPage(pageNum,limit);
08          List<TypeInfo> typeInfoList =  typeInfoMapper.queryTypeInfoAll(name);
09          return new PageInfo<>(typeInfoList);
10      }
11      @Override
12      public void addTypeSubmit(TypeInfo info) {
13          typeInfoMapper.addTypeSubmit(info);
14      }
15      @Override
16      public TypeInfo queryTypeInfoById(Integer id) {
17          return typeInfoMapper.queryTypeInfoById(id);
18      }
19      @Override
20      public void updateTypeSubmit(TypeInfo info) {
21          typeInfoMapper.updateTypeSubmit(info);
22      }
23      @Override
24      public void deleteTypeByIds(List<String> id) {
25          List<Integer> list=new ArrayList<>();
26          for(String cid:id){
27              int id2= Integer.valueOf(cid);
28              list.add(id2);
29          }
30          typeInfoMapper.deleteTypeByIds(list);
31      }
32  }
```

4. 实现 Coutroller 类

在 src 目录下创建一个 com.library.controller 包，在包中创建控制器类 BookInfoController 和 TypeInfoController，代码如文件 18.18 和文件 18.19 所示。

文件 18.18 BookInfoController.java

```
01  @Controller
02  public class BookInfoController {
03      @Autowired
04      private BookInfoService bookInfoService;
05      @Autowired
06      private TypeInfoService typeInfoService;
07      @RequestMapping("/bookAll")
08      @ResponseBody
09      public DataInfo bookAll(BookInfo bookInfo, @RequestParam(defaultValue = "1") Integer
    pageNum, @RequestParam(defaultValue = "15") Integer limit){
10          PageInfo<BookInfo> pageInfo =
11  bookInfoService.queryBookInfoAll(bookInfo,pageNum,limit);
```

```
12            return DataInfo.ok("成功",pageInfo.getTotal(),pageInfo.getList());
13        }
14        // 添加页面的跳转
15        @GetMapping("/bookAdd")
16        public String bookAdd(){
17            return "book/bookAdd";
18        }
19        // 添加类型
20        @RequestMapping("/addBookSubmit")
21        @ResponseBody
22        public DataInfo addBookSubmit(BookInfo info){
23            bookInfoService.addBookSubmit(info);
24            return DataInfo.ok();
25        }
26        //根据 id 查询（修改） 类型
27        @GetMapping("/queryBookInfoById")
28        public String queryTypeInfoById(Integer id, Model model){
29            BookInfo bookInfo= bookInfoService.queryBookInfoById(id);
30            model.addAttribute("info",bookInfo);
31            return "book/updateBook";
32        }
33        @RequestMapping("/updateBookSubmit")
34        @ResponseBody
35        public DataInfo updateBookSubmit(@RequestBody BookInfo info){
36            bookInfoService.updateBookSubmit(info);
37            return DataInfo.ok();
38        }
39        // 删除类型
40        @RequestMapping("/deleteBook")
41        @ResponseBody
42        public DataInfo deleteBook(String ids){
43            List<String> list= Arrays.asList(ids.split(","));
44            bookInfoService.deleteBookByIds(list);
45            return DataInfo.ok();
46        }
47        @RequestMapping("/findAllList")
48        @ResponseBody
49        public List<TypeInfo> findAll(){
50            PageInfo<TypeInfo> pageInfo = typeInfoService.queryTypeInfoAll(null,1,100);
51            List<TypeInfo> lists = pageInfo.getList();
52            return lists;
53        }
54    }
```

文件 18.19　TypeInfoController.java

```
01    @Controller
```

```
02  public class TypeInfoController {
03      @Autowired
04      private TypeInfoService typeInfoService;
05      // 类型管理首页
06      @GetMapping("/typeIndex")
07      public String typeIndex(){
08          return "type/typeIndex";
09      }
10      // 获取 type 数据信息，分页
11      @RequestMapping("/typeAll")
12      @ResponseBody
13      public DataInfo typeAll(String name, @RequestParam(defaultValue = "1") Integer
    pageNum, @RequestParam(defaultValue = "15") Integer limit){
14          PageInfo<TypeInfo> pageInfo =
    typeInfoService.queryTypeInfoAll(name,pageNum,limit);
15          return  DataInfo.ok(" 成 功 ",pageInfo.getTotal(),pageInfo.getList());// 总 条 数
    getTotal，数据封装成 list，以便分页显示，由于加了 ResponseBody，因此会返回一个字符串
16      }
17      // 添加页面的跳转
18      @GetMapping("/typeAdd")
19      public String typeAdd(){
20          return "type/typeAdd";
21      }
22      @PostMapping("/addTypeSubmit")
23      @ResponseBody
24      public DataInfo addTypeSubmit(TypeInfo info){
25          typeInfoService.addTypeSubmit(info);
26          return DataInfo.ok();
27      }
28      //根据 id 查询（修改） 类型
29      @GetMapping("/queryTypeInfoById")
30      public String queryTypeInfoById(Integer id, Model model){
31          TypeInfo info= typeInfoService.queryTypeInfoById(id);
32          model.addAttribute("info",info);
33          return "type/updateType";
34      }
35      @RequestMapping("/updateTypeSubmit")
36      @ResponseBody
37      public DataInfo updateTypeSubmit(@RequestBody TypeInfo info){
38          typeInfoService.updateTypeSubmit(info);
39          return DataInfo.ok();
40      }
41      //删除类型
42      @RequestMapping("/deleteType")
43      @ResponseBody
44      public DataInfo deleteType(String ids){
```

```
45          List<String> list= Arrays.asList(ids.split(","));
46          typeInfoService.deleteTypeByIds(list);
47          return DataInfo.ok();
48      }
49  }
```

5. 实现页面功能

实现页面功能的操作步骤如下：

步骤 01 在项目 src/views/book 目录下创建 Book.vue 页面文件。Book.vue 主要实现图书信息显示、删除、添加和修改的功能，如文件 18.20 所示。

文件 18.20　Book.vue

```
01  <template>
02  <div>
03  <el-card>
04  <div style=" width:300px; display: flex;margin-right: 1%">
05  <el-input v-model="input" prefix-icon="el-icon-search" placeholder="请输入图书名"
    clearable />
06  <el-button>搜索</el-button>
07  </div>
08  <el-row :gutter="10">
09  <el-col :span="4">
10  <el-button type="primary" @click="getOptions">添加图书</el-button>
11  </el-col>
12  </el-row>
13  <!-- 图书列表区域 -->
14  <el-table :data="booklist" border style="width: 100%">
15  <el-table-column fixed prop="isbn" label="编号" width="150">
16  </el-table-column>
17  <el-table-column prop="name" label="名称" width="150">
18  </el-table-column>
19  <el-table-column prop="typeName" label="类型" width="120">
20  </el-table-column>
21  <el-table-column prop="author" label="作者" width="120">
22  </el-table-column>
23  <el-table-column prop="price" label="价格" width="120">
24  </el-table-column>
25  <el-table-column fixed="right" label="操作" >
26  <template slot-scope="scope" >
27  <el-button type="primary" icon="el-icon-edit" @click="showEditDialog(scope.row)">编辑
    </el-button>
28  <el-button type="danger" icon="el-icon-delete"
    @click="removebookById(scope.row.bookId)">删除</el-button>
29  </template>
30  </el-table-column>
```

```
31  </el-table>
32  <el-pagination @size-change="handleSizeChange" @current-change="handleCurrentChange"
33  :current-page="queryInfo.pageNum" :page-sizes="[3, 5,
    10]" :page-size="queryInfo.pageSize"
34  layout="total, sizes, prev, pager, next, jumper" :total="total">
35      </el-pagination>
36  </el-card>
37  <!-- 添加图书的对话框 -->
38  <el-dialog title="添加图书" :visible.sync="addDialogVisible" width="40%"
    @close="addDialogClosed">
39  <!-- 内容主体区域 -->
40  <el-form  :model="addForm"  :rules="addFormRules"  ref="addFormRef"  label-width="40px"
    size="medium"
41  hide-required-asterisk>
42  <el-row :gutter="10" type="flex" align="middle">
43  <el-col :span="12">
44  <el-form-item label="编号" prop="isbn">
45  <el-input v-model="addForm.isbn" placeholder="请输入图书编号"
    @input="change($event)"></el-input>
46  </el-form-item>
47  </el-col>
48  </el-row>
49  <el-row :gutter="10">
50  <el-col :span="12">
51  <el-form-item label="书名" prop="name">
52  <el-input v-model="addForm.name" placeholder="请输入图书名称"></el-input>
53  </el-form-item>
54  </el-col>
55  <el-col :span="12">
56  <el-form-item label="作者" prop="author">
57  <el-input v-model="addForm.author" placeholder="请输入图书作者"></el-input>
58  </el-form-item>
59  </el-col>
60  </el-row>
61  <el-row :gutter="10">
62  <el-col :span="12">
63  <el-form-item label="价格" prop="price">
64  <el-input v-model="addForm.price" placeholder="请输入图书价格"></el-input>
65  </el-form-item>
66  </el-col>
67  <el-col :span="12">
68  <el-form-item label="类型" prop="typeId">
69  <el-select v-model="addForm.typeId" placeholder="图书类别">
70  <el-option v-for="option in optionlist" :label="option.typeName" :value="option.typeId"
71  :key="option.typeId"></el-option>
72  </el-select>
```

```
73  </el-form-item>
74  </el-col>
75  </el-row>
76  </el-form>
77  <!-- 底部区域 -->
78  <span slot="footer" class="dialog-footer">
79  <el-button @click="addDialogVisible = false">取 消</el-button>
80  <el-button type="primary" @click="addbook">确 定</el-button>
81  </span>
82  </el-dialog>
83  <!-- 修改图书的对话框 -->
84  <el-dialog title="修改图书" :visible.sync="editDialogVisible" width="40%"
    @close="editDialogClosed">
85  <!-- 内容主体区域 -->
86  <el-form :model="editForm" :rules="editFormRules" ref="editFormRef" label-width="40px"
    size="medium"
87  hide-required-asterisk>
88  <el-row :gutter="10" type="flex" align="middle">
89  <el-col :span="12">
90  <el-form-item label="编号" prop="isbn">
91  <el-input v-model="editForm.isbn" placeholder="请输入图书编号"></el-input>
92  </el-form-item>
93  </el-col>
94  </el-row>
95  <el-row :gutter="10">
96  <el-col :span="12">
97  <el-form-item label="书名" prop="name">
98  <el-input v-model="editForm.name" placeholder="请输入图书名称"></el-input>
99  </el-form-item>
100 </el-col>
101 <el-col :span="12">
102 <el-form-item label="作者" prop="author">
103 <el-input v-model="editForm.author" placeholder="请输入图书作者"></el-input>
104 </el-form-item>
105 </el-col>
106 </el-row>
107 <el-row :gutter="10">
108 <el-col :span="12">
109 <el-form-item label="价格" prop="price">
110 <el-input v-model="editForm.price" placeholder="请输入图书价格"></el-input>
111 </el-form-item>
112 </el-col>
113 <el-col :span="12">
114 <el-form-item label="类型" prop="typeId">
115 <el-select v-model="editForm.typeId" placeholder="图书类别">
116 <el-option v-for="option in optionlist" :label="option.typeName" :value="option.typeId"
```

```
117 :key="option.typeId"></el-option>
118 </el-select>
119 </el-form-item>
120 </el-col>
121 </el-row>
122 </el-form>
123 <!-- 底部区域 -->
124 <span slot="footer" class="dialog-footer">
125 <el-button @click="editDialogVisible = false">取 消</el-button>
126 <el-button type="primary" @click="editbookInfo">确 定</el-button>
127 </span>
128 </el-dialog>
129 </div>
130 </template>
131 <!--省略部分代码-->
```

进入图书管理模块后，可以对图书进行相关操作，页面如图 18.5、图 18.6 所示。

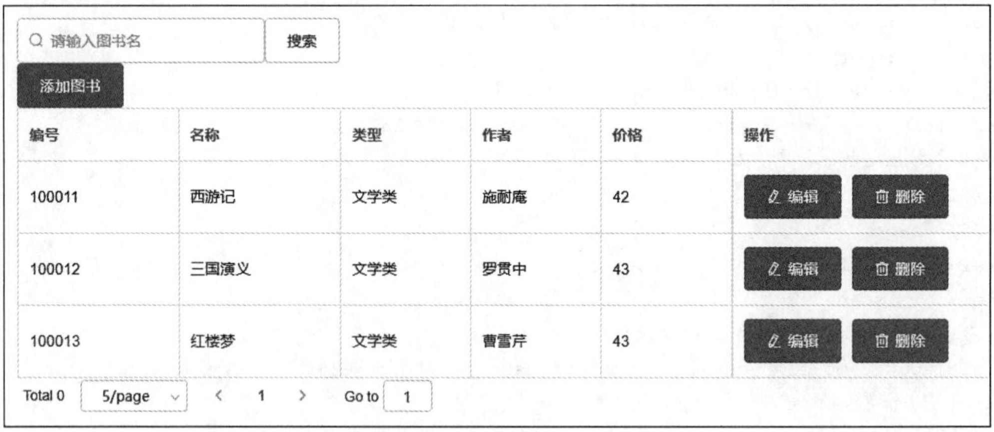

图 18.5　图书信息页面

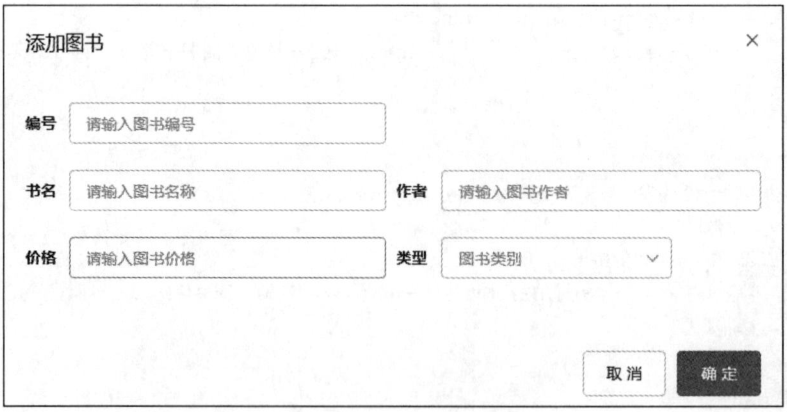

图 18.6　添加图书页面

步骤 02 在项目 src/views/type 目录下创建 type.vue 页面文件。type.vue 主要实现图书信息的显示、删除、添加和修改功能，如文件 18.21 所示。

文件 18.21 type.vue

```
01 <template>
02 <div>
03 <el-card>
04 <!-- 搜索与添加区域 -->
05 <div style=" width:300px; display: flex;margin-right: 1%">
06 <el-input  v-model="name"  prefix-icon="el-icon-search"  placeholder=" 请 输 入 类 型 名 "
   clearable />
07 <el-button>搜索</el-button>
08 </div>
09 <el-row :gutter="10">
10 <el-col :span="4">
11 <el-button type="primary" @click="addDialogVisible = true">添加</el-button>
12 </el-col>
13 </el-row>
14 <el-table :data="userlist" style="width: 100%" border stripe>
15 <el-table-column type="index"></el-table-column>
16 <el-table-column prop="name" label="类型名称" width="120px"></el-table-column>
17 <el-table-column prop="remarks" label="备注" width="150px"></el-table-column>
18 <el-table-column label="操作" width="300px">
19 <template slot-scope="scope">
20 <!-- 修改按钮 -->
21 <el-button type="primary" icon="el-icon-edit" @click="showEditDialog(scope.row)">编辑
   </el-button>
22 <!-- 删除按钮 -->
23 <el-button type="danger" icon="el-icon-delete"
   @click="removeUserById(scope.row.userId)">删除</el-button>
24 </template>
25 </el-table-column>
26 </el-table>
27 <!-- 分页区域 -->
28 <el-pagination @size-change="handleSizeChange" @current-change="handleCurrentChange"
29 :current-page="queryInfo.pageNum" :page-sizes="[1, 2, 5,
   10]" :page-size="queryInfo.pageSize"
30 layout="total, sizes, prev, pager, next, jumper" :total="total">
31 </el-pagination>
32 </el-card>
33 <!-- 添加的对话框 -->
34 <el-dialog title="添加" :visible.sync="addDialogVisible" width="50%"
   @close="addDialogClosed">
35 <!-- 内容主体区域 -->
36 <el-form :model="addForm" :rules="addFormRules" ref="addFormRef" label-width="100px">
37 <el-form-item label="类型名称" prop="name">
38 <el-input v-model="addForm.name"></el-input>
```

```
39  </el-form-item>
40  <el-form-item label="备注" prop="remarks">
41  <el-input v-model="addForm.remarks"></el-input>
42  </el-form-item>
43  </el-form>
44  <!-- 底部区域 -->
45  <span slot="footer" class="dialog-footer">
46  <el-button @click="addDialogVisible = false">取 消</el-button>
47  <el-button type="primary" @click="addUser">确 定</el-button>
48  </span>
49  </el-dialog>
50  <!-- 修改的对话框 -->
51  <el-dialog title="编辑" :visible.sync="editDialogVisible" width="50%"
    @close="editDialogClosed">
52  <!-- 内容主体区域 -->
53  <el-form :model="editForm" :rules="editFormRules" ref="editFormRef" label-width="70px">
54  <el-form-item label="类型名称" prop="name">
55  <el-input v-model="editForm.name"></el-input>
56  </el-form-item>
57  <el-form-item label="备注" prop="remarks">
58  <el-input v-model="editForm.remarks"></el-input>
59  </el-form-item>
60  </el-form>
61  <!-- 底部区域 -->
62  <span slot="footer" class="dialog-footer">
63  <el-button @click="editDialogVisible = false">取 消</el-button>
64  <el-button type="primary" @click="editUserInfo">确 定</el-button>
65  </span>
66  </el-dialog>
67  </div>
68  </template>
69  <!--省略部分代码-->
```

进入类别管理模块后，可以对图书类别进行相关操作，页面如图 18.7、图 18.8 所示。

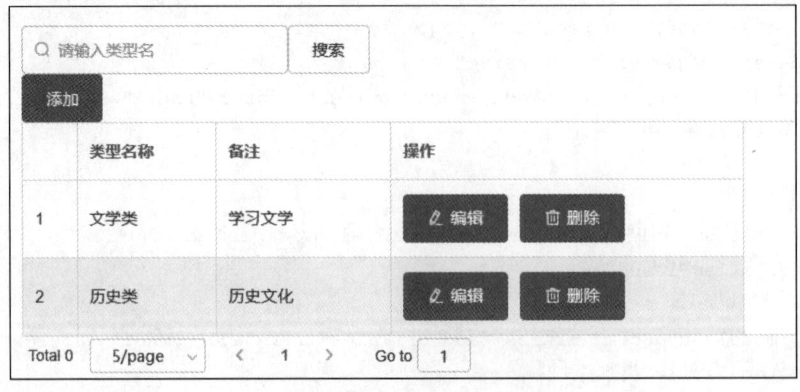

图 18.7 图书类别信息页面

图 18.8　添加图书类别页面

18.4.3　借阅管理模块

1. 创建持久化类

借阅管理模块持久化类是 LendList，代码如文件 18.22 所示。

文件 18.22　LendList.java

```
01 public class LendList implements Serializable {
02     private Integer id;
03     private Integer bookId;
04     private Integer readerId;
05     @DateTimeFormat(pattern = "yyyy-MM-dd HH:mm:ss")//接收页面输入的时间，将它格式化
06     @JSONField(format = "yyyy-MM-dd HH:mm:ss")//后端传的日期格式化
07     private Date lendDate;
08     @DateTimeFormat(pattern = "yyyy-MM-dd HH:mm:ss")
09     @JSONField(format = "yyyy-MM-dd HH:mm:ss")
10     private Date backDate;
11     private Integer backType;
12     private String exceptRemarks;
13     private BookInfo bookInfo;
14     private ReaderInfo readerInfo;
15     private static final long serialVersionUID = 1L;
16     public Integer getId() {
17         return id;
18     }
19     public void setId(Integer id) {
20         this.id = id;
21     }
22     public Integer getBookId() {
23         return bookId;
24     }
25     public void setBookId(Integer bookId) {
26         this.bookId = bookId;
27     }
28     public Integer getReaderId() {
```

```
29            return readerId;
30        }
31        public void setReaderId(Integer readerId) {
32            this.readerId = readerId;
33        }
34        public Date getLendDate() {
35            return lendDate;
36        }
37        public void setLendDate(Date lendDate) {
38            this.lendDate = lendDate;
39        }
40        public Date getBackDate() {
41            return backDate;
42        }
43        public void setBackDate(Date backDate) {
44            this.backDate = backDate;
45        }
46        public Integer getBackType() {
47            return backType;
48        }
49        public void setBackType(Integer backType) {
50            this.backType = backType;
51        }
52        public String getExceptRemarks() {
53            return exceptRemarks;
54        }
55        public void setExceptRemarks(String exceptRemarks) {
56            this.exceptRemarks = exceptRemarks;
57        }
58        public BookInfo getBookInfo() {
59            return bookInfo;
60        }
61        public void setBookInfo(BookInfo bookInfo) {
62            this.bookInfo = bookInfo;
63        }
64        public ReaderInfo getReaderInfo() {
65            return readerInfo;
66        }
67        public void setReaderInfo(ReaderInfo readerInfo) {
68            this.readerInfo = readerInfo;
69        }
70    }
```

2. 实现 DAO 层接口

实现 DAO 层接口的操作步骤如下：

步骤 01 创建DAO层接口。在src目录下的com.library.dao包中创建一个接口LendListMapper，并在接口中编写增、删、改、查等方法，如文件18.23所示。

文件 18.23 LendListMapper.java

```
01  public interface LendListMapper {
02      void deleteByPrimaryKey(Integer id);
03      void insert(LendList record);
04      void insertSelective(LendList record);
05      LendList selectByPrimaryKey(Integer id);
06      void updateByPrimaryKeySelective(LendList record);
07      void updateByPrimaryKey(LendList record);
08      List<LendList> queryLendListAll(LendList lendList);
09      List<LendList> queryLookBookList(@Param("rid") Integer rid,@Param("bid") Integer
    bid);
10      void updateLendListSubmit(LendList lendList);
11  }
```

步骤 02 创建映射文件。在resources目录下的com.library.dao包中创建MyBatis映射文件LendListMapper.xml，并在映射文件中编写增、删、改、查等方法的执行语句，如文件18.24所示。

文件 18.24 LendListMapper.xml

```
01  <?xml version="1.0" encoding="UTF-8"?>
02  <!DOCTYPE mapper PUBLIC "-//mybatis.org//DTD Mapper 3.0//EN"
    "http://mybatis.org/dtd/mybatis-3-mapper.dtd">
03  <mapper namespace="com.library.dao.LendListMapper">
04  <resultMap id="BaseResultMap" type="com.library.po.LendList">
05      <id column="id" jdbcType="INTEGER" property="id" />
06      <result column="bookId" jdbcType="INTEGER" property="bookId" />
07      <result column="readerId" jdbcType="INTEGER" property="readerId" />
08      <result column="lendDate" jdbcType="TIMESTAMP" property="lendDate" />
09      <result column="backDate" jdbcType="TIMESTAMP" property="backDate" />
10      <result column="backType" jdbcType="INTEGER" property="backType" />
11      <result column="exceptRemarks" jdbcType="VARCHAR" property="exceptRemarks" />
12  </resultMap>
13  <sql id="Base_Column_List">
14      id, bookId, readerId, lendDate, backDate, backType, exceptRemarks
15  </sql>
16  <select id="selectByPrimaryKey" parameterType="java.lang.Integer"
    resultMap="BaseResultMap">
17      select
18      <include refid="Base_Column_List" />
19      from lend_list
20      where id = #{id,jdbcType=INTEGER}
21  </select>
22  <delete id="deleteByPrimaryKey" parameterType="java.lang.Integer">
23      delete from lend_list
```

```
24      where id = #{id,jdbcType=INTEGER}
25   </delete>
26   <insert id="insert" parameterType="com.library.po.LendList">
27      insert into lend_list (id, bookId, readerId,
28      lendDate, backDate, backType,
29      exceptRemarks)
30      values (#{id,jdbcType=INTEGER}, #{bookId,jdbcType=INTEGER},
   #{readerId,jdbcType=INTEGER},
31   #{lendDate,jdbcType=TIMESTAMP}, #{backDate,jdbcType=TIMESTAMP},
   #{backType,jdbcType=INTEGER},
32   #{exceptRemarks,jdbcType=VARCHAR})
33   </insert>
34   <insert id="insertSelective" parameterType="com.library.po.LendList">
35      insert into lend_list
36      <trim prefix="(" suffix=")" suffixOverrides=",">
37          <if test="id != null">
38              id,
39          </if>
40          <if test="bookId != null">
41              bookId,
42          </if>
43          <if test="readerId != null">
44              readerId,
45          </if>
46          <if test="lendDate != null">
47              lendDate,
48          </if>
49          <if test="backDate != null">
50              backDate,
51          </if>
52          <if test="backType != null">
53              backType,
54          </if>
55          <if test="exceptRemarks != null">
56              exceptRemarks,
57          </if>
58      </trim>
59      <trim prefix="values (" suffix=")" suffixOverrides=",">
60          <if test="id != null">
61              #{id,jdbcType=INTEGER},
62          </if>
63          <if test="bookId != null">
64              #{bookId,jdbcType=INTEGER},
65          </if>
66          <if test="readerId != null">
67              #{readerId,jdbcType=INTEGER},
```

```
68            </if>
69            <if test="lendDate != null">
70                #{lendDate,jdbcType=TIMESTAMP},
71            </if>
72            <if test="backDate != null">
73                #{backDate,jdbcType=TIMESTAMP},
74            </if>
75            <if test="backType != null">
76                #{backType,jdbcType=INTEGER},
77            </if>
78            <if test="exceptRemarks != null">
79                #{exceptRemarks,jdbcType=VARCHAR},
80            </if>
81        </trim>
82    </insert>
83    <update id="updateByPrimaryKeySelective" parameterType="com.library.po.LendList">
84        update lend_list
85        <set>
86            <if test="bookId != null">
87                bookId = #{bookId,jdbcType=INTEGER},
88            </if>
89            <if test="readerId != null">
90                readerId = #{readerId,jdbcType=INTEGER},
91            </if>
92            <if test="lendDate != null">
93                lendDate = #{lendDate,jdbcType=TIMESTAMP},
94            </if>
95            <if test="backDate != null">
96                backDate = #{backDate,jdbcType=TIMESTAMP},
97            </if>
98            <if test="backType != null">
99                backType = #{backType,jdbcType=INTEGER},
100           </if>
101           <if test="exceptRemarks != null">
102               exceptRemarks = #{exceptRemarks,jdbcType=VARCHAR},
103           </if>
104       </set>
105       where id = #{id,jdbcType=INTEGER}
106   </update>
107   <update id="updateByPrimaryKey" parameterType="com.library.po.LendList">
108       update lend_list
109       set bookId = #{bookId,jdbcType=INTEGER},
110       readerId = #{readerId,jdbcType=INTEGER},
111       lendDate = #{lendDate,jdbcType=TIMESTAMP},
112       backDate = #{backDate,jdbcType=TIMESTAMP},
113       backType = #{backType,jdbcType=INTEGER},
```

```
114     exceptRemarks = #{exceptRemarks,jdbcType=VARCHAR}
115     where id = #{id,jdbcType=INTEGER}
116 </update>
117 
118 <resultMap id="queryLendListAllMap" type="com.library.po.LendList"
    extends="BaseResultMap">
119     <association property="bookInfo" javaType="com.library.po.BookInfo">
120         <id property="id" column="id"></id>
121         <result property="name" column="bookName"></result>
122     </association>
123 
124     <association property="readerInfo" javaType="com.library.po.ReaderInfo">
125         <id property="id" column="id"></id>
126         <result property="realName" column="realName"></result>
127         <result property="readerNumber" column="readerNumber"></result>
128     </association>
129 </resultMap>
130 <!--查询所有记录-->
131 <select id="queryLendListAll" parameterType="com.library.po.LendList"
    resultMap="queryLendListAllMap">
132     SELECT lend.*,
133     book.name as bookName,
134     reader.realName as realName,
135     reader.readerNumber
136     from lend_list lend LEFT JOIN book_info book  on book.id=lend.bookId
137     LEFT JOIN reader_info reader on reader.id=lend.readerId
138     <where>
139     <if test="bookInfo!=null">
140     <if test="bookInfo.name!=null and bookInfo.name!=''">
141         and book.name like '%${bookInfo.name}%'
142     </if>
143     <if test="bookInfo.status!=null and bookInfo.status==1">
144         and book.status=1 and backDate is null
145     </if>
146     <if test="bookInfo.status!=null and bookInfo.status==0">
147         and book.status=0 and backDate is not null
148     </if>
149     </if>
150     <!--根据借阅卡进行查询-->
151     <if test="readerInfo!=null">
152     <if test="readerInfo.readerNumber!=null and readerInfo.readerNumber!=''">
153         and readerNumber like '%${readerInfo.readerNumber}%'
154     </if>
155     </if>
156     <!--根据归还类型进行查询-->
157     <if test="backType!=null">
```

```
158             and backType=#{backType}
159         </if>
160         </where>
161         order by lend.lendDate desc
162     </select>
163
164     <update id="updateLendListSubmit" parameterType="com.library.po.LendList">
165         update lend_list
166         <set>
167             backDate=#{backDate},
168             <if test="backType!=null">
169                 backType=#{backType},
170             </if>
171             <if test="exceptRemarks!=null and exceptRemarks!=''">
172                 exceptRemarks=#{exceptRemarks}
173             </if>
174         </set>
175         where id=#{id}
176     </update>
177
178     <select id="queryLookBookList" resultMap="queryLendListAllMap">
179         SELECT
180         lend.*,
181         reader.readerNumber,
182         bookInfo.name  AS  bookName,
183         reader.realName AS realName
184         FROM
185         lend_list lend
186         LEFT JOIN reader_info reader ON reader.id = lend.readerId
187         LEFT JOIN book_info bookInfo ON bookInfo.id = lend.bookId
188         <where>
189             <if test="bid!=null">and bookInfo.id=#{bid} </if>
190             <if test="rid!=null">and reader.id=#{rid} </if>
191         </where>
192         order by lend.id desc
193     </select>
194 </mapper>
```

3. 实现 Service 层接口

实现 Service 层接口的操作步骤如下：

步骤 01 创建 Service 层接口。在 src 目录下创建一个 com.library.service 包，在包中创建 LendListService 接口，并在该接口中编写相关方法，如文件 18.25 所示。

文件 18.25 LendListService.java

```
01  public interface LendListService {
```

```
02      //分页查询
03      PageInfo<LendList> queryLendListAll(LendList lendList, int page, int limit);
04      //添加借阅记录
05      void addLendListSubmit(LendList lendList);
06      void deleteLendListById(List<String> ids, List<String> bookIds);
07      void updateLendListSubmit(List<String> ids, List<String> bookIds);
08      void backBook(LendList lendList);
09      // 时间线查询
10      List<LendList> queryLookBookList(Integer rid, Integer bid);
11  }
```

步骤02 创建 Service 层接口的实现类。在 src 目录下创建一个 com.library.service.impl 包，并在包中创建 LendListService 接口的实现类 LendListServiceImpl，在类中编辑并实现接口中的方法，如文件 18.26 所示。

文件 18.26 LendListServiceImpl.java

```
01  @Service("lendListService")
02  public class LendListServiceImpl implements LendListService {
03      @Autowired
04      private LendListMapper lendListMapper;
05      @Autowired
06      private BookInfoMapper bookInfoMapper;
07      @Override
08      public PageInfo<LendList> queryLendListAll(LendList lendList,int page, int limit) {
09          PageHelper.startPage(page,limit);
10          List<LendList> list=lendListMapper.queryLendListAll(lendList);
11          PageInfo pageInfo=new PageInfo(list);
12          return pageInfo;
13      }
14      @Override
15      public void addLendListSubmit(LendList lendList) {
16          lendListMapper.insert(lendList);
17      }
18      @Override
19      public void deleteLendListById(List<String> ids, List<String> bookIds) {
20          for(String id:ids){
21              lendListMapper.deleteByPrimaryKey(Integer.parseInt(id));
22          }
23          for(String bid:bookIds){
24              //根据 id 查询图书记录信息
25              BookInfo bookInfo =
    bookInfoMapper.selectByPrimaryKey(Integer.parseInt(bid));
26              bookInfo.setStatus(0);//改为未借出
27              bookInfoMapper.updateByPrimaryKey(bookInfo);
28          }
29      }
```

```
30      @Override
31      public void updateLendListSubmit(List<String> ids, List<String> bookIds) {
32          for(String id:ids){
33              //根据 id 查询借阅记录信息
34              LendList lendList=new LendList();
35              lendList.setId(Integer.parseInt(id));
36              lendList.setBackDate(new Date());
37              lendList.setBackType(0);
38              lendListMapper.updateLendListSubmit(lendList);
39          }
40          for(String bid:bookIds){
41              BookInfo bookInfo =
    bookInfoMapper.selectByPrimaryKey(Integer.parseInt(bid));
42              bookInfo.setStatus(0);//改为未借出
43              bookInfoMapper.updateByPrimaryKey(bookInfo);
44          }
45      }
46      @Override
47      public void backBook(LendList lendList) {
48          LendList lend=new LendList();
49          lend.setId(lendList.getId());
50          lend.setBackType(lendList.getBackType());
51          lend.setBackDate(new Date());
52          lend.setExceptRemarks(lendList.getExceptRemarks());
53          lend.setBookId(lendList.getBookId());
54          lendListMapper.updateLendListSubmit(lend);
55          if(lend.getBackType()==0 || lend.getBackType()==1){
56              BookInfo bookInfo=bookInfoMapper.selectByPrimaryKey(lend.getBookId());
57              bookInfo.setStatus(0);
58              bookInfoMapper.updateByPrimaryKey(bookInfo);
59          }
60      }
61      @Override
62      public List<LendList> queryLookBookList(Integer rid, Integer bid) {
63          return lendListMapper.queryLookBookList(rid, bid);
64      }
65  }
```

4. 实现 Controller 类

在 src 目录下创建一个 com.library.controller 包，在包中创建控制器类 LendListController，如文件 18.27 所示。

文件 18.27　LendListController.java

```
01  @Controller
02  public class LendListController {
03      @Autowired
```

```
04      private LendListService lendListService;
05      @Autowired
06      private ReaderInfoService readerService;
07      @Autowired
08      private BookInfoService bookInfoService;
09      @GetMapping("/lendListIndex")
10      public String lendListIndex(){
11          return "lend/lendListIndex";
12      }
13      @ResponseBody
14      @RequestMapping("/lendListAll")
15      public DataInfo lendListAll(Integer type, String readerNumber, String name, Integer
16  status, @RequestParam(defaultValue = "1")Integer page,@RequestParam(defaultValue =
17  "15")Integer limit){
18          LendList info=new LendList();
19          info.setBackType(type);
20          ReaderInfo reader=new ReaderInfo();
21          reader.setReaderNumber(readerNumber);
22          info.setReaderInfo(reader);
23          BookInfo book=new BookInfo();
24          book.setName(name);
25          book.setStatus(status);
26          info.setBookInfo(book);
27          PageInfo pageInfo=lendListService.queryLendListAll(info,page,limit);
28          return DataInfo.ok("ok",pageInfo.getTotal(),pageInfo.getList());
29      }
30      @GetMapping("/addLendList")
31      public String addLendList(){
32          return "lend/addLendList";
33      }
34      @ResponseBody
35      @RequestMapping("/addLend")
36      public DataInfo addLend(String readerNumber,String ids){
37          //获取图书 id 的集合
38          List<String> list= Arrays.asList(ids.split(","));
39          //判断卡号是否存在
40          ReaderInfo reader=new ReaderInfo();
41          reader.setReaderNumber(readerNumber);
42          PageInfo<ReaderInfo> pageInfo=readerService.queryAllReaderInfo(reader,1,1);
43          if(pageInfo.getList().size()==0){
44              return DataInfo.fail("卡号信息不存在");
45          }else{
46              ReaderInfo readerCard2=pageInfo.getList().get(0);
47              //可借书
48              for(String bid:list) {
49                  LendList lendList = new LendList();
```

```
50                  lendList.setReaderId(readerCard2.getId());//读者 id
51                  lendList.setBookId(Integer.valueOf(bid));//书的 id
52                  lendList.setLendDate(new Date());
53                  lendListService.addLendListSubmit(lendList);
54                  //变更书的状态
55                  BookInfo info =
56  bookInfoService.queryBookInfoById(Integer.valueOf(bid));
57                  //设置书的状态
58                  info.setStatus(1);
59                  bookInfoService.updateBookSubmit(info);
60              }
61          }
62          return DataInfo.ok();
63      }
64      @ResponseBody
65      @RequestMapping("/deleteLendListByIds")
66      public DataInfo deleteLendListByIds(String ids, String bookIds){
67          List list=Arrays.asList(ids.split(","));//借阅记录的 id
68          List blist=Arrays.asList(bookIds.split(","));//图书信息的 id
69          lendListService.deleteLendListById(list,blist);
70          return DataInfo.ok();
71      }
72      @ResponseBody
73      @RequestMapping("/backLendListByIds")
74      public DataInfo backLendListByIds(String ids,String bookIds){
75          List list=Arrays.asList(ids.split(","));//借阅记录的 id
76          List blist=Arrays.asList(bookIds.split(","));//图书信息的 id
77          lendListService.updateLendListSubmit(list,blist);
78          return DataInfo.ok();
79      }
80      @GetMapping("/excBackBook")
81      public String excBackBook(HttpServletRequest request, Model model){
82          //获取借阅记录 id
83          String id=request.getParameter("id");
84          String  bId=request.getParameter("bookId");
85          model.addAttribute("id",id);
86          model.addAttribute("bid",bId);
87          return "lend/excBackBook";
88      }
89      @ResponseBody
90      @RequestMapping("/updateLendInfoSubmit")
91      public DataInfo updateLendInfoSubmit(LendList lendList){
92          lendListService.backBook(lendList);
93          return DataInfo.ok();
94      }
95      // 查阅时间线
```

```
96      @RequestMapping("/queryLookBookList")
97      public String queryLookBookList(String flag,Integer id,Model model){
98          List<LendList> list=null;
99          if(flag.equals("book")){
100             list=lendListService.queryLookBookList(null,id);
101         }else{
102             list=lendListService.queryLookBookList(id,null);
103         }
104         model.addAttribute("info",list);
105         return "lend/lookBookList";
106     }
107     @RequestMapping("/queryLookBookList2")
108     public String queryLookBookList(HttpServletRequest request,Model model){
109         ReaderInfo readerInfo = (ReaderInfo) request.getSession().getAttribute("user");
110         List<LendList> list =
111 list=lendListService.queryLookBookList(readerInfo.getId(),null);
112         model.addAttribute("info",list);
113         return "lend/lookBookList";
114     }
115 }
```

5. 实现页面功能

在项目的 src/views/lend 目录下创建 lend.vue 页面文件。lend.vue 主要实现借阅模块的添加、查看、修改、删除功能，如文件 18.28 所示。

文件 18.28 lend.vue

```
01  <template>
02  <div>
03  <el-card>
04  <!-- 搜索与添加区域 -->
05  <div style=" width:300px; display: flex;margin-right: 1%">
06  <el-input v-model="input" prefix-icon="el-icon-search" placeholder="请输入图书名"
    clearable />
07  <el-button>搜索</el-button>
08  </div>
09  <el-row :gutter="10">
10  <el-col :span="4">
11  <el-button type="primary" @click="addDialogVisible = true">添加借阅信息</el-button>
12  </el-col>
13  </el-row>
14  <!-- 借阅信息列表区域 -->
15  <el-table :data="LendList" style="width: 100%" border stripe>
16  <el-table-column type="index"></el-table-column>
17  <el-table-column prop="readerNumber" label="借书卡号" width="100px"></el-table-column>
18  <el-table-column prop="readerId" label="借阅人"width="100px"></el-table-column>
19  <el-table-column prop="bookId" label="图书" align="center" padding="0px" width="150px">
20  </el-table-column>
21  <el-table-column prop="lendDate" label="借出时间" width="150px"></el-table-column>
```

```
22  <el-table-column prop="backDate" label="归还时间" width="150px"></el-table-column>
23  <el-table-column prop="backType" label="还书类型" width="120px"></el-table-column>
24  <el-table-column label="操作">
25  <template slot-scope="scope">
26  <!-- 修改按钮 -->
27  <el-button type="primary" icon="el-icon-edit"
    @click="showEditDialog(scope.row.bookIds)">编辑</el-button>
28  <!-- 删除按钮 -->
29  <el-button type="danger" icon="el-icon-delete"
    @click="removeBorrowById(scope.row.bookIds)">删除</el-button>
30  </template>
31  </el-table-column>
32  </el-table>
33  <!-- 分页区域 -->
34  <el-pagination @size-change="handleSizeChange" @current-change="handleCurrentChange"
35  :current-page="queryInfo.pageNum" :page-sizes="[1, 3, 5,
    10]" :page-size="queryInfo.pageSize"
36  layout="total, sizes, prev, pager, next, jumper" :total="total">
37  </el-pagination>
38  </el-card>
39  <!-- 添加借阅信息的对话框 -->
40  <el-dialog title="添加借阅信息" :visible.sync="addDialogVisible" width="30%"
    @close="addDialogClosed">
41  <!-- 内容主体区域 -->
42  <el-form :model="addForm" :rules="addFormRules" ref="addFormRef" label-width="auto">
43  <el-form-item label="借阅人" prop="readerId">
44  <el-input v-model="addForm.readerId"></el-input>
45  </el-form-item>
46  <el-form-item label="图书" prop="bookId">
47  <el-input v-model="addForm.bookId"></el-input>
48  </el-form-item>
49  <el-form-item label="借出时间" prop="lendDate">
50  <el-date-picker v-model="addForm.lendDate" type="date" placeholder="选择日期"
    value-format="yyyy-MM-dd"
51  @change="getTime"></el-date-picker>
52  </el-form-item>
53  <el-form-item label="归还时间" prop="backDate">
54  <el-date-picker v-model="addForm.backDate" type="date" placeholder="选择日期"
    value-format="yyyy-MM-dd"
55  @change="getTime"></el-date-picker>
56  </el-form-item>
57  <el-form-item label="还书类型" prop="backType">
58  <el-input v-model="addForm.backType"></el-input>
59  </el-form-item>
60  </el-form>
61  <span slot="footer" class="dialog-footer">
62  <el-button @click="addDialogVisible = false">取 消</el-button>
63  <el-button type="primary" @click="addBorrow">确 定</el-button>
64  </span>
65  </el-dialog>
66  <!-- 修改借阅信息的对话框 -->
67  <el-dialog title="修改借阅信息" :visible.sync="editDialogVisible" width="30%"
```

```
    @close="editDialogClosed">
68  <!-- 内容主体区域 -->
69  <el-form :model="editForm" :rules="editFormRules" ref="editFormRef" label-width="auto">
70  <el-form-item label="借书卡号" prop="readerNumber">
71  <el-input v-model="editForm.readerNumber"></el-input>
72  </el-form-item>
73  <el-form-item label="借阅人" prop="readerId">
74  <el-input v-model="editForm.readerId"></el-input>
75  </el-form-item>
76  <el-form-item label="图书" prop="bookId">
77  <el-input v-model="editForm.bookId"></el-input>
78  </el-form-item>
79  <el-form-item label="借出时间" prop="lendDate">
80  <el-date-picker  v-model="editForm.lendDate"  type="date"  placeholder=" 选 择 日 期 "
    value-format="yyyy-MM-dd">
81  </el-date-picker>
82  </el-form-item>
83  <el-form-item label="归还时间" prop="backDate">
84  <!-- <el-input v-model="editForm.backTime"></el-input> -->
85  <el-date-picker  v-model="editForm.backDate"  type="date"  placeholder=" 选 择 日 期 "
    value-format="yyyy-MM-dd">
86  </el-date-picker>
87  </el-form-item>
88  <el-form-item label="还书类型" prop="backType">
89  <el-input v-model="editForm.bookId"></el-input>
90  </el-form-item>
91  </el-form>
92  <!-- 底部区域 -->
93  <span slot="footer" class="dialog-footer">
94  <el-button @click="editDialogVisible = false">取 消</el-button>
95  <el-button type="primary" @click="editBorrowInfo">确 定</el-button>
96  </span>
97  </el-dialog>
98  </div>
99  </template>
100 <!--省略部分代码-->
```

进入借阅管理模块后，可以对图书的借阅信息进行相关操作，页面如图 18.9、图 18.10 所示。

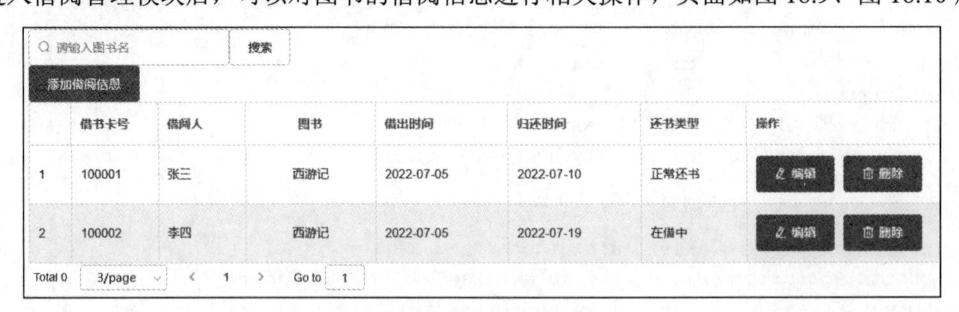

图 18.9　借阅信息页面

添加借阅信息

借阅人

图书

借出时间　选择日期

归还时间　选择日期

还书类型

取消　确定

图 18.10　添加借阅信息页面

18.4.4　读者（管理员）管理模块

1. 创建持久化类

读者（管理员）模块持久化类有 Admin 和 ReaderInfo，具体代码如文件 18.29 和文件 18.30 所示。

文件 18.29　Admin.java

```
01  public class Admin implements Serializable {
02      private Integer id;
03      private String username;
04      private String password;
05      private Integer adminType;
06      private static final long serialVersionUID = 1L;
07      public Integer getId() {
08          return id;
09      }
10      public void setId(Integer id) {
11          this.id = id;
12      }
13      public String getUsername() {
14          return username;
15      }
16      public void setUsername(String username) {
17          this.username = username == null ? null : username.trim();
18      }
```

```
19      public String getPassword() {
20          return password;
21      }
22      public void setPassword(String password) {
23          this.password = password == null ? null : password.trim();
24      }
25      public Integer getAdminType() {
26          return adminType;
27      }
28      public void setAdminType(Integer adminType) {
29          this.adminType = adminType;
30      }
31  }
```

文件 18.30 ReaderInfo.java

```
01  public class ReaderInfo implements Serializable {
02      private Integer id;
03      private String username;
04      private String password;
05      private String realName;
06      private String sex;
07      private Date birthday;
08      private String address;
09      private String tel;
10      private String email;
11      private Date registerDate;
12      private String readerNumber;
13      private static final long serialVersionUID = 1L;
14      public Integer getId() {
15          return id;
16      }
17      public void setId(Integer id) {
18          this.id = id;
19      }
20      public String getUsername() {
21          return username;
22      }
23      public void setUsername(String username) {
24          this.username = username == null ? null : username.trim();
25      }
26      public String getPassword() {
27          return password;
28      }
29      public void setPassword(String password) {
30          this.password = password == null ? null : password.trim();
31      }
```

```
32      public String getRealName() {
33          return realName;
34      }
35      public void setRealName(String realName) {
36          this.realName = realName == null ? null : realName.trim();
37      }
38      public String getSex() {
39          return sex;
40      }
41      public void setSex(String sex) {
42          this.sex = sex == null ? null : sex.trim();
43      }
44
45      public Date getBirthday() {
46          return birthday;
47      }
48      public void setBirthday(Date birthday) {
49          this.birthday = birthday;
50      }
51      public String getAddress() {
52          return address;
53      }
54
55      public void setAddress(String address) {
56          this.address = address == null ? null : address.trim();
57      }
58      public String getTel() {
59          return tel;
60      }
61      public void setTel(String tel) {
62          this.tel = tel == null ? null : tel.trim();
63      }
64      public String getEmail() {
65          return email;
66      }
67      public void setEmail(String email) {
68          this.email = email == null ? null : email.trim();
69      }
70      public Date getRegisterDate() {
71          return registerDate;
72      }
73      public void setRegisterDate(Date registerDate) {
74          this.registerDate = registerDate;
75      }
76      public String getReaderNumber() {
77          return readerNumber;
```

```
78      }
79      public void setReaderNumber(String readerNumber) {
80          this.readerNumber = readerNumber == null ? null : readerNumber.trim();
81      }
82  }
```

2. 实现 DAO 层接口

实现 DAO 层接口的操作步骤如下：

步骤 01 创建 DAO 层接口。在 src 目录下的 com.library.dao 包中创建一个接口 AdminMapper 和 ReaderInfoMapper，并在接口中编写增、删、改、查等方法，如文件 18.31 和文件 18.32 所示。

文件 18.31　AdminMapper.java

```
01  public interface AdminMapper {
02      int deleteByPrimaryKey(Integer id);
03      int insert(Admin record);
04      int insertSelective(Admin record);
05      Admin selectByPrimaryKey(Integer id);
06      int updateByPrimaryKeySelective(Admin record);
07      int updateByPrimaryKey(Admin record);
08      // 管理员查询
09      List<Admin> queryAdminInfoAll(Admin admin);
10      // 根据用户名和密码查询用户信息
11      Admin queryUserByNameAndPassword(@Param("username") String
12  username,@Param("password") String password);
13  }
```

文件 18.32　ReaderInfoMapper.java

```
01  public interface ReaderInfoMapper {
02      int deleteByPrimaryKey(Integer id);
03      int insert(ReaderInfo record);
04      int insertSelective(ReaderInfo record);
05      ReaderInfo selectByPrimaryKey(Integer id);
06      int updateByPrimaryKeySelective(ReaderInfo record);
07      int updateByPrimaryKey(ReaderInfo record);
08      List<ReaderInfo> queryAllReaderInfo(ReaderInfo readerInfo);
09      ReaderInfo queryUserInfoByNameAndPassword(@Param("username") String username,
10  @Param("password") String password);
11  }
```

步骤 02 创建映射文件。在 resources 目录下的 com.library.dao 包中创建 MyBatis 映射文件 AdminMapper.xml 和 ReaderInfoMapper.xml，并在映射文件中编写增、删、改、查等方法的执行语句，如文件 18.33 和文件 18.34 所示。

文件 18.33　AdminMapper.xml

```
01  <?xml version="1.0" encoding="UTF-8"?>
```

```
02  <!DOCTYPE mapper PUBLIC "-//mybatis.org//DTD Mapper 3.0//EN"
    "http://mybatis.org/dtd/mybatis-3-mapper.dtd">
03  <mapper namespace="com.library.dao.AdminMapper">
04  <resultMap id="BaseResultMap" type="com.library.po.Admin">
05  <id column="id" jdbcType="INTEGER" property="id" />
06  <result column="username" jdbcType="VARCHAR" property="username" />
07  <result column="password" jdbcType="VARCHAR" property="password" />
08  <result column="adminType" jdbcType="INTEGER" property="adminType" />
09  </resultMap>
10  <sql id="Base_Column_List">
11      id, username, password, adminType
12  </sql>
13  <select id="selectByPrimaryKey" parameterType="java.lang.Integer"
    resultMap="BaseResultMap">
14      select
15      <include refid="Base_Column_List" />
16          from admin
17          where id = #{id,jdbcType=INTEGER}
18  </select>
19  <delete id="deleteByPrimaryKey" parameterType="java.lang.Integer">
20      delete from admin
21      where id = #{id,jdbcType=INTEGER}
22  </delete>
23  <insert id="insert" parameterType="com.library.po.Admin">
24      insert into admin (id, username, password,
25      adminType)
26      values (#{id,jdbcType=INTEGER}, #{username,jdbcType=VARCHAR},
        #{password,jdbcType=VARCHAR},
27      #{adminType,jdbcType=INTEGER})
28  </insert>
29  <insert id="insertSelective" parameterType="com.library.po.Admin">
30      insert into admin
31      <trim prefix="(" suffix=")" suffixOverrides=",">
32      <if test="id != null">
33          id,
34      </if>
35      <if test="username != null">
36          username,
37      </if>
38      <if test="password != null">
39          password,
40      </if>
41      <if test="adminType != null">
42          adminType,
43      </if>
44      </trim>
```

```
45      <trim prefix="values (" suffix=")" suffixOverrides=",">
46      <if test="id != null">
47          #{id,jdbcType=INTEGER},
48      </if>
49      <if test="username != null">
50          #{username,jdbcType=VARCHAR},
51      </if>
52      <if test="password != null">
53          #{password,jdbcType=VARCHAR},
54      </if>
55      <if test="adminType != null">
56          #{adminType,jdbcType=INTEGER},
57      </if>
58      </trim>
59  </insert>
60  <update id="updateByPrimaryKeySelective" parameterType="com.library.po.Admin">
61      update admin
62      <set>
63      <if test="username != null">
64          username = #{username,jdbcType=VARCHAR},
65      </if>
66      <if test="password != null">
67          password = #{password,jdbcType=VARCHAR},
68      </if>
69      <if test="adminType != null">
70          adminType = #{adminType,jdbcType=INTEGER},
71      </if>
72      </set>
73      where id = #{id,jdbcType=INTEGER}
74  </update>
75  <update id="updateByPrimaryKey" parameterType="com.library.po.Admin">
76      update admin
77      set username = #{username,jdbcType=VARCHAR},
78      password = #{password,jdbcType=VARCHAR},
79      adminType = #{adminType,jdbcType=INTEGER}
80      where id = #{id,jdbcType=INTEGER}
81  </update>
82  <select id="queryAdminInfoAll" parameterType="com.library.po.Admin"
    resultType="com.library.po.Admin">
83      SELECT * from admin
84      <where>
85      <if test="username!=null">
86          and username like '%${username}%'
87      </if>
88      <if test="adminType!=null">
89          and adminType like '%${adminType}%'
```

```
90         </if>
91     </where>
92 </select>
93 <select id="queryUserByNameAndPassword" resultType="com.library.po.Admin">
94     select * from admin where username=#{username} and password=#{password}
95 </select>
96 </mapper>
```

文件 18.34　ReaderInfoMapper.xml

```
01 <?xml version="1.0" encoding="UTF-8"?>
02 <!DOCTYPE mapper PUBLIC "-//mybatis.org//DTD Mapper 3.0//EN"
   "http://mybatis.org/dtd/mybatis-3-mapper.dtd">
03 <mapper namespace="com.library.dao.ReaderInfoMapper">
04 <resultMap id="BaseResultMap" type="com.library.po.ReaderInfo">
05 <id column="id" jdbcType="INTEGER" property="id" />
06 <result column="username" jdbcType="VARCHAR" property="username" />
07 <result column="password" jdbcType="VARCHAR" property="password" />
08 <result column="realName" jdbcType="VARCHAR" property="realName" />
09 <result column="sex" jdbcType="VARCHAR" property="sex" />
10 <result column="birthday" jdbcType="DATE" property="birthday" />
11 <result column="address" jdbcType="VARCHAR" property="address" />
12 <result column="tel" jdbcType="VARCHAR" property="tel" />
13 <result column="email" jdbcType="VARCHAR" property="email" />
14 <result column="registerDate" jdbcType="TIMESTAMP" property="registerDate" />
15 <result column="readerNumber" jdbcType="VARCHAR" property="readerNumber" />
16 </resultMap>
17 <sql id="Base_Column_List">
18 id, username, password, realName, sex, birthday, address, tel, email, registerDate,
19 readerNumber
20 </sql>
21 <select id="selectByPrimaryKey" parameterType="java.lang.Integer"
   resultMap="BaseResultMap">
22 select
23 <include refid="Base_Column_List" />
24 from reader_info
25 where id = #{id,jdbcType=INTEGER}
26 </select>
27 <delete id="deleteByPrimaryKey" parameterType="java.lang.Integer">
28 delete from reader_info
29 where id = #{id,jdbcType=INTEGER}
30 </delete>
31 <insert id="insert" parameterType="com.library.po.ReaderInfo">
32 insert into reader_info (id, username, password, realName, sex,
33 birthday, address, tel,
34 email, registerDate, readerNumber
35 )
```

```
36  values (#{id,jdbcType=INTEGER}, #{username,jdbcType=VARCHAR},
    #{password,jdbcType=VARCHAR},
37  #{realName,jdbcType=VARCHAR}, #{sex,jdbcType=VARCHAR},
38  #{birthday,jdbcType=DATE}, #{address,jdbcType=VARCHAR}, #{tel,jdbcType=VARCHAR},
39  #{email,jdbcType=VARCHAR}, #{registerDate,jdbcType=TIMESTAMP},
    #{readerNumber,jdbcType=VARCHAR}
40  )
41  </insert>
42  <insert id="insertSelective" parameterType="com.library.po.ReaderInfo">
43  insert into reader_info
44  <trim prefix="(" suffix=")" suffixOverrides=",">
45  <if test="id != null">
46  id,
47  </if>
48  <if test="username != null">
49  username,
50  </if>
51  <if test="password != null">
52  password,
53  </if>
54  <if test="realName != null">
55  realName,
56  </if>
57  <if test="sex != null">
58  sex,
59  </if>
60  <if test="birthday != null">
61  birthday,
62  </if>
63  <if test="address != null">
64  address,
65  </if>
66  <if test="tel != null">
67  tel,
68  </if>
69  <if test="email != null">
70  email,
71  </if>
72  <if test="registerDate != null">
73  registerDate,
74  </if>
75  <if test="readerNumber != null">
76  readerNumber,
77  </if>
78  </trim>
79  <trim prefix="values (" suffix=")" suffixOverrides=",">
```

```
80  <if test="id != null">
81  #{id,jdbcType=INTEGER},
82  </if>
83  <if test="username != null">
84  #{username,jdbcType=VARCHAR},
85  </if>
86  <if test="password != null">
87  #{password,jdbcType=VARCHAR},
88  </if>
89  <if test="realName != null">
90  #{realName,jdbcType=VARCHAR},
91  </if>
92  <if test="sex != null">
93  #{sex,jdbcType=VARCHAR},
94  </if>
95  <if test="birthday != null">
96  #{birthday,jdbcType=DATE},
97  </if>
98  <if test="address != null">
99  #{address,jdbcType=VARCHAR},
100 </if>
101 <if test="tel != null">
102 #{tel,jdbcType=VARCHAR},
103 </if>
104 <if test="email != null">
105 #{email,jdbcType=VARCHAR},
106 </if>
107 <if test="registerDate != null">
108 #{registerDate,jdbcType=TIMESTAMP},
109 </if>
110 <if test="readerNumber != null">
111 #{readerNumber,jdbcType=VARCHAR},
112 </if>
113 </trim>
114 </insert>
115 <update id="updateByPrimaryKeySelective" parameterType="com.library.po.ReaderInfo">
116 update reader_info
117 <set>
118 <if test="username != null">
119 username = #{username,jdbcType=VARCHAR},
120 </if>
121 <if test="password != null">
122 password = #{password,jdbcType=VARCHAR},
123 </if>
124 <if test="realName != null">
125 realName = #{realName,jdbcType=VARCHAR},
```

```
126 </if>
127 <if test="sex != null">
128 sex = #{sex,jdbcType=VARCHAR},
129 </if>
130 <if test="birthday != null">
131 birthday = #{birthday,jdbcType=DATE},
132 </if>
133 <if test="address != null">
134 address = #{address,jdbcType=VARCHAR},
135 </if>
136 <if test="tel != null">
137 tel = #{tel,jdbcType=VARCHAR},
138 </if>
139 <if test="email != null">
140 email = #{email,jdbcType=VARCHAR},
141 </if>
142 <if test="registerDate != null">
143 registerDate = #{registerDate,jdbcType=TIMESTAMP},
144 </if>
145 <if test="readerNumber != null">
146 readerNumber = #{readerNumber,jdbcType=VARCHAR},
147 </if>
148 </set>
149 where id = #{id,jdbcType=INTEGER}
150 </update>
151 <update id="updateByPrimaryKey" parameterType="com.library.po.ReaderInfo">
152 update reader_info
153 set username = #{username,jdbcType=VARCHAR},
154 password = #{password,jdbcType=VARCHAR},
155 realName = #{realName,jdbcType=VARCHAR},
156 sex = #{sex,jdbcType=VARCHAR},
157 birthday = #{birthday,jdbcType=DATE},
158 address = #{address,jdbcType=VARCHAR},
159 tel = #{tel,jdbcType=VARCHAR},
160 email = #{email,jdbcType=VARCHAR},
161 registerDate = #{registerDate,jdbcType=TIMESTAMP},
162 readerNumber = #{readerNumber,jdbcType=VARCHAR}
163 where id = #{id,jdbcType=INTEGER}
164 </update>
165 <select id="queryAllReaderInfo" resultType="com.library.po.ReaderInfo"
    parameterType="com.library.po.ReaderInfo">
166 select * from reader_info
167 <where>
168 <if test="readerNumber!=null">
169 and readerNumber like '%${readerNumber}%'
170 </if>
```

```
171 <if test="username!=null">
172 and username like '%${username}%'
173 </if>
174 <if test="tel!=null">
175 and tel like '%${tel}%'
176 </if>
177 </where>
178 </select>
179 <select id="queryUserInfoByNameAndPassword" resultType="com.library.po.ReaderInfo">
180 select * from reader_info where username=#{username} and password=#{password}
181 </select>
182 </mapper>
```

3. 实现 Service 层接口

实现 Service 层接口的操作步骤如下：

步骤 01 创建 Service 层接口。在 src 目录下创建一个 com.library.service 包，在包中创建 AdminService 接口和 ReaderInfoService，并在该接口中编写相关方法，如文件 18.35 和文件 18.36 所示。

文件 18.35　AdminService.java

```
01  public interface AdminService {
02      // 查询所有管理员（分页）
03      PageInfo<Admin> queryAdminAll(Admin admin,Integer pageNum,Integer limit);
04      // 添加提交
05      void addAdminSubmit(Admin admin);
06      // 根据 id 查询（修改）
07      Admin queryAdminById(Integer id);
08      // 修改提交
09      void updateAdminSubmit(Admin admin);
10      // 删除
11      void deleteAdminByIds(List<String> ids);
12      // 根据用户名和密码查询用户信息
13      Admin queryUserByNameAndPassword(String username,String password);
14  }
```

文件 18.36　ReaderInfoService.java

```
01  public interface ReaderInfoService {
02      PageInfo<ReaderInfo> queryAllReaderInfo(ReaderInfo readerInfo,Integer
03  pageNum,Integerlimit);
04      void addReaderInfoSubmit(ReaderInfo readerInfo);
05      ReaderInfo queryReaderInfoById(Integer id);
06      void updateReaderInfoSubmit(ReaderInfo readerInfo);
07      void deleteReaderInfoByIds(List<String> ids);
08      // 根据用户名和密码查询用户信息
09      ReaderInfo queryUserInfoByNameAndPassword(String username,String password);
10  }
```

步骤 02 创建 Service 层接口的实现类。在 src 目录下创建一个 com.library.service.impl 包，并在包中创建 AdminService 接口的实现类 AdminServiceImpl 和 ReaderInfoService 接口的实现类 ReaderInfoServiceImpl，在类中编辑并实现接口中的方法，如文件 18.37 和文件 18.38 所示。

文件 18.37　AdminServiceImpl.java

```
01  @Service("adminService")
02  public class AdminServiceImpl implements AdminService {
03      @Autowired
04      private AdminMapper adminMapper;
05      @Override
06      public PageInfo<Admin> queryAdminAll(Admin admin, Integer pageNum, Integer limit) {
07          PageHelper.startPage(pageNum,limit);
08          List<Admin> adminList = adminMapper.queryAdminInfoAll(admin);
09          return new PageInfo<>(adminList) ;
10      }
11      @Override
12      public void addAdminSubmit(Admin admin) {
13          adminMapper.insert(admin);
14      }
15      @Override
16      public Admin queryAdminById(Integer id) {
17          return adminMapper.selectByPrimaryKey(id);
18      }
19      @Override
20      public void updateAdminSubmit(Admin admin) {
21          adminMapper.updateByPrimaryKey(admin);
22      }
23      @Override
24      public void deleteAdminByIds(List<String> ids) {
25          for (String id : ids){
26              adminMapper.deleteByPrimaryKey(Integer.parseInt(id));
27          }
28      }
29      @Override
30      public Admin queryUserByNameAndPassword(String username, String password) {
31          return adminMapper.queryUserByNameAndPassword(username,password);
32      }
33  }
```

文件 18.38　ReaderInfoServiceImpl.java

```
01  @Service("readerInfoService")
02  public class ReaderInfoServiceImpl implements ReaderInfoService {
03      @Autowired
04      private ReaderInfoMapper readerInfoMapper;
05      @Override
06      public PageInfo<ReaderInfo> queryAllReaderInfo(ReaderInfo readerInfo, Integer
```

```
07  pageNum, Integer limit) {
08          PageHelper.startPage(pageNum,limit);
09          List<ReaderInfo> readerInfoList =
10  readerInfoMapper.queryAllReaderInfo(readerInfo);
11          return new PageInfo<>(readerInfoList);
12      }
13      @Override
14      public void addReaderInfoSubmit(ReaderInfo readerInfo) {
15          readerInfoMapper.insert(readerInfo);
16      }
17      @Override
18      public ReaderInfo queryReaderInfoById(Integer id) {
19          return readerInfoMapper.selectByPrimaryKey(id);
20      }
21      @Override
22      public void updateReaderInfoSubmit(ReaderInfo readerInfo) {
23          readerInfoMapper.updateByPrimaryKey(readerInfo);
24      }
25      @Override
26      public void deleteReaderInfoByIds(List<String> ids) {
27          for (String id : ids){
28              readerInfoMapper.deleteByPrimaryKey(Integer.parseInt(id));
29          }
30      }
31      @Override
32      public ReaderInfo queryUserInfoByNameAndPassword(String username,String password) {
33          return readerInfoMapper.queryUserInfoByNameAndPassword(username, password);
34      }
35  }
```

4. 实现 Controller 类

在 src 目录下创建一个 com.library.controller 包，在包中创建控制器类 AdminController 和 ReaderInfoController，相应代码如文件 18.39 和文件 18.40 所示。

文件 18.39　AdminController.java

```
01  @Controller
02  public class AdminController {
03      @Autowired
04      private AdminService adminService;
05      @GetMapping("/adminIndex")
06      public String adminIndex(){
07          return "admin/adminIndex";
08      }
09      @RequestMapping("/adminAll")
10      @ResponseBody
11      public DataInfo queryAdminAll(Admin admin, @RequestParam(defaultValue = "1") Integer
```

```
12  pageNum, @RequestParam(defaultValue = "15") Integer limit){
13          PageInfo<Admin> pageInfo = adminService.queryAdminAll(admin,pageNum,limit);
14          return DataInfo.ok("成功",pageInfo.getTotal(),pageInfo.getList());
15      }
16      // 添加页面的跳转
17      @GetMapping("/adminAdd")
18      public String adminAdd(){
19          return "admin/adminAdd";
20      }
21      // 添加提交
22      @RequestMapping("/addAdminSubmit")
23      @ResponseBody
24      public DataInfo addBookSubmit(Admin admin){
25          adminService.addAdminSubmit(admin);
26          return DataInfo.ok();
27      }
28      // 根据 id 进行查询
29      @GetMapping("/queryAdminById")
30      public String queryAdminById(Integer id, Model model){
31          model.addAttribute("id",id);
32          return "admin/updateAdmin";
33      }
34      // 修改提交
35      @RequestMapping("/updatePwdSubmit")
36      @ResponseBody
37      public DataInfo updatePwdSubmit(Integer id,String oldPwd,String newPwd){
38          Admin admin = adminService.queryAdminById(id);//根据 id 查询对象
39          if (!oldPwd.equals(admin.getPassword())){
40              return DataInfo.fail("输入的旧密码错误");
41          }else{
42              admin.setPassword(newPwd);
43              adminService.updateAdminSubmit(admin);           //修改数据库
44              return DataInfo.ok();
45          }
46      }
47      // 删除
48      @RequestMapping("/deleteAdminByIds")
49      @ResponseBody
50      public DataInfo deleteAdminByIds(String ids){
51          List<String> list = Arrays.asList(ids.split(","));
52          adminService.deleteAdminByIds(list);
53          return DataInfo.ok();
54      }
55  }
```

文件 18.40　ReaderInfoController.java

```
01	@Controller
02	public class ReaderInfoController {
03	    @Autowired
04	    private ReaderInfoService readerInfoService;
05	    @Autowired
06	    private AdminService adminService;
07	
08	    @GetMapping("/readerIndex")
09	    public String readerIndex(){
10	        return "reader/readerIndex";
11	    }
12	    @RequestMapping("/readerAll")
13	    @ResponseBody
14	    public DataInfo queryReaderAll(ReaderInfo readerInfo, @RequestParam(defaultValue =
15	"1") Integer pageNum, @RequestParam(defaultValue = "15") Integer limit){
16	        PageInfo<ReaderInfo> pageInfo =
17	readerInfoService.queryAllReaderInfo(readerInfo,pageNum,limit);
18	        return DataInfo.ok("成功",pageInfo.getTotal(),pageInfo.getList());
19	    }
20	    @RequestMapping("/readerAdd")
21	    public String readerAdd(){
22	        return "reader/readerAdd";
23	    }
24	    @RequestMapping("/addReaderSubmit")
25	    @ResponseBody
26	    public DataInfo addReaderSubmit(@RequestBody ReaderInfo readerInfo){
27	        readerInfo.setPassword("123456");//设置默认密码
28	        readerInfoService.addReaderInfoSubmit(readerInfo);
29	        return DataInfo.ok();
30	    }
31	    @GetMapping("/queryReaderInfoById")
32	    public String queryReaderInfoById(Integer id, Model model){
33	        ReaderInfo readerInfo = readerInfoService.queryReaderInfoById(id);
34	        model.addAttribute("info",readerInfo);
35	        return "reader/updateReader";
36	    }
37	    @RequestMapping("/updateReaderSubmit")
38	    @ResponseBody
39	    public DataInfo updateReaderSubmit(@RequestBody ReaderInfo readerInfo){
40	        readerInfoService.updateReaderInfoSubmit(readerInfo);
41	        return DataInfo.ok();
42	    }
43	    // 删除
44	    @RequestMapping("/deleteReader")
45	    @ResponseBody
46	    public DataInfo deleteReader(String ids){
47	        List<String> list= Arrays.asList(ids.split(","));
48	        readerInfoService.deleteReaderInfoByIds(list);
```

```
49            return DataInfo.ok();
50        }
51        @RequestMapping("/updatePwdSubmit2")
52        @ResponseBody
58        public DataInfo updatePwdSubmit(HttpServletRequest request, String oldPwd, String
54    newPwd){
55            HttpSession session = request.getSession();
56            if(session.getAttribute("type")=="admin"){
57                //管理员
58                Admin admin = (Admin)session.getAttribute("user");
59                Admin admin1 = (Admin)adminService.queryAdminById(admin.getId());
60                if (!oldPwd.equals(admin1.getPassword())){
61                    return DataInfo.fail("输入的旧密码错误");
62                }else{
63                    admin1.setPassword(newPwd);
64                    adminService.updateAdminSubmit(admin1);// 修改数据库
65                }
66            }else{
67                //读者
68                ReaderInfo readerInfo = (ReaderInfo) session.getAttribute("user");
69                ReaderInfo readerInfo1 =
70    readerInfoService.queryReaderInfoById(readerInfo.getId());//根据 id 查询对象
71                if (!oldPwd.equals(readerInfo1.getPassword())){
72                    return DataInfo.fail("输入的旧密码错误");
73                }else{
74                    readerInfo1.setPassword(newPwd);
75                    readerInfoService.updateReaderInfoSubmit(readerInfo1);// 修改数据库
76                }
77            }
78            return DataInfo.ok();
79        }
80    }
```

5. 实现页面功能

实现页面功能的操作步骤如下：

步骤 01 在项目的 src/views/Reader 目录下创建 Reader.vue 页面文件。Reader.vue 主要实现对读者进行管理的功能，代码如文件 18.41 所示。

文件 18.41 Reader.vue

```
01  <template>
02  <div>
03  <el-card>
04  <!-- 搜索与添加区域 -->
05  <div style=" width:300px; display: flex;margin-right: 1%">
06  <el-input v-model="username" prefix-icon="el-icon-search" placeholder="请输入用户名"
    clearable />
07  <el-button>搜索</el-button>
08  </div>
```

```
09  <el-row :gutter="10">
10  <el-col :span="4">
11  <el-button type="primary" @click="addDialogVisible = true">添加读者</el-button>
12  </el-col>
13  </el-row>
14  <el-table :data="userlist" style="width: 100%" border stripe>
15  <el-table-column type="index"></el-table-column>
16  <!-- <el-table-column prop="userId" label="用户 id" type="hidden"></el-table-column> -->
17  <el-table-column prop="readerNumber" label="读者卡号" ></el-table-column>
18  <el-table-column prop="username" label="用户名"></el-table-column>
19  <el-table-column prop="realName" label="姓名"></el-table-column>
20  <el-table-column prop="sex" label="性别"></el-table-column>
21  <el-table-column prop="tel" label="电话"></el-table-column>
22  <el-table-column prop="registerDate" label="办卡时间"></el-table-column>
23  <el-table-column prop="email" label="邮箱"></el-table-column>
24  <el-table-column label="操作" width="300px">
25  <template slot-scope="scope">
26  <!-- 修改按钮 -->
27  <el-button type="primary" icon="el-icon-edit" @click="showEditDialog(scope.row)">编辑
    </el-button>
28  <!-- 删除按钮 -->
29  <el-button type="danger" icon="el-icon-delete"
    @click="removeUserById(scope.row.userId)">删除</el-button>
30  </template>
31  </el-table-column>
32  </el-table>
33  <!-- 分页区域 -->
34  <el-pagination @size-change="handleSizeChange" @current-change="handleCurrentChange"
35  :current-page="queryInfo.pageNum" :page-sizes="[1, 2, 5,
    10]" :page-size="queryInfo.pageSize"
36  layout="total, sizes, prev, pager, next, jumper" :total="total">
37  </el-pagination>
38  </el-card>
39  <!-- 添加用户的对话框 -->
40  <el-dialog title="添加读者" :visible.sync="addDialogVisible" width="50%"
    @close="addDialogClosed">
41  <!-- 内容主体区域 -->
42  <el-form :model="addForm" :rules="addFormRules" ref="addFormRef" label-width="70px">
43  <el-form-item label="用户名" prop="username">
44  <el-input v-model="addForm.username"></el-input>
45  </el-form-item>
46  <el-form-item label="密码" prop="password">
47  <el-input v-model="addForm.password"></el-input>
48  </el-form-item>
49  <el-form-item label="姓名" prop="realName">
50  <el-input v-model="addForm.realName"></el-input>
51  </el-form-item>
52  <el-form-item label="性别" prop="sex">
53  <el-input v-model="addForm.sex"></el-input>
54  </el-form-item>
```

```
55  <el-form-item label="电话" prop="tel">
56  <el-input v-model="addForm.tel"></el-input>
57  </el-form-item>
58  <el-form-item label="邮箱" prop="email">
59  <el-input v-model="addForm.email"></el-input>
60  </el-form-item>
61  </el-form>
62  <!-- 底部区域 -->
63  <span slot="footer" class="dialog-footer">
64  <el-button @click="addDialogVisible = false">取 消</el-button>
65  <el-button type="primary" @click="addUser">确 定</el-button>
66  </span>
67  </el-dialog>
68  <!-- 修改用户的对话框 -->
69  <el-dialog title="修改用户" :visible.sync="editDialogVisible" width="50%"
    @close="editDialogClosed">
70  <!-- 内容主体区域 -->
71  <el-form :model="editForm" :rules="editFormRules" ref="editFormRef" label-width="70px">
72  <el-form-item label="用户名" prop="username">
73  <el-input v-model="editForm.username"></el-input>
74  </el-form-item>
75  <el-form-item label="姓名" prop="realName">
76  <el-input v-model="editForm.realName"></el-input>
77  </el-form-item>
78  <el-form-item label="性别" prop="sex">
79  <el-input v-model="editForm.sex"></el-input>
80  </el-form-item>
81  <el-form-item label="生日" prop="birthday">
82  <el-input v-model="editForm.birthday"></el-input>
83  </el-form-item>
84  <el-form-item label="住址" prop="address">
85  <el-input v-model="editForm.address"></el-input>
86  </el-form-item>
87  <el-form-item label="电话" prop="tel">
88  <el-input v-model="editForm.tel"></el-input>
89  </el-form-item>
90  <el-form-item label="邮箱" prop="email">
91  <el-input v-model="editForm.email"></el-input>
92  </el-form-item>
93  </el-form>
94  <!-- 底部区域 -->
95  <span slot="footer" class="dialog-footer">
96  <el-button @click="editDialogVisible = false">取 消</el-button>
97  <el-button type="primary" @click="editUserInfo">确 定</el-button>
98  </span>
99  </el-dialog>
100 </div>
101 </template>
102 <!--省略部分代码-->
```

进入读者管理模块后，可以对读者信息进行相关操作，页面如图 18.11、图 18.12 所示。

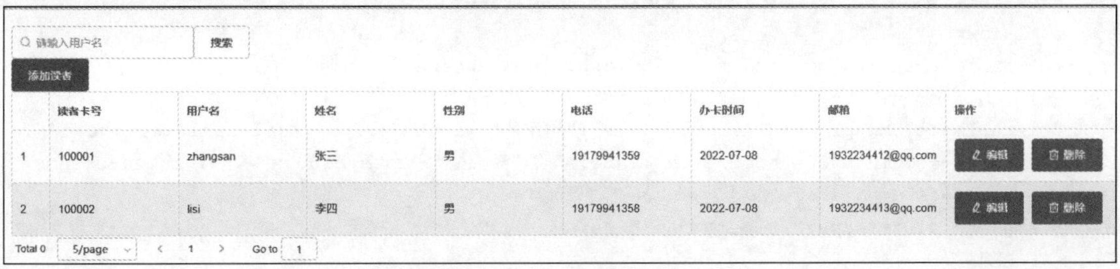

图 18.11　读者信息页面

图 18.12　添加读者页面

步骤 02 在项目 src/views/admin 目录下创建 admin.vue 页面文件。admin.vue 主要实现了对管理员进行添加、删除等操作，代码如文件 18.42 所示。

文件 18.42　admin.vue

```
01	<template>
02	<div>
03	<el-card>
04	<!-- 搜索与添加区域 -->
05	<div style=" width:300px; display: flex;margin-right: 1%">
06	<el-input v-model="username" prefix-icon="el-icon-search" placeholder="请输入用户名"
	clearable />
07	<el-button>搜索</el-button>
08	</div>
09	<el-row :gutter="10">
10	<el-col :span="4">
```

```
11  <el-button type="primary" @click="addDialogVisible = true">添加</el-button>
12  </el-col>
13  </el-row>
14  <el-table :data="userlist" style="width: 100%" border stripe>
15  <el-table-column type="index"></el-table-column>
16  <el-table-column prop="username" label="用户名" width="120px"></el-table-column>
17  <el-table-column prop="adminType" label="管理员类型" width="150px"></el-table-column>
18  <el-table-column label="操作" width="300px">
19  <template slot-scope="scope">
20  <!-- 修改按钮 -->
21  <el-button type="primary" icon="el-icon-edit" @click="showEditDialog(scope.row)">修改密
    码</el-button>
22  <!-- 删除按钮 -->
23  <el-button type="danger" icon="el-icon-delete"
    @click="removeUserById(scope.row.userId)">删除</el-button>
24  </template>
25  </el-table-column>
26  </el-table>
27  <!-- 分页区域 -->
28  <el-pagination @size-change="handleSizeChange" @current-change="handleCurrentChange"
29  :current-page="queryInfo.pageNum" :page-sizes="[1, 2, 5,
    10]" :page-size="queryInfo.pageSize"
30  layout="total, sizes, prev, pager, next, jumper" :total="total">
31  </el-pagination>
32  </el-card>
33  <!-- 添加的对话框 -->
34  <el-dialog title="添加" :visible.sync="addDialogVisible" width="50%"
    @close="addDialogClosed">
35  <!-- 内容主体区域 -->
36  <el-form :model="addForm" :rules="addFormRules" ref="addFormRef" label-width="100px">
37  <el-form-item label="用户名" prop="username">
38  <el-input v-model="addForm.username"></el-input>
39  </el-form-item>
40  <el-form-item label="密码" prop="password">
41  <el-input v-model="addForm.password"></el-input>
42  </el-form-item>
43  <el-form-item label="管理员类型" prop="adminType">
44  <el-input v-model="addForm.adminType"></el-input>
45  </el-form-item>
46  </el-form>
47  <!-- 底部区域 -->
48  <span slot="footer" class="dialog-footer">
49  <el-button @click="addDialogVisible = false">取 消</el-button>
50  <el-button type="primary" @click="addUser">确 定</el-button>
51  </span>
52  </el-dialog>
53  <!-- 修改的对话框 -->
54  <el-dialog title="修改密码" :visible.sync="editDialogVisible" width="50%"
    @close="editDialogClosed">
55  <!-- 内容主体区域 -->
```

```
56  <el-form :model="editForm" :rules="editFormRules" ref="editFormRef" label-width="70px">
57  <el-form-item label="旧密码" prop="oldPassword">
58  <el-input v-model="editForm.oldPassword"></el-input>
59  </el-form-item>
60  <el-form-item label="新密码" prop="newPassword">
61  <el-input v-model="editForm.newPassword"></el-input>
62  </el-form-item>
63  </el-form>
64  <!-- 底部区域 -->
65  <span slot="footer" class="dialog-footer">
66  <el-button @click="editDialogVisible = false">取 消</el-button>
67  <el-button type="primary" @click="editUserInfo">确 定</el-button>
68  </span>
69  </el-dialog>
70  </div>
71  </template>
72  <!--省略部分代码-->
```

进入管理员管理模块后，可以对管理员信息进行相关操作，页面如图 18.13~图 18.15 所示。

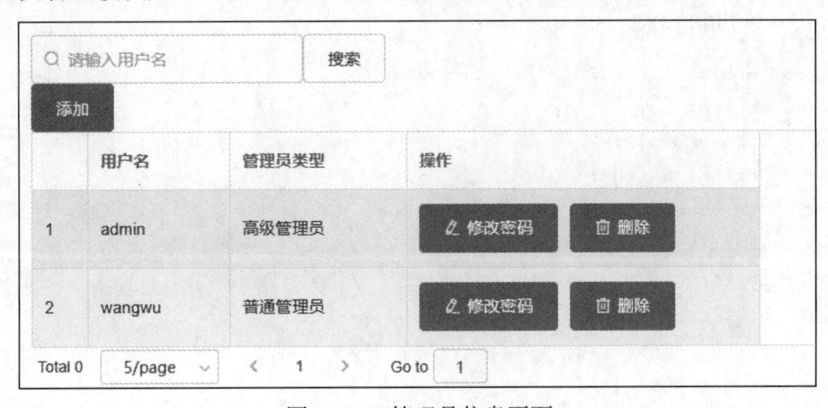

图 18.13　管理员信息页面

图 18.14　添加管理员页面

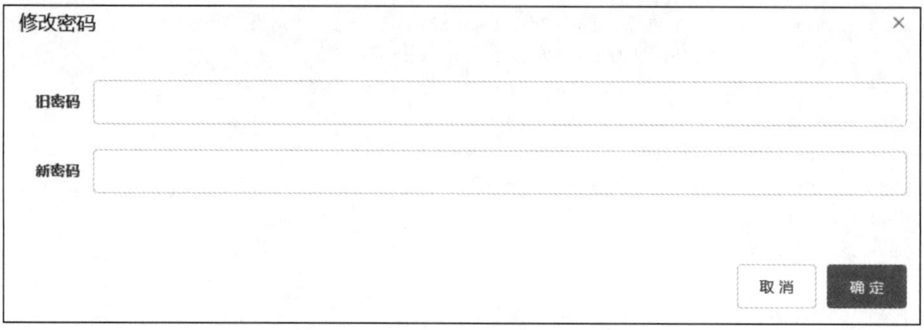

图 18.15　修改密码页面

18.4.5　公告管理模块

1. 创建持久化类

公告管理模块持久化类有 Notice，具体代码如文件 18.43 所示。

文件 18.43　Notice.java

```
01  public class Notice implements Serializable {
02      private Integer id;
03      private String topic;
04      private String content;
05      private String author;
06      @DateTimeFormat(pattern = "yyyy-MM-dd HH:mm:ss")//接收页面传来的时间格式
07      @JSONField(format="yyyy-MM-dd HH:mm:ss")//对返回的时间对象用 fastjson 格式化时间
08      private Date createDate;
09      private static final long serialVersionUID = 1L;
10      public Integer getId() {
11          return id;
12      }
13      public void setId(Integer id) {
14          this.id = id;
15      }
16      public String getTopic() {
17          return topic;
18      }
19      public void setTopic(String topic) {
20          this.topic = topic == null ? null : topic.trim();
21      }
22      public String getContent() {
23          return content;
24      }
25      public void setContent(String content) {
26          this.content = content == null ? null : content.trim();
27      }
```

```
28      public String getAuthor() {
29          return author;
30      }
31      public void setAuthor(String author) {
32          this.author = author == null ? null : author.trim();
33      }
34      public Date getCreateDate() {
35          return createDate;
36      }
37      public void setCreateDate(Date createDate) {
38          this.createDate = createDate;
39      }
40  }
```

2. 实现 DAO 层接口

实现 DAO 层接口的操作步骤如下：

步骤 01 创建 DAO 层接口。在 src 目录下的 com.library.dao 包中创建一个接口 NoticeMapper，并在接口中编写增、删、改、查等方法，如文件 18.44 所示。

文件 18.44 NoticeMapper.java

```
01  public interface NoticeMapper {
02      int deleteByPrimaryKey(Integer id);
03      int insert(Notice record);
04      int insertSelective(Notice record);
05      Notice selectByPrimaryKey(Integer id);
06      int updateByPrimaryKeySelective(Notice record);
07      int updateByPrimaryKey(Notice record);
08      // 查询所有公告信息
09      List<Notice> queryNoticeAll(Notice notice);
10  }
```

步骤 02 创建映射文件。在 resources 目录下的 com.library.dao 包中创建 MyBatis 映射文件 NoticeMapper.xml，并在映射文件中编写增、删、改、查等方法的执行语句，如文件 18.45 所示。

文件 18.45 NoticeMapper.xml

```
01  <?xml version="1.0" encoding="UTF-8"?>
02  <!DOCTYPE mapper PUBLIC "-//mybatis.org//DTD Mapper 3.0//EN"
    "http://mybatis.org/dtd/mybatis-3-mapper.dtd">
03  <mapper namespace="com.library.dao.NoticeMapper">
04  <resultMap id="BaseResultMap" type="com.library.po.Notice">
05  <id column="id" jdbcType="INTEGER" property="id" />
06  <result column="topic" jdbcType="VARCHAR" property="topic" />
07  <result column="content" jdbcType="VARCHAR" property="content" />
08  <result column="author" jdbcType="VARCHAR" property="author" />
09  <result column="createDate" jdbcType="TIMESTAMP" property="createDate" />
```

```
10 </resultMap>
11 <sql id="Base_Column_List">
12 id, topic, content, author, createDate
13 </sql>
14 <select id="selectByPrimaryKey" parameterType="java.lang.Integer"
   resultMap="BaseResultMap">
15 select
16 <include refid="Base_Column_List" />
17 from notice
18 where id = #{id,jdbcType=INTEGER}
19 </select>
20 <delete id="deleteByPrimaryKey" parameterType="java.lang.Integer">
21 delete from notice
22 where id = #{id,jdbcType=INTEGER}
23 </delete>
24 <insert id="insert" parameterType="com.library.po.Notice">
25 insert into notice (id, topic, content,
26 author, createDate)
27 values (#{id,jdbcType=INTEGER}, #{topic,jdbcType=VARCHAR}, #{content,jdbcType=VARCHAR},
28 #{author,jdbcType=VARCHAR}, #{createDate,jdbcType=TIMESTAMP})
29 </insert>
30 <insert id="insertSelective" parameterType="com.library.po.Notice">
31 insert into notice
32 <trim prefix="(" suffix=")" suffixOverrides=",">
33 <if test="id != null">
34 id,
35 </if>
36 <if test="topic != null">
37 topic,
38 </if>
39 <if test="content != null">
40 content,
41 </if>
42 <if test="author != null">
43 author,
44 </if>
45 <if test="createDate != null">
46 createDate,
47 </if>
48 </trim>
49 <trim prefix="values (" suffix=")" suffixOverrides=",">
50 <if test="id != null">
51 #{id,jdbcType=INTEGER},
52 </if>
53 <if test="topic != null">
54 #{topic,jdbcType=VARCHAR},
```

```
55  </if>
56  <if test="content != null">
57  #{content,jdbcType=VARCHAR},
58  </if>
59  <if test="author != null">
60  #{author,jdbcType=VARCHAR},
61  </if>
62  <if test="createDate != null">
63  #{createDate,jdbcType=TIMESTAMP},
64  </if>
65  </trim>
66  </insert>
67  <update id="updateByPrimaryKeySelective" parameterType="com.library.po.Notice">
68  update notice
69  <set>
70  <if test="topic != null">
71  topic = #{topic,jdbcType=VARCHAR},
72  </if>
73  <if test="content != null">
74  content = #{content,jdbcType=VARCHAR},
75  </if>
76  <if test="author != null">
77  author = #{author,jdbcType=VARCHAR},
78  </if>
79  <if test="createDate != null">
80  createDate = #{createDate,jdbcType=TIMESTAMP},
81  </if>
82  </set>
83  where id = #{id,jdbcType=INTEGER}
84  </update>
85  <update id="updateByPrimaryKey" parameterType="com.library.po.Notice">
86  update notice
87  set topic = #{topic,jdbcType=VARCHAR},
88  content = #{content,jdbcType=VARCHAR},
89  author = #{author,jdbcType=VARCHAR},
90  createDate = #{createDate,jdbcType=TIMESTAMP}
91  where id = #{id,jdbcType=INTEGER}
92  </update>
93  <select id="queryNoticeAll" parameterType="com.library.po.Notice"
    resultType="com.library.po.Notice">
94  select * from notice
95  <where>
96  <if test="topic!=null and topic!=''">
97  and topic like '%${topic}%'
98  </if>
99  </where>
```

```
100 order by createDate desc
101 </select>
102 </mapper>
```

3. 实现 Service 层接口

实现 Service 层接口的操作步骤如下：

步骤01 创建 Service 层接口。在 src 目录下创建一个 com.library.service 包，在包中创建 NoticeService 接口，并在该接口中编写相关方法，如文件 18.46 所示。

文件 18.46 NoticeService.java

```
01 public interface NoticeService {
02     // 查询所有公告
03     PageInfo<Notice> queryAllNotice(Notice notice,Integer pageNum,Integer limit);
04     void addNotice(Notice notice);
05     Notice queryNoticeById(Integer id);
06     void deleteNoticeByIds(List<String> ids);
07 }
```

步骤02 创建 Service 层接口的实现类。在 src 目录下创建一个 com.library.service.impl 包，并在包中创建 NoticeService 接口的实现类 NoticeServiceImpl，在类中编辑并实现接口中的方法，如文件 18.47 所示。

文件 18.47 NoticeServiceImpl.java

```
01 @Service("noticeService")
02 public class NoticeServiceImpl implements NoticeService {
03     @Autowired
04     private NoticeMapper noticeMapper;
05     @Override
06     public PageInfo<Notice> queryAllNotice(Notice notice,Integer pageNum,Integer limit)
07     {
08         PageHelper.startPage(pageNum,limit);
09         List<Notice> noticeList = noticeMapper.queryNoticeAll(notice);
10         return new PageInfo<>(noticeList);
11     }
12     @Override
13     public void addNotice(Notice notice) {
14         noticeMapper.insert(notice);
15     }
16     @Override
17     public Notice queryNoticeById(Integer id) {
18         return noticeMapper.selectByPrimaryKey(id);
19     }
20     @Override
21     public void deleteNoticeByIds(List<String> ids) {
22         for (String id : ids){
```

```
23              noticeMapper.deleteByPrimaryKey(Integer.parseInt(id));
24          }
25      }
26  }
```

4. 实现 Controller 类

在 src 目录下创建一个 com.library.controller 包，在包中创建控制器类 NoticeController，代码如文件 18.48 所示。

文件 18.48　NoticeController.java

```
01  public interface NoticeMapper {
02      int deleteByPrimaryKey(Integer id);
03      int insert(Notice record);
04      int insertSelective(Notice record);
05      Notice selectByPrimaryKey(Integer id);
06      int updateByPrimaryKeySelective(Notice record);
07      int updateByPrimaryKey(Notice record);
08      // 查询所有公告信息
09      List<Notice> queryNoticeAll(Notice notice);
10  }
```

5. 实现页面功能

在项目 src/views/notice 目录下创建 notice.vue 页面文件。notice.vue 主要实现用户登录到后台的功能，代码如文件 18.49 所示。

文件 18.49　notice.vue

```
01  <template>
02  <div>
03  <el-card>
04  <!-- 搜索与添加区域 -->
05  <div style=" width:300px; display: flex;margin-right: 1%">
06  <el-input v-model="topic" prefix-icon="el-icon-search" placeholder="请输入公告主题名"
    clearable />
07  <el-button>搜索</el-button>
08  </div>
09  <el-row :gutter="10">
10  <el-col :span="4">
11  <el-button type="primary" @click="addDialogVisible = true">添加</el-button>
12  </el-col>
13  </el-row>
14  <el-table :data="userlist" style="width: 100%" border stripe>
15  <el-table-column type="index"></el-table-column>
16  <el-table-column prop="topic" label="公告主题" width="120px"></el-table-column>
17  <el-table-column prop="content" label="公告内容" width="500px" ></el-table-column>
18  <el-table-column prop="author" label="发布人" width="100px" ></el-table-column>
19  <el-table-column prop="createDate" label="发布时间" width="120px" ></el-table-column>
20  <el-table-column label="操作" width="300px">
```

```
21  <template slot-scope="scope">
22  <!-- 修改按钮 -->
23  <el-button type="primary" icon="el-icon-edit" @click="showEditDialog(scope.row)">编辑
    </el-button>
24  <!-- 删除按钮 -->
25  <el-button type="danger" icon="el-icon-delete"
    @click="removeUserById(scope.row.userId)">删除</el-button>
26  </template>
27  </el-table-column>
28  </el-table>
29  <!-- 分页区域 -->
30  <el-pagination @size-change="handleSizeChange" @current-change="handleCurrentChange"
31  :current-page="queryInfo.pageNum" :page-sizes="[1, 2, 5,
    10]" :page-size="queryInfo.pageSize"
32  layout="total, sizes, prev, pager, next, jumper" :total="total">
33  </el-pagination>
34  </el-card>
35  <!-- 添加的对话框 -->
36  <el-dialog title="添加" :visible.sync="addDialogVisible" width="50%"
    @close="addDialogClosed">
37  <!-- 内容主体区域 -->
38  <el-form :model="addForm" :rules="addFormRules" ref="addFormRef" label-width="100px">
39  <el-form-item label="公告主题" prop="topic">
40  <el-input v-model="addForm.topic"></el-input>
41  </el-form-item>
42  <el-form-item label="公告内容" prop="content">
43  <el-input v-model="addForm.content"></el-input>
44  </el-form-item>
45  <el-form-item label="发布人" prop="author">
46  <el-input v-model="addForm.author"></el-input>
47  </el-form-item>
48  <el-form-item label="发布时间" prop="createDate">
49  <el-input v-model="addForm.createDate"></el-input>
50  </el-form-item>
51  </el-form>
52  <!-- 底部区域 -->
53  <span slot="footer" class="dialog-footer">
54  <el-button @click="addDialogVisible = false">取 消</el-button>
55  <el-button type="primary" @click="addUser">确 定</el-button>
56  </span>
57  </el-dialog>
58  <!-- 修改的对话框 -->
59  <el-dialog title="编辑" :visible.sync="editDialogVisible" width="50%"
    @close="editDialogClosed">
60  <!-- 内容主体区域 -->
61  <el-form :model="editForm" :rules="editFormRules" ref="editFormRef" label-width="70px">
62  <el-form-item label="公告主题" prop="topic">
63  <el-input v-model="editForm.topic"></el-input>
64  </el-form-item>
65  <el-form-item label="公告内容" prop="content">
```

```
66  <el-input v-model="editForm.content"></el-input>
67  </el-form-item>
68  <el-form-item label="发布人" prop="author">
69  <el-input v-model="editForm.author"></el-input>
70  </el-form-item>
71  <el-form-item label="发布时间" prop="createDate">
72  <el-input v-model="editForm.createDate"></el-input>
73  </el-form-item>
74  </el-form>
75  <!-- 底部区域 -->
76  <span slot="footer" class="dialog-footer">
77  <el-button @click="editDialogVisible = false">取 消</el-button>
78  <el-button type="primary" @click="editUserInfo">确 定</el-button>
79  </span>
80  </el-dialog>
81  </div>
82  </template>
83  <!--省略部分代码-->
```

进入公告模块后，可以对公告信息进行相关操作，页面如图 18.16、图 18.17 所示。

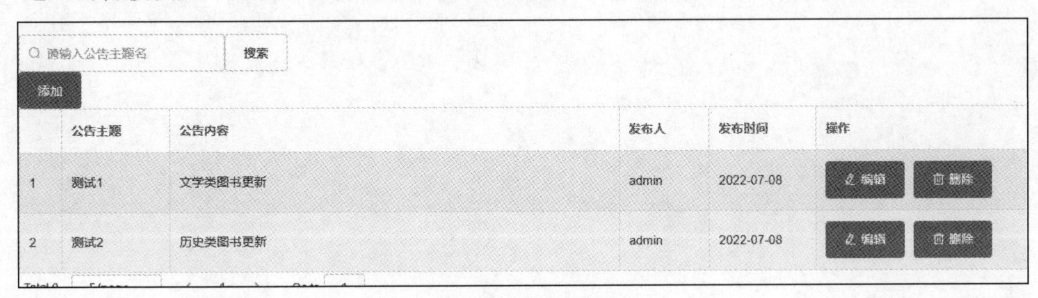

图 18.16　公告信息页面

添加

公告主题

公告内容

发布人

发布时间

取 消　确 定

图 18.17　添加公告页面

18.5 项目小结

本章主要通过一个较为完整的图书管理系统的设计与实现，系统介绍了 SSM+Vue.js 框架的整合和实战应用。首先，对系统的功能、结构等进行简单介绍；其次，对系统数据库表进行分析和设计；再次，详细讲解了系统的开发环境的搭建；最后，详细介绍了图书管理系统中的主要功能模块的设计与实现。通过本章的学习，读者可以熟练地掌握 SSM+Vue.js 框架的使用，并能熟练地使用 SSM+Vue.js 框架实现系统功能模块的开发工作。